Optimization Techniques and their Applications to Mine Systems

This book describes the fundamental and theoretical concepts of optimization algorithms in a systematic manner, along with their potential applications and implementation strategies in mining engineering. It explains basics of systems engineering, linear programming, and integer linear programming, transportation and assignment algorithms, network analysis, dynamic programming, queuing theory and their applications to mine systems. Reliability analysis of mine systems, inventory management in mines, and applications of non-linear optimization in mines are discussed as well. All the optimization algorithms are explained with suitable examples and numerical problems in each of the chapters.

Features include:

- Integrates operations research, reliability, and novel computerized technologies in single volume, with a modern vision of continuous improvement of mining systems.
- Systematically reviews optimization methods and algorithms applied to mining systems including reliability analysis.
- Gives out software-based solutions such as MATLAB®, AMPL, LINDO for the optimization problems.
- All discussed algorithms are supported by examples in each chapter.
- Includes case studies for performance improvement of the mine systems.

This book is aimed primarily at professionals, graduate students, and researchers in mining engineering.

Optimization Techniques and their Applications to Mine Systems

Amit Kumar Gorai and Snehamoy Chatterjee

CRC Press is an imprint of the
Taylor & Francis Group, an **informa** business

First edition published 2023
by CRC Press
6000 Broken Sound Parkway NW, Suite 300, Boca Raton, FL 33487-2742

and by CRC Press
4 Park Square, Milton Park, Abingdon, Oxon, OX14 4RN

CRC Press is an imprint of Taylor & Francis Group, LLC

Library of Congress Cataloging-in-Publication Data
Names: Gorai, Amit Kumar, author. | Chatterjee, Snehamoy, editor.
Title: Optimization techniques and their applications to mine systems / Amit Kumar Gorai, Snehamoy Chatterjee.
Description: Boca Raton : CRC Press, 2022. | Includes bibliographical references and index.
Identifiers: LCCN 2022002805 | ISBN 9781032060989 (hardback) | ISBN 9781032060996 (paperback) |
ISBN 9781003200703 (ebook)
Subjects: LCSH: Mining engineering.
Classification: LCC TN153.G575 2022 | DDC 622.0285–dc23/eng/20220408
LC record available at https://lccn.loc.gov/2022002805

ISBN: 9781032060989 (hbk)
ISBN: 9781032060996 (pbk)
ISBN: 9781003200703 (ebk)

DOI: 10.1201/9781003200703

Typeset in Times
by Newgen Publishing UK

Contents

Preface

Mining is one of the oldest industries and was discovered almost 20,000 years ago. Today, the mining industry generates more than US$700 bn revenue worldwide, only by the top 40 mining companies. Although mining companies are generating a significant amount of revenue, the net profit margin decreased from 25 per cent in 2010 to 10 per cent in 2018. As time passes, mining is becoming more challenging due to greater depth, low-grade, limited resources, and complex geo-mining conditions. Therefore, mine system optimization will play an important role in maximizing profit by satisfying several constraints. Moreover, today's mining industry uses complex and sophisticated systems whose reliability has become a critical issue. A significant amount of research is going on around the globe to address these challenges; however, to the best of the authors' knowledge, there is no single book available that covers system engineering and optimization from mining industry prospects. Students, researchers, and engineers need to consult with multiple sources to find reliable information related to this subject that causes serious difficulty. This book combines different systems engineering and optimization concepts in the light of mining engineering to make a one-stop-shop for all information seekers. This book covers almost every aspect of systems engineering and optimizations and is presented so that the readers don't require previous knowledge about the subject to understand the contents. The book describes the fundamentals and theoretical concepts of optimization algorithms and their potential applications and implementation strategies in mines. This book includes chapters on the basics of systems engineering, linear programming, and integer linear programming and their applications in mines, transportation and assignment algorithms, network analysis, dynamic programming, queuing theory and its applications to mine systems, reliability analysis of mine systems, inventory management in mines, and applications of non-linear optimization in mines. The book contains example problems and their solutions, and at the end of each chapter, there are various problems to provide readers the opportunity to comprehend their knowledge and understanding about the topics. A wide-ranging list of references is provided to give readers a view of developments in the area over the years. The book is composed of 11 chapters.

This book will be valuable to many individuals, including graduate and undergraduate students, researchers, academicians in mining engineering, mining engineering professionals, and associated professionals concerned with mining equipment.

We have a tremendous debt of gratitude to many individuals and organizations, especially to those companies around the globe who have shared their data to use in this book. The quality of this book is also substantially improved from the reviewers' suggestions (Julian M. Ortiz of Queen's University, Mustafa Kumral of McGill University, Victor Octavio Tenorio of the University of Arizona) and my colleagues in academia and industry.

Author biographies

Dr Amit Kumar Gorai is an Associate Professor in the Department of Mining Engineering at the National Institute of Technology, Rourkela, India. Prior to joining at NIT Rourkela, Dr Gorai had worked at Birla Institute of Technology Mesra, Ranchi, for over seven years. He has published over 60 research articles in the area of reliability analysis of mining systems, machine learning applications for quality monitoring of ores/coal, remote sensing applications for environmental management in mines, and so on. Dr Gorai has also written one guidebook, *A Complete Guide for Mining Engineers*, and one edited book, *Sustainable Mining Practices*.

Dr. Gorai received his PhD from the Indian School of Mines, Dhanbad, India. He is the recipient of the Endeavour Executive Fellowship from the Australian Government for working at the University of New South Wales, Sydney, Australia, and Raman Postdoctoral Fellowship from University Grants Commission, New Delhi for working at Jackson State University, MS, USA. He has been teaching Mine Systems Engineering at NIT Rourkela for the last few years.

He has completed several sponsored research projects in the field of environmental modelling. His current research area is systems optimization, machine learning, GIS, and remote sensing. Dr Gorai is a member of the Institution of Engineers India (IEI), The Mining, Geological & Metallurgical Institute (MGMI) of India, and the International Associate of Mathematical Geosciences (IAMG).

Dr Snehamoy Chatterjee is an Associate Professor and Witte Family Endowed Faculty Fellow in Mining Engineering in the Geological and Mining Engineering and Sciences Department at Michigan Technological University. Before joining Michigan Tech, Dr Chatterjee worked as an Assistant Professor at the National Institute of Technology, Rourkela, India. Dr Chatterjee specializes in ore reserve estimation, short- and long-range mine planning, mining machine reliability analysis, mine safety evaluation, and the application of image analysis and artificial intelligence in mining problems. He received his PhD in Mining Engineering from the Indian Institute of Technology Kharagpur, India. Dr Chatterjee worked as a Post-Doctoral Fellow at the University of Alaska Fairbanks and as a research associate at COSMO Stochastic Mine Planning Laboratory, McGill University, Canada, where he focused on mine planning optimization and ore-body modelling under uncertainty. Presently, Dr Chatterjee is actively involved in research work in resource modelling, production planning, online quality monitoring, and machine learning. He teaches courses and advises students on topics related to mine planning, mineral resource modelling, mining machine reliability, and vision-based online quality monitoring. He has completed several sponsored research and industry projects for different government organizations and mining companies in India and the USA.

Dr Chatterjee is an active member of the International Associate of Mathematical Geosciences (IAMG), the Society for Mining, Metallurgy, and Exploration, Inc. (SME), the American Geophysical Union (AGU). He has served as a co-convener and a technical committee member for several international mining conferences. He is also a reviewer for more than 30 journals and has received The Editor's Best Reviewer Awards 2014 from Mathematical Geosciences Journal. He is the 2015 APCOM Young Professional Award recipient at the 37th APCOM in Fairbanks, Alaska. He is an editorial board member of the *International Journal of Mining, Reclamation and Environment* and *Journal of Artificial Intelligence Research*, and associate editor of *Results in Geophysical Sciences*.

1 Introduction to mine systems

1.1 DEFINITION OF A SYSTEM

A system can be defined as a device or scheme that accepts single or multiple inputs and provides single or multiple outputs. According to Dooge (1973), a system is 'any structure, device, scheme, or procedure, real or abstract, that interrelates an input and output or cause and effect or any other things/information in a given time reference'. Systems theory views the mines as a complex system of interconnected subsystems. A mining system can be classified as an open system or a closed system depending on the demarcation and definition of the boundary. Depending on the demarcation of the boundary, one can check which entities are inside the system and which are outside. In an open system, materials/mass and energy both can flow through the boundary of the system; whereas, in a closed system, energy can flow through the boundary of the system, but materials/mass remain fixed and cannot flow through the boundary of the system. One can make a simplified representation of the mine systems in order to understand it and predict its future behaviour. A typical example of a mining system is represented in Figure 1.1.

1.2 TYPES OF SYSTEM

Any system can be defined or classified in multiple ways. These are as follows:

- **Simple and Complex Systems:** If the input has a direct relation with the output, the system is said to be a simple system. It may be linear or non-linear in nature.

  On the other hand, a complex system is a combination of several simple systems. All these simple systems can be termed a sub-system. Each subsystem has a distinct relation between input and output. It may be linear or non-linear in nature.

  Typical examples of a simple and a complex system are presented here.

  Example of a simple system

$$y = a_1 x_1$$

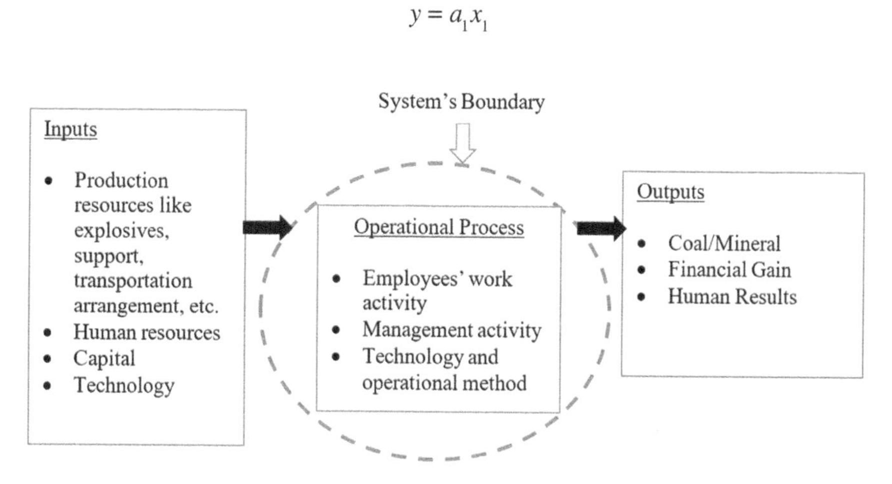

FIGURE 1.1 A typical example of a mining system (mine as an open system).

DOI: 10.1201/9781003200703-1

Example of a complex system

$$y = a_1 x_1 + a_2 x_2 + \cdots a_n x_n$$

$$\bullet\, y = y_1 + y_2 + \cdots y_n$$

where y is the output from the system, $x_1, x_2, \ldots, x_n$ are inputs to the system, and a_1, $a_2, \ldots, a_n$ are weights associated with the inputs. Both the above simple and complex systems are linear. In the simple system, the output is directly proportional to the input, but the same does not hold true for the complex system. Thus, the complex system represents n number of sub-systems or simple systems.

- **Linear and Nonlinear Systems**: A linear system is a type of system in which the output from the system varies directly with respect to the inputs of the system. Also, a linear system satisfies superposition and homogenate principles. In a linear system, the output of a combination of inputs is equal to the sum of the outputs from each of the individual inputs.

For example, the inputs of two systems are $x_1(t)$ and $x_2(t)$, and the corresponding outputs are respectively $y_1(t)$ and $y_2(t)$. Then, for a linear system, superposition and homogenate principles will hold good, and mathematically, this can be represented as:

$$f\left[a_1\, x_1(t) + a_2\, x_2(t)\right] = a_1\, f\left[x_1(t)\right] + a_2\, f\left[x_2(t)\right] = a_1\, y_1(t) + a_2\, y_2(t)$$

It is evident from the above relationship that the output of the overall system is equal to the output of the individual system.

On the other hand, if the condition of superposition and homogenate principles is not satisfied, then the system is called a non-linear system. That is, in a non-linear system, the above equation does not hold.

$$f\left[a_1 x_1(t) + a_2 x_2(t)\right] \neq a_1 f\left[x_1(t)\right] + a_2 f\left[x_2(t)\right]$$

$$f\left[a_1 x_1(t) + a_2 x_2(t)\right] \neq a_1 y_1(t) + a_2 y_2(t)$$

- **Time-Variant and Time-Invariant Systems**: If any time shifts in the input in the system cause the same amount of time shift in the output, the system is said to be a time-invariant system.

If the above condition is not satisfied, the system is said to be a time-variant system

Example of a time-invariant system

$$y(t) = k + x(t)$$

where $y(t)$ is output and $x(t)$ is input of the system. k is any constant.
Let there is delay input by Δt and output is $y_1(t)$
We have, $y_1(t) = k + x(t - \Delta t)$
Again, assuming for the delay in output by Δt and output is $y_2(t)$
We have, $y_2(t) = y(t - \Delta t) = k + x(t - \Delta t)$

Therefore, the above two equations indicate that

$$y_1(t) = y_2(t)$$

Thus, the above system is a time-invariant system.

Example of a time-variant system

$$y(t) = tx(t)$$

Let there is delay input by Δt and output is $y_1(t)$
We have, $y_1(t) = tx(t - \Delta t)$
Again, assuming for the delay in output by Δt and output is $y_2(t)$
We have, $y_2(t) = y(t - \Delta t) = (t - \Delta t)x(t - \Delta t)$

Therefore, the above two equations indicate that

$$y_1(t) \neq y_2(t)$$

Thus, the above system is a time-variant system.

- **Continuous and discrete changes systems**: A system is said to be continuous if the variable(s) are subjected to change continuously over time.

On the other hand, if the variable(s) are subjected to change at a discrete interval of time.

For example, a system in mine showing the number of dump trucks waiting in the queue for being loaded is a discrete number. That is, for any moment, the number of dump trucks is an integer number.

On the other hand, the strata pressure on the roof is continuous in nature. That is, every moment, the pressure is subjected to change.

The graphical representation of the characteristics of the discrete and continuous data is shown in Figure 1.2.

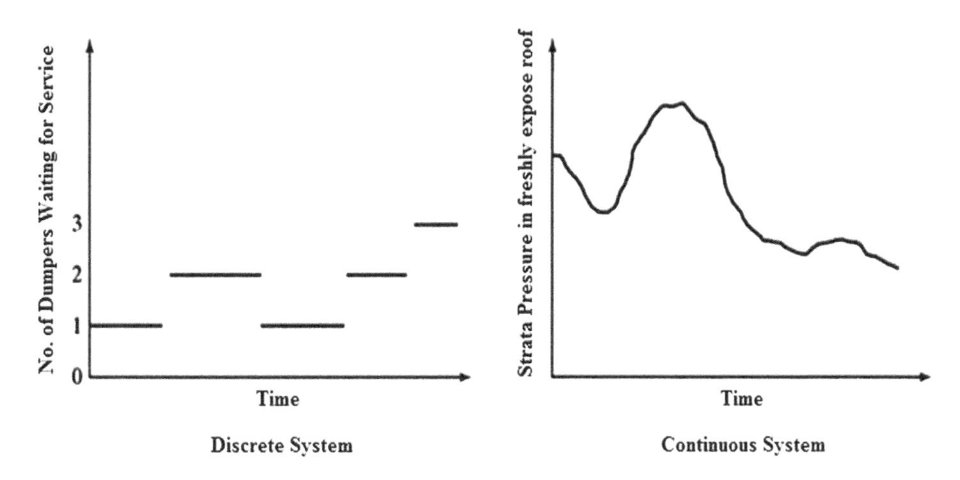

FIGURE 1.2 Discrete vs. continuous system.

- **Lumped Parameter and Distributed Parameter Systems**: If the dependent variables are a function of time alone and the variation in space is either non-existent or ignored, the system is said to be a lumped parameter system. This type of system can be represented by ordinary differential equations.
  On the other hand, if all the dependent variables are functions of time and at least one variable has spatial variation, then the system is said to be a distributed parameter system. This type of system is governed by a partial differential equation.
- **Static and Dynamic Systems:** If the output of the system doesn't depend on the time-dependent input variable, the system is said to be a static system. But, if the output of the system depends on the time-dependent input, then the system is said to be a dynamic system. A static system has memory-less property, but the dynamic system does not have the same.

Example of a static system: $y(t) = c\,x(t)$.

At $t = 0$, the output of the system is given by

$$y(0) = c * x(0)$$

It is clear that the output depends on the current input only, and hence the system represents a static, memory-less property.

Example of a dynamic system: $y(t) = c_1 x(t) + c_2\, x(t-n)$.

At $t = 0$, the output of the system is given by

$$y(0) = c_1 x(0) + c_2 x(-n)$$

In the above equation, $x(-n)$ represents the past value for the current input, and thus the system needs memory to get this current output. This type of system is called a dynamic system.

- **Deterministic and Probabilistic Systems**: If the occurrence of all events is known with complete certainty, the system is said to be a deterministic system. In a deterministic system, the output remains the same for constant input.

But, if the occurrence of an event cannot be perfectly known, the system is said to be probabilistic. In a probabilistic system, the output may not be constant for the same input, and thus the input-output relationship in this type of system is probabilistic in nature.

For example, the failure time of a mining machine is probabilistic, but the capacity of the same machine is deterministic.

1.3 SYSTEM APPROACH

In any mining system, the relationship between the input–output is controlled by multiple factors like the characteristic of the deposit (shape, size, depth, and strength), mining method, type of machinery, workforce, etc., and the physical laws governing the system. In many of the mining systems, the characteristics of the deposit and the laws governing the system are very complex. Thus modelling of those complex systems requires considering simplifying assumptions and transformation functions to determine the output corresponding to input. The system analysis process requires defining and formulation of the system by constructing a mathematical model, wherein the input-output relationships are estimated through existing operating conditions of the system.

In general, the objective of the system approach is to break down a complex system into multiple simple sub-systems for a better understanding of the different components of the complex system. Most of the mining systems are open as most of the sub-systems are linked to each other. To analyse and understand a mining system, it needs to be identified the different sub-systems by defining the boundaries.

The system approach should focus on the common objects of the system without neglecting the sub-systems. Major characteristics of a system approach are:

1. **Holism:** It tells that a change in any sub-system of a system directly or indirectly affects the entire properties of the system (Boulding, 1985; Litterer, 1973; von Bertalanffy, 1968).
2. **Specialization:** The entire system can be divided into different subsystems for easy understanding of the typical role of each sub-system in the system.
3. **Non-summation:** Every subsystem is of importance to the entire system, and hence it is of utmost importance to understand the role of individual sub-systems to get the complete perspective of the system (Boulding, 1985; Litterer, 1973).
4. **Grouping:** The process of grouping may lead to its own complexity by more specialized sub-systems. Therefore, it is desired to make a group with related disciplines or sub-disciplines.
5. **Coordination:** The grouped components in a sub-system need coordination and control in the study of systems. It is difficult to maintain a unified, holistic concept without proper coordination.
6. **Emergent properties:** This property tells that the group of interrelated components has a common property instead of properties of any individual component. This is the general view of a system.

1.4 SYSTEM ANALYSIS

System analysis is a process of collecting information, identifying problems, and decomposing a system into smaller sub-systems. It is usually done using a standard optimization technique based on the formulated mathematical equations. It should be noted that systems analysis is not simply to solve a mathematical model but requires decision-making for designing a system. The system analysis techniques can be used for solving both descriptive and prescriptive models. A descriptive model explains how a system works; whereas, a prescriptive model offers a solution for optimal operation of the system for achieving the desired objectives.

1.5 ELEMENTS OF A MINING SYSTEM

The elements of a mining system are represented in a diagram, as shown in Figure 1.3.

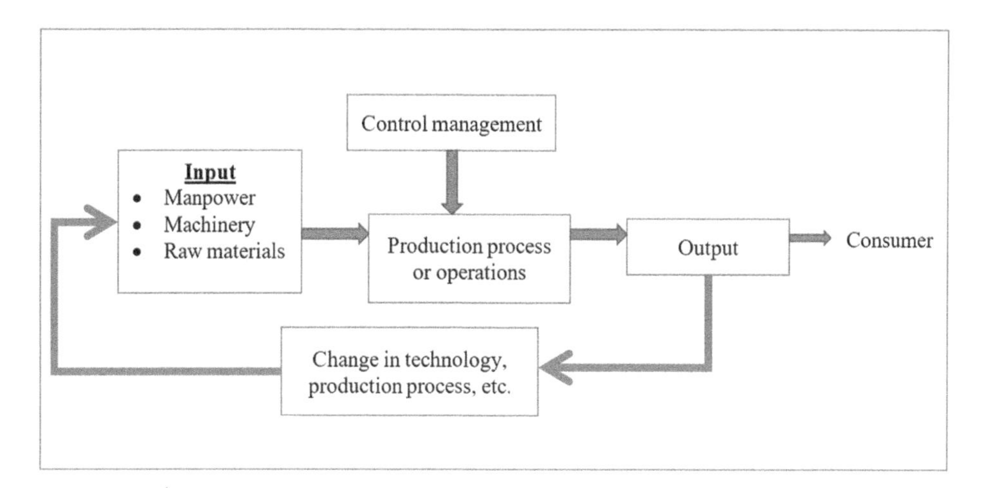

FIGURE 1.3 Elements of a mining system.

- **Inputs and outputs**
  The main objective of a mining system is to maximize (or minimize) outputs from the system for a given input. The inputs to a typical mining system are workforce, heavy earth moving machinery, explosives, supports, transportation equipment, etc. The outputs of any mining system are production, productivity, quality, accident and incident rates, etc.
- **Production operation**
  The production operation in a mining system involves the actual transformation of input (in-situ coal or ore) to output (extracted coal or mineral) for supplying to the consumer. The production process involves many operations like drilling and blasting, loading into dump trucks/haul trucks or conveyor belts, transportation, etc. Therefore, the operational process needs to modify by changing the input either totally or partially, depending on the desired productions.
- **Control management**
  The control management guides the mining system by analysing the pattern of activities governing input, operations, and output. The decision-making played a significant role in optimizing the productions operations based on the available resources and desired outputs. The production behaviour of a mining system is controlled by the operational process. To optimize the system, management should know how much input is needed to obtain the desired output.
- **Feedback**
  Feedback of the production process requires decision-making on the system's alteration to achieve the desired output. Positive feedback is routine that encourages the performance of the system; whereas, negative feedback provides the information to management for action.
- **Environment**
  The output of the system depends on the environment facilitates the mining systems. It includes any external factors like political interference and market conditions that affect the actual performance of the system.
- **Boundaries and interface**
  Each system has boundaries that determine its periphery of influence and control. Boundaries are the limits that identify its sub-system, processes, and interrelationship among them. Thus, the definition of the boundaries of any mining system is crucial in determining the nature of its interface with other systems for successful design.

1.6 DEFINITION AND CLASSIFICATION OF OPTIMIZATION PROBLEM

Optimization is a process of searching for the most cost-effective or most efficient solution of any system or sub-systems under the given constraints of any defined problem by maximizing desired factors and minimizing undesired ones. In other words, optimization problems are represented by a mathematical equation of the objective function and constraints, and decision variables are estimated based on a formal search procedure for optimizing the objective function. An optimization problem can be classified in multiple ways like the nature of constraints, characteristics of decision variables, type of equations used, and several objective functions. A brief description of the classification is given below.

1.6.1 BASED ON THE EXISTENCE OF CONSTRAINTS

- **Constrained optimization problem**: Any problem is said to be a constrained optimization problem if it has one or more constraints. The generalized form of a constrained optimization problem is shown below:

$$\begin{cases} Maximize\,(Minimize)\ Z = \sum\limits_{\substack{i=1\ to\ m \\ k=1\ to\ m}} c_i x_i^{p_k} & p_k \in Real\ number \\ Subject\ to\ \ \sum\limits_{j=1\ to\ n} \sum\limits_{i=1\ to\ m} a_{ji} x_i^{p_k} \le b_i & \\ x_i \ge 0 \qquad i = 1\ to\ m & \end{cases}$$

In the above optimization problem, Z is called the objective function that needs to be optimized. The variables $x_1, x_2, \ldots, x_m$ are called decision variables. The decision variables are subject to n number of constraints along with the non-negativity constraints. The value of p_k can be any real number.

- **Unconstrained optimization problem**: Any problem is said to be an unconstrained optimization problem if it has no constraints. The generalized form of an unconstrained optimization problem is shown below:

$$Maximize\,(Minimize)\ Z = \sum_{\substack{i=1\ to\ m \\ k=1\ to\ m}} c_i x_i^{p_k} \qquad p_k \in Real\ number$$

In the above optimization problem, Z is called the objective function that needs to be optimized. The variables x_i are called decision variables. The decision variables are not subjected to any constraints. The value of p_k can be any real number.

1.6.2 Based on the nature of the equations involved

- **Linear programming problem (LPP)**: If the objective function and all the constraints of a problem are linear, then the problem is said to be a linear programming problem (LPP). It is probably the single-most applied optimization technique in engineering decision-making. Thus, the generalized form of the LPP is

$$\begin{cases} \text{Maximize}(\text{Minimize})Z = \sum\limits_{i=1\ \text{to}\ m} c_i x_i \\ \text{Subject to}\ \ \sum\limits_{i=1\ \text{to}\ m} a_{ji} x_i \le b_j \\ \qquad x_i \ge 0 \qquad i = 1\ \text{to}\ m \text{ and } j = 1\,\text{to}\,n \end{cases}$$

If the decision variables are restricted to integer values, the same problem represents an integer-programming problem.

- **Quadratic programming problem:** In a quadratic programming problem, the objective function is a quadratic function, and all constraint functions are linear. The general form of a quadratic problem is as follows:

$$\text{Maximize}\ (\text{Minimize})Z = q^T x + \frac{1}{2} x^T Q x$$

$$\text{Subject to } Ax = a$$

$$Bx \le b$$

$$x \ge 0$$

The objective function is arranged such that the vector q contains all of the (single-differentiated) linear terms and Q contains all of the (twice-differentiated) quadratic terms. The constants contained in the objective function are left out of the general formulation.

As for the constraints, the matrix equation $Ax = a$ contains all of the linear equality constraints, and $Bx \le b$ are the linear inequality constraints.

- **Non-linear programming problem**: In non-linear programming problem, the objective function and/or one or more constraints are non-linear. An unconstrained problem can be represented as a non-linear programming problem if at least one of the values of p_k is not equal to 1.

1.6.3 Based on the permissible values of the decision variables

- **Integer programming problem:** Integer programming, which is a variant of LPP, the decision variables are integer. It is a special case of LPP in which all the decision variables are restricted to take only integer (or discrete) values. The general form of an integer programming problem is given by

$$\begin{cases} \text{Maximize}(\text{Minimize})Z = \sum\limits_{i=1 \text{ to } m} c_i x_i \\ \text{Subject to } \sum\limits_{i=1 \text{ to } m} a_{ji} x_i \le b_j \\ x_i \in \text{Integer} \qquad i = 1 \text{ to } m \text{ and } j = 1 \text{ to } n \end{cases}$$

If some of the variables are integer values and the remaining are real values, the problem is a mixed integer programming problem. The binary programming problem is a special case of an integer programming problem, where decision variables are all binary, that is, 0 or 1.

1.6.4 Based on the number of the objective function

- **Single-objective problem:** If the optimization problem has only one objective function, then the problem is said to be a single-objective problem.
- **Multi-objective problem:** If the optimization problem has more than one objective function, then the problem is said to be a multi-objective problem.

The mathematical form of a multi-objective problem can be represented as follows:

$$\text{Minimize } f_1(x), f_2(x), \ldots, f_k(x)$$

$$\text{Subject to } g_j(x) \le 0, \qquad j = 1, 2, \ldots, m$$

In the given problem, the objective functions, $f_1(x), f_2(x), \ldots, f_k(x)$ need to be minimized simultaneously. The problem has m number of constraints.

1.7 SOLVING OPTIMIZATION PROBLEMS

In the last few decades, many algorithms have been developed for solving different types of optimization problems. The selection of a suitable algorithm for solving an optimization problem depends on the nature of the problem. The major advancement in optimization happened after the development of digital computers. Currently, many options are available to solve complex optimization problems. A few techniques are explained below.

1.7.1 Classical optimization techniques

The classical optimization techniques are useful for solving constrained and unconstrained single- and multi-variable optimization problems that involve continuous and differentiable functions. The followings are the few popular classical optimization techniques.

1.7.1.1 Direct methods

Direct methods are the simple searching approach to obtain the optimum solution. These methods do not require any derivatives at any points for finding the optimal solution. The golden-section search (Press et al., 2007) can be used to solve the one-dimensional problem, whereas univariate search or random search method (Rastrigin, 1963) can be used for solving multi-dimensional problems.

1.7.1.2 Gradient methods

In the gradient method, the derivative information of the optimization function is used to locate the solution. The first derivatives of the function offer slopes of the function that become zero at the optimal point. The steepest slope or gradient of the function tells the optimal solution. For a one-dimensional problem, Newton's method (Avriel, 2003) can be used to find the optimum solution of the function.

1.7.1.3 Linear programming methods

In this method, linear mathematical functions are formulated to represent both the objective functions and constraints to derive the optimal solution. For single and two-variable linear programming problems, the graphical solution method can be used. For more than two variable problems, the simplex method (Murty, 1983) can be used to determine the optimal solution.

1.7.1.4 Interior point method

Interior point method (IPM), also referred to as barrier method, is generally used to solve linear and nonlinear convex optimization problems. The method was first proposed by Soviet mathematician Dikin in 1967 and reinvented in the USA in the mid-1980s. In this method, the violations of inequality constraints are prevented by shifting the objective function with a barrier term that causes the optimal unconstrained value to be in the feasible space.

1.7.2 Advanced optimization techniques

Most of the real mining optimization problems involve complexities like a large number of variables, non-linearity, and multiple conflicting objectives. Furthermore, it is difficult to find the global optimum solution due to the large search space. In general, if the system cannot be solved using the classical optimization solving methods, the evolutionary algorithms (EAs) can be used. EAs are applied to a large-scale optimization problem for obtaining near-optimum solutions. This type of algorithm can be easily applied to an optimization problem with many decision variables and non-linear objective functions and constraints.

Goldberg (1989) has developed the first evolutionary optimization technique, called a genetic algorithm (GA). GA algorithm was designed based on the Darwinian principle 'the survival of the fittest and the natural process of evolution through reproduction'. In the last few decades, many other evolutionary algorithms like Particle Swarm Optimization (PSO) (Eberhart and Kennedy, 1995), estimation of distribution algorithm (EDA) (Pelikan, 2005), Tabu search (Glover, 1986), Ant Colony (Colorni et al., 1991), etc., have been developed.

In this type of algorithm, the solving process starts from a population of possible random solutions and moved towards the optimal by incorporating generation and selection.

2 Basics of probability and statistics

2.1 DEFINITION OF PROBABILITY

Probability is defined as the chances of occurrence of any event in an experiment. The sum of all the possible outcomes is called the sample space, and a subset of sample space is known as an event.

If there are S exhaustive (i.e., at least one of the events must occur), mutually exclusive (i.e., only one event occurs at a time), and equally likely outcomes of a random experiment (i.e., equal chances of occurrence of each event) and r of them are favourable to the occurrence of an event A, the probability of the occurrence of the event A is given by

$$P(A) = \frac{r}{S} \tag{2.1}$$

It is sometimes expressed as 'odds in favour of A' or the 'odds against A'. The 'odds in favour of A' is defined as the ratio of occurrence of event A to the non-occurrence of event A. On the other hand, 'odds against A' is defined as the ratio of non-occurrence of event A to the occurrence of event A

$$\text{Odds in favour of A} = \frac{\text{Probability of occurrence of event } A}{\text{Probability of non-occurrence of event } A} = \frac{r/S}{(S-r)/S} = \frac{r}{S-r}$$

Again,

$$\text{Odds against A} = \frac{\text{Probability of non-occurrence of event } A}{\text{Probability of occurrence of event } A} = \frac{(S-r)/S}{r/S} = \frac{S-r}{r}$$

The total probability of the occurrence of any event ranges from 0 to 1.

$$\text{i.e.,} 0 \le P(A) \le 1$$

$P(A) = 0$ indicates event A is impossible, and $P(A) = 1$ indicates the event is certain.

In the above discussion, the discrete sample space was considered. But, the probability can also be determined for a continuous sample space. For a continuous sample space, the probability of occurrence is measured as a probability density function. The **probability density function** of any continuous random variable gives the relative likelihood of any outcome in a specific range. Therefore, for a continuous random variable, the probability of an outcome of any single or discrete outcome is zero.

Example 2.1: Two detonators are picked at random from a detonator box that has 12 detonators, of which four are defective. Determine the probability that both the detonators have chosen are defective.

DOI: 10.1201/9781003200703-2

Solution: Let A be the event of picking two defective detonators.

The probability of occurrence of event A is given by

$$P(A) = \frac{\text{Number of ways of selection 2 defective detonators out of 4}}{\text{Number of ways of selection 2 detonators out of 12.}}$$

$$\Rightarrow P(A) = \frac{4C_2}{12C_2} = \frac{6}{66} = \frac{1}{11} = 0.09$$

where $4C_2$ represents the number of ways two items can be picked from four items at a time, and $12C_2$ represents the number of ways two items can be picked from 12 items at a time.

The probability of occurrence of event A is 0.09. Therefore, the probability of the event that both the picked detonators are defective is 0.09.

Example 2.2: From open-pit coal mine, 500 workers were randomly chosen. It was found that 30 workers experienced an injury in the year 2020. The distribution of injury, based on the younger age group $(age \leq 35\, years)$ and older age group $(age > 35\, years)$, generates the following cross-classification table.

	Number of workers		
Age group	**Injured**	**Non-Injured**	**Row total**
Younger age group	10	120	130
Older age group	20	350	370
Column total	30	470	500

Determine the odds of injury for the younger age group compared to the older age group.

Solution

We have,

Number of workers injured in younger age group = N_{YI} = 10
Number of workers non-injured in younger age group = N_{YNI} = 120
Number of workers injured in older age group = N_{OI} = 20
Number of workers non-injured in older age group = N_{ONI} = 350

$$\text{Odds of injury for the younger group} = P(Y) = \frac{N_{YI}}{N_{YNI}} = \frac{10}{120}$$

$$\text{Odds of injury for the older group} = P(O) = \frac{N_{OI}}{N_{ONI}} = \frac{20}{350}$$

$$\therefore \text{Odds of injury for younger compared to older age group} = \frac{P(Y)}{P(O)}$$

$$= \frac{10}{120} * \frac{350}{20} = 1.45$$

Therefore, the odds of the injury for the younger age group as compared to the older age group is 1.45.

2.2 ADDITIONAL THEORY OF PROBABILITY

If two events A and B are mutually exclusive, the probability of occurrence of either A or B is given by

$$P(A \cup B) = P(A) + P(B) \tag{2.2}$$

where $P(A)$ is the probability of even A and $P(B)$ is the probability of event B.

If events A and B are not mutually exclusive, then

$$P(A \cup B) = P(A) + P(B) - P(A \cap B) \tag{2.3}$$

$P(A \cap B)$ represents the probability of occurrence of both events simultaneously.

For n number of mutually exclusive events $(A_1, A_2, \ldots, A_n)$, the probability of occurrence of either of the $A_1, A_2, \ldots, A_n$ events can be presented as:

$$P(A_1 \cup A_2 \cup \cdots \cup A_n) = P(A_1) + P(A_2) + \cdots = P(A_n) \tag{2.4}$$

In the case of three non-mutually exclusive events:

$$P(A \cup B \cup C) = P(A) + P(B) + P(C) - P(A \cap B) - P(B \cap C) - P(A \cap C) + P(A \cap B \cap C) \tag{2.5}$$

where

$P(A)$ = probability of occurring the event A
$P(B)$ = probability of occurring the event B
$P(C)$ = probability of occurring the event C
$P(A \cap B)$ = probability of occurring both the events A and B simultaneously
$P(B \cap C)$ = probability of occurring both the events C and B simultaneously
$P(A \cap C)$ = probability of occurring both the events A and C simultaneously
$P(A \cap B \cap C)$ = probability of occurring all the three events A, B, and C simultaneously

Example 2.3: The probability of failure of a dump truck A is 0.7, and that of dump truck B is 0.2. If the probability of failure of both the dump trucks is 0.3 simultaneously, determine the probability that neither of the dump trucks fails.

Solution

We have,

$$\text{Probability of failure of a dumper } A = P(A) = 0.7$$

$$\text{Probability of failure of a dumper } B = P(B) = 0.2$$

$$\text{Probability of failure of both the dump trucks A and B} = P(A \cap B) = 0.3$$

$$\text{Probability of failure of either of the dump trucks A or B} = P(A \cap B) = P(A) + P(B) - P(A \cap B)$$

$$\Rightarrow P(A \cap B) = 0.7 + 0.2 - 0.3 = 0.6$$

$$\text{The probability that neither of the dump trucks fails} = P(\bar{A} \cap \bar{B}) = 1 - P(A \cap B) = 1 - 0.6 = 0.4$$

Therefore, the probability that neither the dump truck fails is 0.4.

2.3 PROBABILITY DISTRIBUTIONS

The probability distribution for a discrete random variable is referred to as the probability mass function (PMF); whereas, the same is referred to as probability density function (PDF) for the continuous random variable. A PMF, $P(X)$ or PDF, $f(x)$ must be non-negative, and the sum of the probability of the entire sample space must be equal to 1. The probability distribution can be represented by a discrete or continuous function, as explained in subsequent subsections.

The probability distribution for a discrete random variable is represented by spikes of probability values correspond to the random variable.

An important probability measure is the cumulative distribution function (CDF). The CDF of a discrete random variable, $F_X(k)$ represents the probability that the random variable X is less than or equal to k. It can be represented mathematically, as

$$P(X \leq k) = F_X(k) = \sum_{x=0}^{k} P(x) \tag{2.6}$$

For example, a mine worker takes leave randomly on any one day of the week. The probabilities of taking the leave each day are given in Table 2.1. The probability of taking the leave on Sunday and Saturday are 0.3 and 0.2, respectively. For the rest of the days, the probability of taking leave is 0.1. The spikes of PMF and CDF for the given data (Table 2.1) are shown in Figures 2.1(a) and 2.1(b), respectively. The CDF values can be directly determined from the PMF values by cumulating the values. The CDF and PMF values for the first day of the week (Monday) are the same. In the subsequent days, the CDF values are calculated by taking the sum of the current day PMF value and the previous day's CDF value. The CDF value for the last day is 1 [Figure 2.1(b)], which is equal to the sum of the PMF for each day.

On the other hand, the probability distribution of a continuous random variable is represented by smooth curves, as shown in Figure 2.2(a). The CDF of a continuous random variable is a non-decreasing function with a maximum value of 1, as shown in Figure 2.2(b). Mathematically, it can be represented as

$$P(X \leq k) = F_X(k) = \int_{0}^{k} f(x)dx \tag{2.7}$$

where $f(x)$ is a probability density function of a continuous random variable, X. $F_X(k)$ represents the cumulative distribution function for the random variable $X \leq k$, which is measured by the ordinate of the probability curve.

The shaded area of Figure 2.2(a) represents the probability that $X \leq k$. It is also represented by CDF value, $F_X(k)$ [Figure 2(b)]. The total probability for any PDF is equal to 1, as shown in Eq. (2.1), and can be presented as:

TABLE 2.1
Probability Mass Function

	Monday	Tuesday	Wednesday	Thursday	Friday	Saturday	Sunday
Day (X)	1	2	3	4	5	6	7
$P(X=k)$	1/10	1/10	1/10	1/10	1/10	2/10	3/10
$F_X(k)$	1/10	2/10	3/10	4/10	5/10	7/10	1

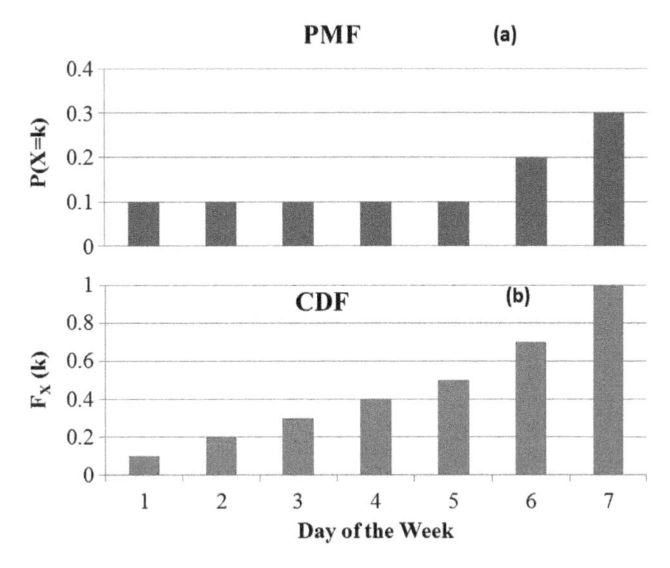

FIGURE 2.1 Probability distribution of a discrete random variable: (a) probability mass function (PMF) and (b) cumulative distribution function (CDF).

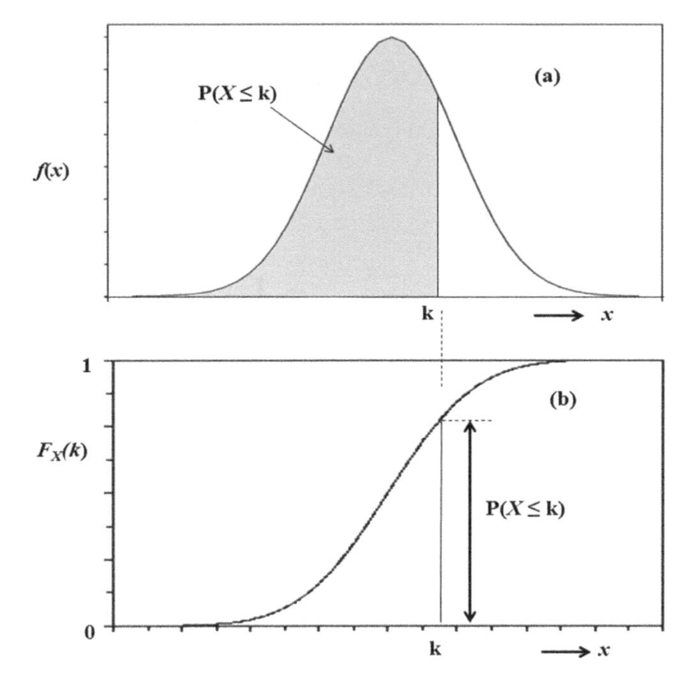

FIGURE 2.2 Probability density function of a continuous random variable: (a) shaded area represents $P(X \le k)$ and (b) cumulative distribution function (CDF), showing the probability vs. k.

$$\int_{-\infty}^{+\infty} f(x)dx = 1 \tag{2.8}$$

The probability of a continuous random variable taking a value exactly equal to a given value is zero. This can be proved from the following derivation:

$$P(X = m) = P(m < X \leq m) = \int_{m}^{m} f(x)dx = 0$$

2.4 COMMON PROBABILITY DISTRIBUTION FUNCTIONS

2.4.1 Uniform distribution

For discrete uniform distribution, a finite number of values are equally likely to be observed, as shown in Figure 2.3(a). The PMF for a discrete random variable in the interval [a, b] is given by

$$f(x) = \frac{1}{n}, \qquad \text{for } a \leq x \leq b, \ n = b - a + 1$$

The cumulative distribution function for discrete uniform distribution in the interval [a,b] can be determined as

$$F(x) = \frac{i}{n}, \qquad i = 1, 2, \ldots, n$$

The CDF for a discrete uniform distribution is shown in Figure 2.3(b).

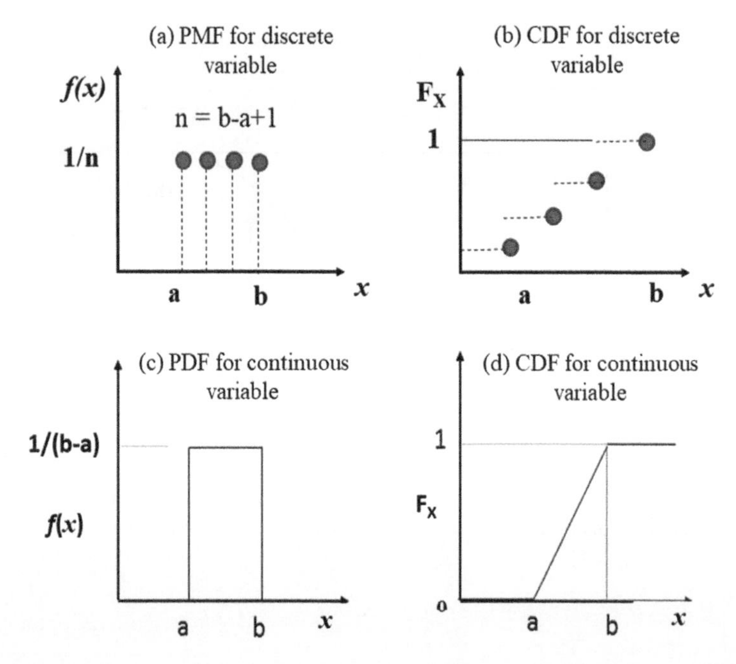

FIGURE 2.3 Characteristics of uniform distribution: (a) PMF of a discrete random variable, (b) CDF of a discrete random variable, (c) PDF of a continuous random variable, and (d) CDF of a continuous random variable.

On the other hand, the probability density function for continuous uniform distribution in the interval $[a, b]$ is given by

$$f(x)=\begin{cases}0, & x<a \\ \dfrac{1}{b-a}, & a\le x\le b \\ 0, & x>b\end{cases} \tag{2.9}$$

The cumulative distribution function for a continuous uniform distribution on the interval [a,b] can be determined as

$$P(X\ge a)=F(x)=\int_a^x f(x)\,dx$$

$$F_X(x)=\begin{cases}0, & x<a \\ \dfrac{x-a}{b-a}, & a\le x\le b \\ 1, & x>b\end{cases} \tag{2.10}$$

A uniform distribution, also called a rectangular distribution, is a distribution that has a constant probability, as shown in Figure 2.3(c). The nature of the CDF of a uniform distribution function is shown in Figure 2.3(d). The CDF of uniform distribution is a straight line that intercepts the x-axis at a value a, and it has a slope of $\dfrac{1}{b-a}$.

Example 2.4: The daily explosives demand in an opencast mine is uniformly distributed between 2500 and 3250 kg. The explosive tank, which has a storage capacity of 3000 kg, is refilled daily after the end of the last shift. What is the probability that the tank will be empty before the end of the last shift?

Solution

For uniform distribution of demand in the range of a (=2550) to b (=3250), the probability density function can be represented as

$$f(x)=\frac{1}{b-a}=\frac{1}{3250-2550}=\frac{1}{700}$$

$$P(k<x\le b)=\int_k^b f(x)\,dx=\int_{3000}^{3250}\frac{1}{700}dx=\frac{1}{700}\int_{3000}^{3250}dx=\frac{1}{700}[x]_{3000}^{3250}=0.355$$

Therefore, the probability that the explosive tank will be empty before the last shift is 0.355.

Example 2.5: The daily coal production from a mine follows a continuous uniform distribution with a range [3000, 3500] tons. Find the probability that the production in a randomly selected day has greater than 3200 tons.

Solution

The probability that the production in a randomly selected day is greater than 3200 tonnes is given by

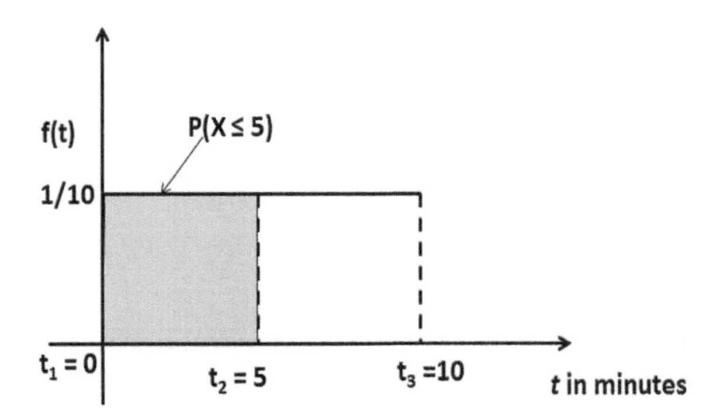

FIGURE 2.4 PDF of the given uniform distribution function.

$$P(3200 < x \leq 3500) = \int_{3200}^{3500} f(x)dx = \int_{3200}^{3500} \frac{1}{3500-3000}dx = \frac{1}{500}\int_{3200}^{3500} dx = \frac{1}{500}[x]_{3200}^{3500} = \frac{300}{500} = 0.6$$

Therefore, the probability of coal production greater than 3200 tonnes is 0.6.

Example 2.6: A mine worker arrives at pit-bottom at a random time (i.e., no prior knowledge of the scheduled start time) to ride on the cage on the next trip. Cage starts at pit-bottom every 10 minutes without fail, and hence the next trip will start any time during the next 10 minutes with evenly distributed probability (a uniform distribution). Find the probability that the cage will start within the next 5 minutes after the arrival of the worker at the pit bottom.

Solution

The probability density function (Figure 2.4) represents a horizontal line above the interval from 0 to 10 minutes for a uniform distribution. As the total probability is one, the total area under the curve must be one, and hence the height of the horizontal line is 1/10 with the bin size of 1 minute.

The probability that the cage will start within the next 5 minutes after the arrival of the worker at the pit-bottom is $P(0 \leq X \leq 5)$. This represents the shaded region in Figure 2.4. Its area is, (5)*(1/10) =1/2.

Thus, the probability that the cage will start within the next 5 minutes after the arrival of the worker at the pit bottom is

$$P(0 \leq X \leq 5) = \frac{1}{2}$$

2.4.2 Normal distribution

Although many distribution functions are developed and applied for different purposes, the **Gaussian distribution**, also known as the **normal distribution**, is the most widely used distribution function across all disciplines. The probability density function of the normal distribution is given by

$$f(x) = \frac{1}{\sqrt{2\pi\sigma}} e^{-\frac{(x-\mu)^2}{2\sigma^2}} \tag{2.11}$$

where

μ = mean of the distribution
σ^2 = variance of the distribution
$x \in (-\infty, \infty)$

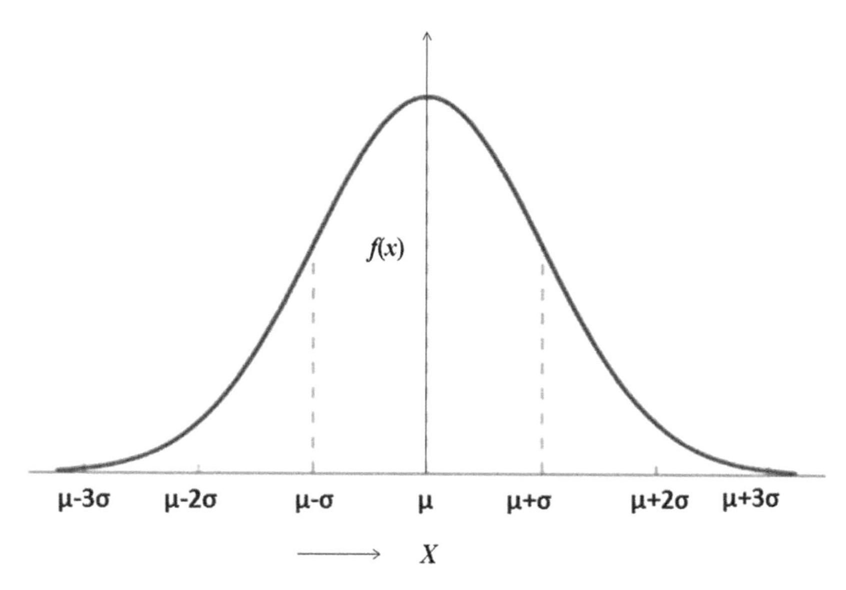

FIGURE 2.5 The probability density function of a normal distribution.

The nature of the PDF for a normal distribution function is shown in Figure 2.5. The PDF of the normal distribution is a bell-shaped curve and is symmetric around mean μ. Therefore, the probability of both the right-hand and left-hand sides of mean μ is equal, and that is 0.5.

A normal distribution with arbitrary data range can be converted into a standardized normal density by putting $\frac{(x-\mu)}{\sigma} = z$, which referred to ***z*-score** or **standard score.** Taking derivative of $\frac{(x-\mu)}{\sigma} = z$ with respect to z, we get $dz = dx / \sigma$. Thus, Eq. 2.11 can be rewritten as

$$f(z) = \frac{1}{\sqrt{2\pi}} e^{-\frac{Z^2}{2}}$$

In the normalized PDF, $\mu = 0$ and $\sigma = 1$, as shown in Figure 2.6. The CDF of a standardized normal density function is given by:

$$F(z) = \int_0^Z f(z)dz = \frac{1}{\sqrt{2\pi}} \int_0^Z e^{-\frac{Z^2}{2}} dz$$

The peak of the curve (at the mean) in a normalized PDF of the normal distribution function is approximately 0.399.

Furthermore, the distribution can easily be normalized to adapt to the particular mean and standard deviation of interest. It can be demonstrated using the probability, $P(X \le b)$, written as

$$P(X \le b) = \int_{-\infty}^{b} f(x)dx = \int_{-\infty}^{b} \frac{1}{\sigma\sqrt{2\pi}} e^{-\frac{(x-\mu)^2}{2\sigma^2}} dx$$

The above expression determines the area under the curve from the extreme left ($-\infty$) to $x = b$. This can be represented as the shaded region, as shown in Figure 2.7.

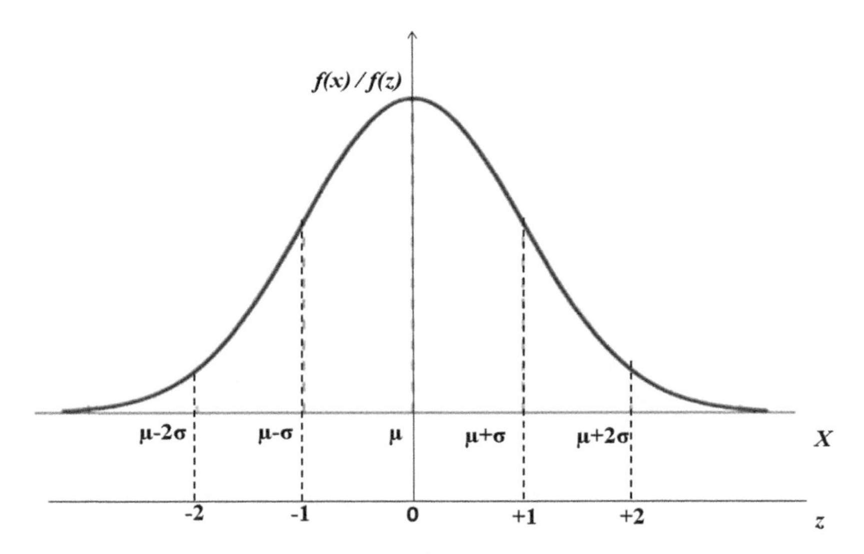

FIGURE 2.6 The probability density function of the standardized normal distribution curve.

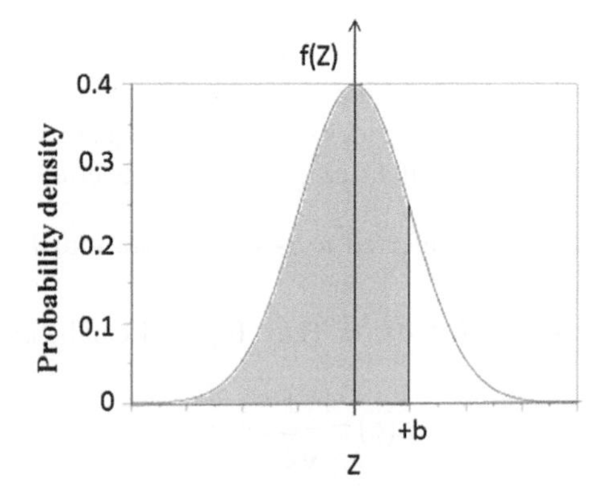

FIGURE 2.7 Shaded area represents the probability of $X \leq b$.

Thus, by replacing $\frac{(x-\mu)}{\sigma} = z$, and performing derivative with respect to z, the probability expression can be written as:

$$P(X \leq \mathrm{b}) = \int_{-\infty}^{b} f(x)dx = \int_{-\infty}^{b} \frac{1}{\sigma\sqrt{2\pi}} e^{-\frac{(x-\mu)^2}{2\sigma^2}} dx = \int_{-\infty}^{b} \frac{1}{\sqrt{2\pi}} e^{-\frac{Z^2}{2}} dz$$

We can transform any data that follows a normal distribution to a standardized normal distribution. This transformation will help to determine the probabilities for *any* normally distributed data using the standardized normal distribution (Table 2.2) without any application of integral calculus. The beauty of normal distribution is that we can easily calculate the probability within a specific range by knowing the standard deviation value. For example, if the mean is μ and the standard deviation is σ, then about 68% of the values lie within $\mu \pm \sigma$, and 95% of values lie within $\mu \pm 2\sigma$ in a normal distribution, as shown in Figure 2.8.

TABLE 2.2
Standardized Normal Distribution Table

z	0.00	0.01	0.02	0.03	0.04	0.05	0.06	0.07	0.08	0.09
0.0	0.5000	0.5010	0.5080	0.5120	0.5160	0.5199	0.5239	0.5279	0.5319	0.5359
0.1	0.5398	0.5438	0.5478	0.5517	0.5557	0.5596	0.5636	0.5675	0.5714	0.5753
0.2	0.5793	0.5832	0.5871	0.5910	0.5918	0.5987	0.6026	0.6064	0.6103	0.6141
0.3	0.6179	0.6217	0.6255	0.6293	0.6331	0.6368	0.6106	0.6143	0.6180	0.6517
0.4	0.6554	0.6591	0.6628	0.6664	0.6700	0.6736	0.6772	0.6808	0.6844	0.6879
0.5	0.6915	0.6950	0.6985	0.7019	0.7054	0.7088	0.7123	0.7157	0.7190	0.7224
0.6	0.7257	0.7291	0.7324	0.7357	0.7389	0.7422	0.7454	0.7186	0.7517	0.7549
0.7	0.7580	0.7611	0.7642	0.7673	0.7704	0.7734	0.7764	0.7794	0.7823	0.7852
0.8	0.7881	0.7910	0.7939	0.7967	0.7995	0.8023	0.8051	0.8078	0.8106	0.8133
0.9	0.8159	0.8186	0.8212	0.8238	0.8264	0.8289	0.8315	0.8310	0.8365	0.8389
1.0	0.8413	0.8438	0.8161	0.8185	0.8508	0.8531	0.8554	0.8577	0.8599	0.8621
1.1	0.8643	0.8665	0.8686	0.8708	0.8729	0.8749	0.8770	0.8790	0.8810	0.8830
1.2	0.8849	0.8869	0.8888	0.8907	0.8925	0.8944	0.8962	0.8980	0.8997	0.9015
1.3	0.9032	0.9049	0.9066	0.9082	0.9099	0.9115	0.9131	0.9147	0.9162	0.9177
1.4	0.9192	0.9207	0.9222	0.9236	0.9251	0.9265	0.9279	0.9292	0.9306	0.9319
1.5	0.9332	0.9345	0.9357	0.9370	0.9382	0.9394	0.9106	0.9418	0.9429	0.9441
1.6	0.9452	0.9163	0.9474	0.9184	0.9495	0.9505	0.9515	0.9525	0.9535	0.9545
1.7	0.9554	0.9564	0.9573	0.9582	0.9591	0.9599	0.9608	0.9616	0.9625	0.9633
1.8	0.9641	0.9649	0.9656	0.9664	0.9671	0.9678	0.9686	0.9693	0.9699	0.9706
1.9	0.9713	0.9719	0.9726	0.9732	0.9738	0.9714	0.9750	0.9756	0.9761	0.9767
2.0	0.9772	0.9778	0.9783	0.9788	0.9793	0.9798	0.9803	0.9808	0.9812	0.9817
2.1	0.9821	0.9826	0.9830	0.9834	0.9838	0.9842	0.9816	0.9850	0.9854	0.9857
2.2	0.9861	0.9864	0.9868	0.9871	0.9875	0.9878	0.9881	0.9884	0.9887	0.9890
2.3	0.9893	0.9896	0.9898	0.9901	0.9904	0.9906	0.9909	0.9911	0.9913	0.9916
2.4	0.9918	0.9920	0.9922	0.9924	0.9927	0.9929	0.9931	0.9932	0.9934	0.9936
2.5	0.9938	0.9910	0.9941	0.9943	0.9945	0.9916	0.9918	0.9949	0.9951	0.9952
2.6	0.9953	0.9955	0.9956	0.9957	0.9958	0.9960	0.9961	0.9962	0.9963	0.9964
2.7	0.9965	0.9966	0.9967	0.9968	0.9969	0.9970	0.9971	0.9972	0.9973	0.9974
2.8	0.9974	0.9975	0.9976	0.9977	0.9977	0.9978	0.9979	0.9979	0.9980	0.9981
2.9	0.9981	0.9982	0.9982	0.9983	0.9984	0.9984	0.9985	0.9985	0.9986	0.9986

The probability value of a normally distributed function can be calculated from the standardized normal distribution. The label for rows contains the integer part and the first decimal place of z. The label for columns contains the second decimal place of z. The values within the table are the probabilities corresponding to the different z-value. These probabilities are calculations of the area under the normal curve from the starting point (negative infinity) to a specified value with the maximum up to positive infinity.

For example, to find the value for $z \leq 0.72$, the value corresponding to the row with 0.7 and the column with 0.02 gives a probability of 0.7611 for a cumulative from $-\infty$. Thus, four different cases can be observed when the z-score value is used for probability calculation.

Case 1: $P(z \leq a)$

The probability of z is less than any specific positive value 'a' can be determined using the standard normal distribution table data, as given in Table 2.2. This is generally represented as Φ.

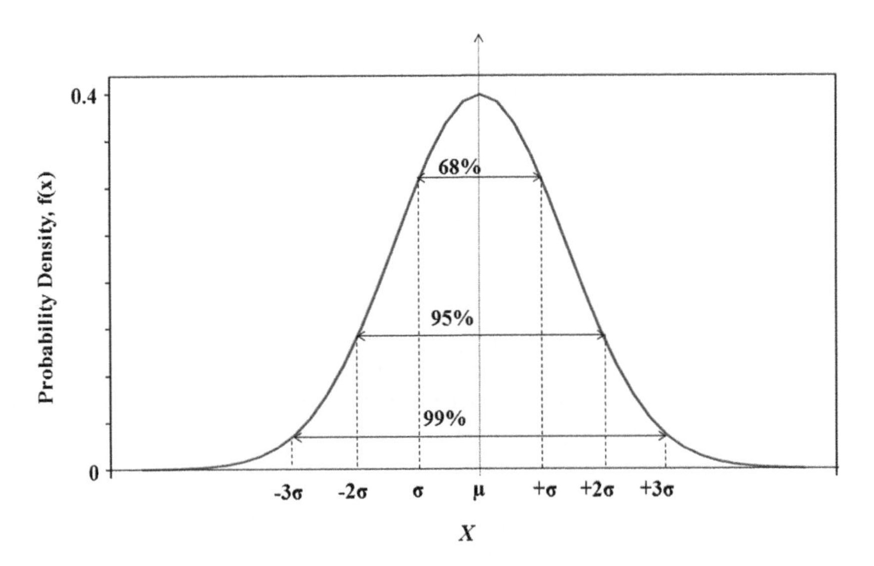

FIGURE 2.8 Probability density function of a normal distribution with σ, 2σ, and 3σ limits.

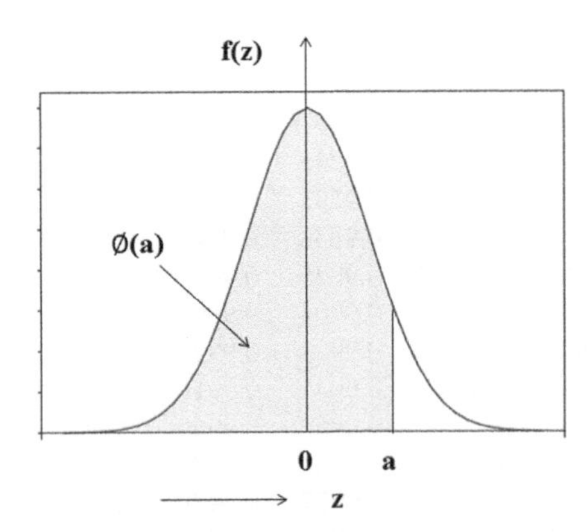

FIGURE 2.9 Shaded area represents $P(z \leq \text{a})$.

Therefore, for any positive value, a, the probability of z less than acan be determined as

$$P(z \leq a) = \phi(\text{a}) \qquad \text{for } a > 0$$

Graphically, this can be represented as shown in Figure 2.9. The shaded region (Figure 2.9) represents the probability of z less and equal to a.

Case 2: $P(z \leq -a)$

Since the standard normal distribution table only provides the probability for values less than a positive z-value (i.e., z-values on the right-hand side of the mean), the probability for z less than a negative value can be determined using an indirect method. The standard normal distribution has a total area (probability) equal to 1, and it is also symmetrical around the mean. Thus, the probability for $z \leq -a$ [Figure 2.10(a)] will be the same as the probability of $z > a$ [Figure 2.10(b)]. The probability of $z \leq a$ is shown in Figure 2.10(c).

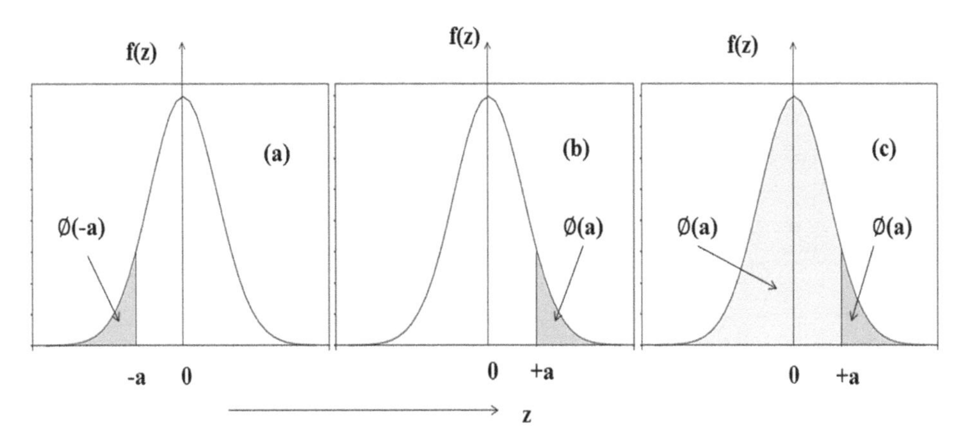

FIGURE 2.10 Graphical representation of the probability: (a) shaded area representing $P(z \le -a)$, (b) the shaded area representing $P(z \ge a)$, and (c) the shaded area representing $P(z \le a)$.

We have,

$$P(z \le -a) = \phi(-a) \quad \text{for } a > 0$$

Also, we know

$$P(z > a) + P(z \le a) = 1$$

$$\Rightarrow P(z > a) = 1 - P(z \le a) = 1 - \phi(a)$$

$$\Rightarrow P(z \le -a) = 1 - P(z \le a) = 1 - \phi(a)$$

$$\textbf{Note: } P(z \le -a) = P(z > a)$$

Therefore, $\phi(-a) = 1 - \phi(a)$.

Case 3: $P(a < z \le b)$

The shaded area in Figure 2.11 represents the probability of $\mathbf{a} < z \le \mathbf{b}$.

Probability for $a < z \le b$ can be given by

$$P(a < z \le b) = P(z \le b) - P(z \le a) = \phi(b) - \phi(a)$$

Case 4: $-P(-a < z \le b)$

The shaded area in Figure 2.12 represents the probability of $-\mathbf{a} < z \le \mathbf{b}$.

Probability for $-a < z \le b$ can be given by

$$\begin{aligned} P(-a < z \le b) &= P(z \le b) - P(z \le -a) = P(z \le b) - P(z > a) \\ &= P(z \le b) - \{1 - P(z \le a)\} = \phi(b) - \{1 - \phi(a)\} \end{aligned}$$

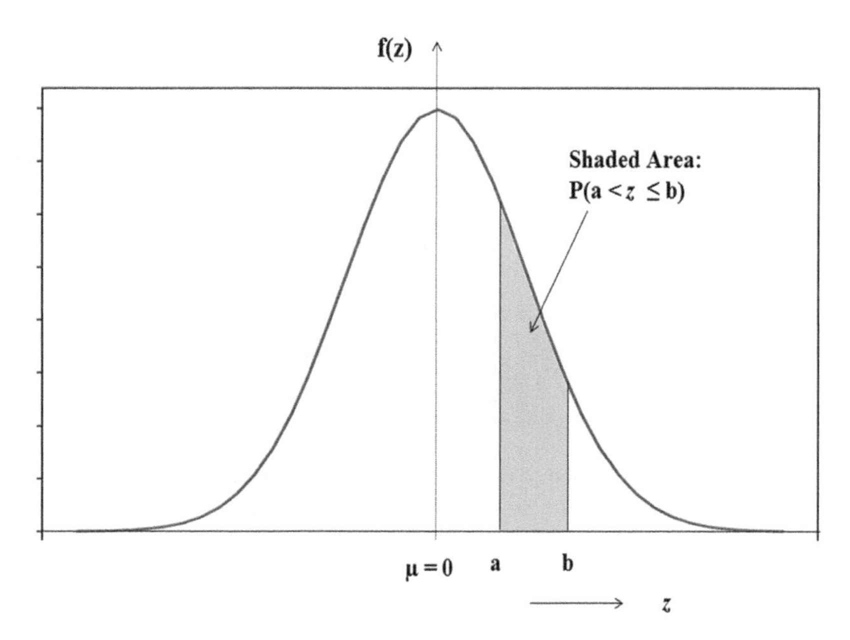

FIGURE 2.11 Shaded area represents $P(a < z \leq b)$.

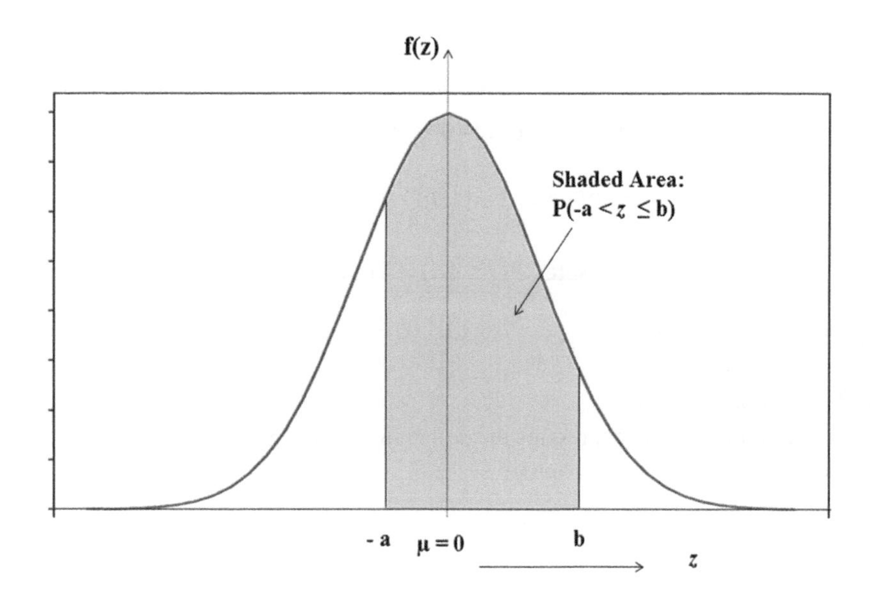

FIGURE 2.12 Shaded area represents $P(-a < z \leq b)$.

Example 2.7: A mine management is investigating the time the workers take to complete a specific task, using an individual learning approach. The mine management determined that the time to complete a task follows a normal distribution with a mean μ of 60 minutes and a standard deviation σ of 6 minutes. Determine the probability that a randomly selected mine worker will perform the task in

a) Less than 71 minutes.
b) Greater than 66 minutes.
c) Less than 54 minutes.
d) Greater than 66 minutes and less than 71 minutes.

Solution

(a) The probability to complete the task in less than 71 minutes can be calculated as:

$$P(X<t)=P\left(\frac{X-\mu}{\sigma}<\frac{t-\mu}{\sigma}\right)=P\left(z<\frac{t-\mu}{\sigma}\right)=P\left(z<\frac{71-60}{6}\right)=P(z<1.83)$$

To calculate the probability, we need to find the z-score value corresponding to 1.83 from Table 2.2. First, find '1.8' on the left side (z-column) and move across the table to '0.03' on the top or bottom, and record the value in the cell, which is **0.96442**.

$$P(z<1.83)=0.9644$$

It reveals that the randomly selected mine worker finished the work in less than 71 minutes is **96.44%**.

(b) The probability to complete the task in more than 66 minutes can be calculated as:

$$P(X>t)=P\left(\frac{X-\mu}{\sigma}>\frac{t-\mu}{\sigma}\right)=P\left(z>\frac{t-\mu}{\sigma}\right)=P\left(z>\frac{66-60}{6}\right)=P(z>1)=1-P(z\leq 1)$$

$$=1-0.8413=0.1587$$

Find '1.0' on the left side (z-column) and move across the table to '0.00' on the top or bottom, and record the value in the cell: **0.8413.** Thus the probability that the randomly selected mine worker finishes the work in more than 66 minutes is **15.87%**.

(c) The probability to complete the task within 54 minutes can be calculated as:

$$P\left(z<\frac{t-\mu}{\sigma}\right)=P\left(z<\frac{54-60}{6}\right)=P(z<-1)=P(z>1)$$

$$=0.1587 \text{ [Refer sub-problem (b)]}$$

Thus the probability that the randomly selected mine worker finishes the work in less than 54 minutes is **15.87%**.

(d) The probability to complete the task in greater than 66 minutes and less than 71 minutes can be calculated as:

$$P(t_1<X<t_2)=P\left(\frac{t_1-\mu}{\sigma}<z<\frac{t_2-\mu}{\sigma}\right)=P\left(\frac{66-60}{6}<z<\frac{71-60}{6}\right)$$

$$\Rightarrow P(1<z<1.833)=P(z<1.833)-P(z<1)=0.9644-0.8413=0.1231$$

Therefore, the probability that the randomly selected mine worker finishes the work in between 66 and 71 minutes is **12.31%**.

Example 2.8: The mean grade of the iron ore sample in a particular mine is 65%, with a standard deviation of 8%. Determine the probability that the mean grade of 25 random iron ore samples is 68%, assuming the grade is normally distributed.

Solution
Given data:

Mean grade of the random sample ($\bar{x}$) = 68%
Number of random sample (n) = 25
Mean grade of the iron samples (μ) = 65%
Standard deviation of the iron samples (σ) = 8%

The z-score can be determined as

$$z = \frac{\bar{x} - \mu}{\sigma / \sqrt{n}} = \frac{68 - 65}{8 / \sqrt{25}} = 1.87$$

The z-score value corresponds to 1.87 is 0.9697. This reveals that the probability of randomly selected iron ore samples, having a mean grade value of 68% is 0.9697. This also indicates that the probability that the random samples are taken from the same iron ore mine is 0.9697.

2.4.3 Poisson distribution

Poisson distribution, a discrete probability distribution, tells the number of occurrences of a specific event within a particular time interval. In Poisson distribution, the events occur with a known constant mean rate, and the time interval between two consecutive events is independent. Poisson distribution is a count distribution and is thus represented by a probability mass function.

The probability mass function of Poisson distribution is written as:

$$f(X = x) = \frac{\lambda^x e^{-\lambda}}{x!} \qquad x = 0, 1, 2, \ldots \tag{2.12}$$

where
λ = mean of the number of occurrences within a specific time interval (or mean of the distribution).
x = number of occurrences of a given event.
$x!$ = factorial of x.

The expected number of occurrences $E(X)$ per unit time is equal to the mean number of arrival (λ) per unit time. It can be derived as follows:

$$E(X) = \sum_{x=0}^{\infty} xf(x) = \sum_{x=0}^{\infty} x \frac{\lambda^x e^{-\lambda}}{x!} = \lambda e^{-\lambda} \sum_{x=1}^{\infty} \frac{\lambda^{x-1}}{(x-1)!} = \lambda e^{-\lambda} * e^{\lambda} = \lambda$$

Example 2.9: If the failure occurs of a shovel according to a Poisson distribution with an average of four failures in every 20 weeks. Determine the probability that there will not be more than one failure during a particular week.

Solution
The average number of failures per week is given by

$$\lambda = \frac{4}{20} = 0.2$$

The probability mass function for a Poisson distribution is given by

$$P(X = x) = \frac{\lambda^x e^{-\lambda}}{x!}$$

The probability that there will not be more than one failure during a particular week is given by

$$P(X \le 1) = P(X=0) + P(X=1) = \frac{\lambda^1 e^{-\lambda}}{!1} + \frac{\lambda^0 e^{-\lambda}}{!0} = \frac{0.2^1 e^{-0.2}}{!1} + \frac{0.2^0 e^{-0.2}}{!0}$$

$$\Rightarrow P(X \le 1) = \frac{0.2 * 0.82}{1} + \frac{1 * 0.82}{1} = 0.983$$

Therefore, the probability that there will not be more than one shovel failure during that week is 98.3%.

Example 2.10: Trucks arrive on the production bench at an average rate of 30 per hour. It was observed that the arrival pattern follows the Poisson distribution.

a. Find the probability that none of the trucks arrives in a given minute.
b. Determine the expected number of trucks that arrive in two minutes.

(a) The average number of truck arrives per minute is

$$\lambda = \frac{30}{60} = 0.5 \text{ per minute}$$

The probability that zero trucks arrive in a given minute is given by

$$P(X=0) = \frac{\lambda^0 e^{-\lambda}}{0!} = \frac{0.5^0 e^{-2}}{0!} = 0.367$$

(b) Expected number of trucks arrive in 4 minutes = $E(X) = \lambda * t = 0.5 * 4 = 2$

2.4.4 Exponential distribution

Exponential distribution is a probability distribution function for the continuous random variable within the range of zero to positive infinity.

The probability density function of the exponential distribution is represented in Eq. (2.13).

$$f(x) = \lambda e^{-\lambda x} = \frac{1}{\mu} e^{-\frac{x}{\mu}} \qquad x \ge 0 \tag{2.13}$$

where λ is the rate parameter, and μ represents the mean of the interval between two successive events.

The characteristics of the exponential function with different λ (0.5, 1.0, 1.5) are represented in Figure 2.13.

The cumulative distribution of the exponential distribution is

$$F_X(x) = P(X \le x) = \int_0^x f(x)\,dx = \int_0^x \lambda e^{-\lambda x}\,dx = \lambda \left[\frac{e^{-\lambda x}}{-\lambda}\right]_0^x = 1 - e^{-\lambda x}$$

Thus the generalized form of the cumulative density function can be written as:

$$F_X(x) = P(X \le x) = \begin{cases} 1 - e^{-\lambda x} & x \ge 0 \\ 0 & \text{otherwise} \end{cases} \tag{2.14}$$

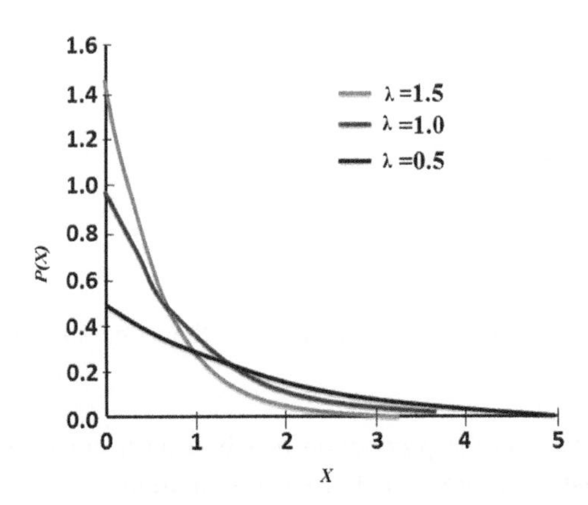

FIGURE 2.13 Probability density function of an exponential distribution function for different occurrence rate.

The expected value, $E(X)$ or the mean value (μ) of the exponential distribution can be derived as follows:

$$E(X) = \mu = \int_0^\infty xf(x) = \int_0^\infty x\lambda e^{-\lambda x} = \left[-xe^{-\lambda x} - \frac{e^{-\lambda x}}{-\lambda} \right]_0^\infty = \frac{1}{\lambda} \tag{2.15}$$

The variance of an exponential distribution is $1/\lambda^2$

Example 2.11: The average life of a shovel deployed in a mine is ten years. The duration of time the shovel effectively working in the mine is exponentially distributed.

a. Determine the probability that the shovel successfully operates for more than seven years.
b. What is the probability that the shovel lasts between 9 and 11 years?

Solution

Let X = the amount of time (in years) the shovel successfully operates.

Mean life of the shovel (μ) = 10 years

a. The probability that the shovel effectively worked in the mine for more than seven years is given by

$$P(X > 7) = 1 - P(X \leq 7) = 1 - \left(1 - e^{-\frac{x}{\mu}}\right) = e^{-\frac{7}{10}} = e^{-0.7} = 0.496$$

Thus the probability that the shovel effectively works for more than 7 years is 0.496. This can also be represented by the area of the shaded region, shown in Figure 2.14.

(b) $P(9 < X \leq 11) = P(X \leq 11) - P(X \leq 9)$

$$= \left(1 - e^{-11/10}\right) - \left(1 - e^{-9/10}\right) = e^{-0.9} - e^{-1.1} = 0.406 - 0.332 = 0.074$$

Therefore, the probability that the shovel effectively works between the 9th and 11th year is 0.074. This can also be represented by the area of the shaded region, shown in Figure 2.15.

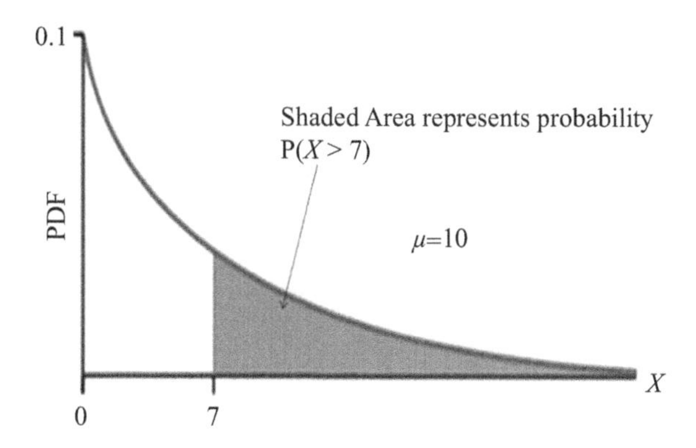

FIGURE 2.14 Shaded region representing the probability of effective work of shovel for more than 7 years.

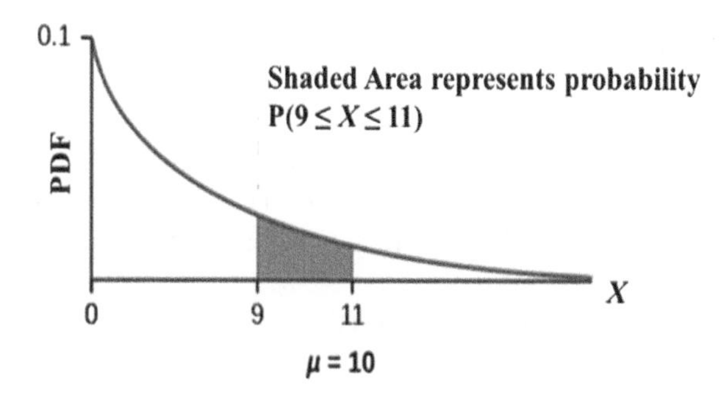

FIGURE 2.15 Shaded region representing the probability of effective work of shovel between 9 and 11 years.

Example 2.12: Suppose the loading time of a mining machine (in minutes) follows an exponential distribution with a mean loading rate of 1/6 per minute. Determine the probability that the machine will take more than 6 minutes to complete loading.

Solution

Let X = the length of a loading time, in minutes.
Average loading rate = λ = 1/6 per minute
Average loading time = $\mu = 1/\lambda$ = 6 minutes
Therefore,

$$P(X>6)=1-P(X<6)=1-\left(1-e^{-x/\mu}\right)=e^{-\frac{6}{6}}=e^{-1}=0.44$$

The probability that the machine takes more than 6 minutes for loading is 0.44.

2.5 CONDITIONAL PROBABILITY

Conditional probability is the probability of occurrence of one event with some relationship to (conditioning to) one or more other events. For two events A and B with $P(B)>0$, the conditional probability of A and B, $P(A \mid B)$ is defined as

$$P(A \mid B)=\frac{P(A\cap B)}{P(B)}, \qquad P(B)>0 \tag{2.16}$$

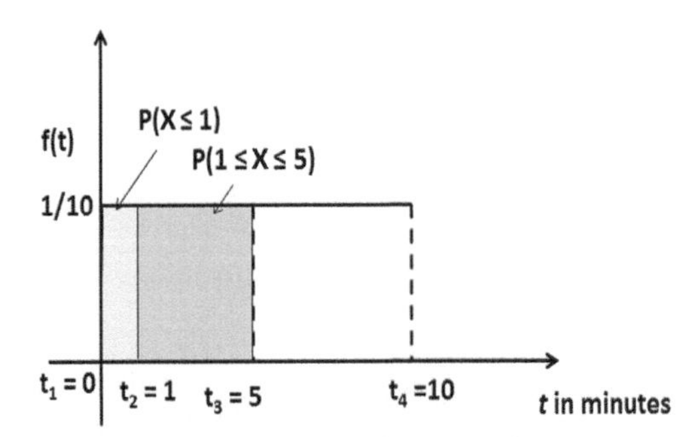

FIGURE 2.16 Probability density function of the uniform distribution.

Example 2.13: In **Example 2.6**, determine the probability that the cage will start within the next 4 minutes after the arrival of the worker at the pit-bottom if it is known that the cage has not started 1 minute before the arrival of the worker.

Solution

Let A be the event that the cage will start within 4 minutes of arrival and B be the event that the cage has not started within 1 minute before the arrival of the worker. The shaded area in Figure 2.16 shows the probability of different conditions. We are interested in determining $P(A|B)$.

$$P(B) = P(T > 1) = 1 - P(T \le 1) = 1 - 1 * \frac{1}{10} = 1 - 0.1 = 0.9$$

$$P(A) = P(T \le 5) = 5 * \frac{1}{10} = 0.5$$

Since $A \subset B$, we have, $A \cap B = A$

Therefore, $P(A|B) = \frac{P(A \cap B)}{P(B)} = \frac{P(A)}{P(B)} = \frac{0.5}{0.9} = 0.555$

The probability that the cage will start within the next 4 minutes after the arrival of the worker at the pit-bottom if it is known that the cage has not started 1 minute before the arrival of the worker is 0.555.

2.6 MEMORYLESS PROPERTY OF THE PROBABILITY DISTRIBUTION

The memoryless property indicates that a given probability distribution is independent of its historical occurrences. If a probability density has the memoryless property, the chance of occurrence of any event in the future is independent of the occurrence of the event in the past. The exponential distribution is one of the probability distribution functions that has memoryless properties.

For example, the inter-arrival time between trucks in a production bench is exponentially distributed with a mean arrival time of T minutes. Assuming that the time elapsed since the last truck arrived is t ($t > T$) minutes. Since a significant length of time has now elapsed, there is a high chance that a truck will arrive within the next minute. But for the exponential distribution, this does not hold true. The additional time spent waiting for the next truck to arrive does not depend on the time that has already elapsed since the last truck arrived. This is referred to as the **memoryless property**. Specifically, the **memoryless property** says that

$$P(X > T + t \mid X > T) = P(X > t) \quad \text{for all } T \ge 0 \text{ and } t \ge 0 \tag{2.17}$$

Proof:
Let X be the variable representing the random inter-arrival time between two successive events at time T and $(T+t)$.

From conditional probability property, we have

$$P(X > T+t | X > T) = \frac{P(X > T+t,\ X > T)}{P(X > T)} = \frac{P(X > T+t) \cap P(X > T)}{P(X > T)}$$
$$= \frac{P(X > T+t)}{P(X > T)} = \frac{1 - F_X(T+t)}{1 - F_X(T)}$$

For an exponential distribution, the CDF, $F_X(x)$ is given by

$$F_X(x) = 1 - e^{-\lambda x}$$

Therefore,

$$P(X > T+t | X > T) = \frac{e^{-\lambda(T+t)}}{e^{-\lambda T}} = e^{-\lambda t} = P(X > t)$$

Example 2.14: The time a shovel takes to load a truck follows an exponential distribution with a mean of four minutes. If the truck has already spent four minutes in a queue before loading, determine the probability that the truck will spend at least an additional three minutes for completion of the loading operation?

Solution
Mean loading time (λ) = 4 minutes.

$$P(X < t) = \left(1 - e^{-t/\lambda}\right) = 1 - e^{-0.25t}$$

The probability that the truck will spend at least an additional three minutes with the shovel, which already spent 4 minutes, is given by

$$P(X > 7 | X > 4) = P(X > 3) = 1 - P(X \le 3) = \left(1 - e^{-0.25t}\right)$$

$$\Rightarrow P(X > 7 | X > 4) = \left(1 - e^{-0.25*3}\right) = 0.4724$$

Therefore, the probability that the truck will spend at least an additional three minutes after waiting four minutes for completion of the loading operation is 0.4724.

Example 2.15: The successful operational time, t of a dragline, is defined as the amount of time (in years) it works until it breaks down, satisfies the following probability density function.

$$P(T \ge t) = e^{-t/5}, \qquad t \ge 0$$

If the dragline successfully works more than or equal to 2 years, determine the probability of breaking down in the third year.

Solution: Let A be the event that the dragline breakdown in the third year and B be the event that the dragline does not break down in the first two years.

$$P(B) = P(T \ge 2) = e^{-2/5} = 0.67$$

$$P(A)=P(2\le T\le 3)=P(T\ge 2)-P(T\ge 3)=e^{-\frac{2}{5}}-e^{-\frac{3}{5}}=0.67-0.5488=0.1215$$

Since $A\subset B$, we have, $A\cap B=A$

Therefore, the probability that the dragline fail in the third year after a successful operation of two years is given by

$$P(A|B)=\frac{P(A\cap B)}{P(B)}=\frac{P(A)}{P(B)}=\frac{0.1215}{0.67}=0.1813$$

That is, there is an 18.13% chance that the dragline fails in the third year after a successful run of the first two years.

2.7 THEOREM OF TOTAL PROBABILITY FOR COMPOUND EVENTS

Let events $A_1, A_2, \ldots, A_n$, form partitions of the sample space *S*, where all the events have a non-zero probability of occurrence. For any event, *A* associated with *S*, according to the total probability theorem,

$$P(A)=\sum_{i=1}^{n}P(A_i)P\left(\frac{A}{A_i}\right)=\sum_{i=1}^{n}P(A\cap A_i),\ \mathrm{P}(\mathrm{A}_i)\ne 0,\quad \mathrm{i}=1,2,3,\ldots,n \tag{2.18}$$

Proof:

The collection of events $A_1, A_2, \ldots, A_n$ are said to partition a sample space *S* (Figure 2.17) if the following conditions hold.

a. $A_1\cup A_2\cup\ldots\cup A_n=S$
b. $A_i\cap A_j=\phi \qquad i,j\in 1 \text{ to } n$, and $i\ne j$
c. $A_i\ne\phi \qquad i\in 1 \text{ to } n$

If *A* is any event within *S*, it can be expressed as a union of subsets as

$$A=(A\cap A_1)\cup(A\cap A_2)\cup\ldots\cup(A\cap A_n)$$

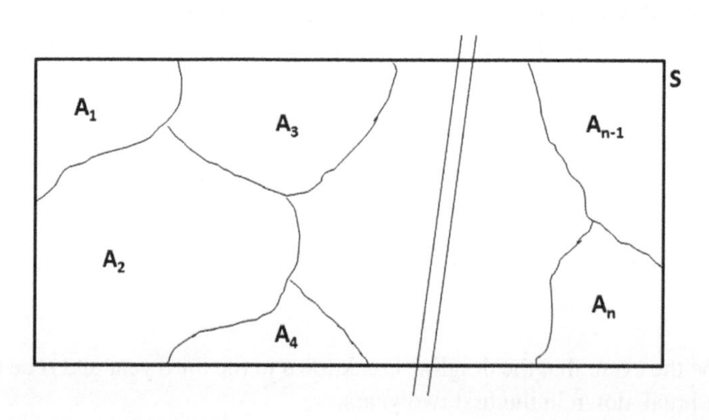

FIGURE 2.17 Partition of sample space by *n* events.

The bracketed events $(A\cap A_1),(A\cap A_2),\ldots,(A\cap A_n)$ are mutually exclusive (if one occurs, then none of the others can occur). Thus, from the law of probability for mutually exclusive events, we have

$$P(A)=P(A\cap A_1)+P(A\cap A_2)+\cdots+P(A\cap A_n)$$

$$\Rightarrow P(A)=P(A\cap A_1)+P(A\cap A_2)+\cdots+P(A\cap A_n)$$

$$\Rightarrow P(A)=P(A\mid A_1)*P(A_1)+P(A\mid A_2)*P(A_2)+\cdots+P(A\mid A_n)*P(A_n)$$

$$\Rightarrow P(A)=\sum_{i=1}^{n}P(A\mid A_i)*P(A_i)$$

2.8 BAYES' RULE

The most useful property of conditional probability is Bayes' rule. If the value of conditional probability $P(B|A)$ is known, the value of conditional probability $P(A|B)$ can be easily determined using Bayes' rule. The theorem can be derived as follows:

The conditional probability, $P(A|B)$ for two events, A and B can be presented as:

$$P(A|B)=\frac{P(A\cap B)}{P(B)},\quad \text{where } P(B)\neq 0$$

Similarly, the conditional probability $P(B|A)$ for two events, A and B can be written as:

$$P(B|A)=\frac{P(B\cap A)}{P(A)},\quad \text{where } P(A)\neq 0$$

Since the joint probabilities of two events are equal, we have

$$P(B\cap A) = P(A\cap B)$$

Therefore,

$$P(A\cap B) = P(A\mid B)\,P(B) = P(B\mid A)\,P(A)$$

$$\Rightarrow P(A\mid B)=\frac{P(B\mid A)P(A)}{P(B)},\quad P(B)\neq 0\quad \text{[Bayes'Rule]} \tag{2.19}$$

In order to find $P(B)$ in Bayes' formula, we need to use the law of total probability, as demonstrated in Eq. 2.7. The Bayes' rule can be reduced to

$$P(A_k\mid B) = \frac{P(B\mid A_k)\,P(A_k)}{\sum_i P(B\mid A_i)P(A_i)} \tag{2.20}$$

where, $A_1, A_2, \ldots, A_n$ form a partition of the sample space.

Example 2.16: Iron ore samples are collected from three mines, A, B, and C. In mine A, 60% of samples have grade values greater than cut-off grade, in mine B, 50% of the samples have grades greater than cut-off grade, and in mine C, 40% of the samples have grades greater than the cut-off grade. Of the total sample from three mines, 40% are from mine A, 25% from mine B, and 35% from mine C. Given that a particular sample has a grade greater than the cut-off grade determines the probability that the sample is collected from mine B.

Solution
The probability that the sample is from mine B if the grade value (G) is higher than the cut-off grade (CoG) is given by

$$P(\text{sample} \in \text{mine B} | G > \text{CoG}) = \frac{P(G > \text{CoG} | \text{sample} \in \text{mine B}) * P(\text{sample} \in \text{mine B})}{\begin{array}{c} P(G > \text{CoG} | \text{sample} \in \text{mine A}) * P(\text{sample} \in \text{mine A}) + \\ P(G > \text{CoG} | \text{sample} \in \text{mine B}) * P(\text{sample} \in \text{mine B}) + \\ P(G > \text{CoG} | \text{sample} \in \text{mine C}) * P(\text{sample} \in \text{mine C}) \end{array}}$$

$$\Rightarrow P(\text{sample} \in \text{mine B} | G > \text{CoG}) = \frac{0.5*0.25}{0.6*0.4+0.5*0.25+0.40*0.35} = \frac{0.125}{0.24+0.125+0.14}$$

$$= \frac{0.125}{0.505} = 0.247$$

Therefore, the probability that the sample is from mine B if the grade value is higher than the cut-off grade is 0.247.

2.9 DEFINITION OF STATISTICS

Statistics is a branch of mathematics that deals with data collection, data analysis, and interpretation. Data are information, which are collected through experiment or observation. In statistics, a **random variable or stochastic variable** can be defined as a variable whose values depend on the outcomes of a random process (Blitzstein and Hwang, 2014). For example, the failure occurrence time of heavy earth-moving machinery deployed in a mine is a random variable, as the same cannot be estimated deterministically through an algebraic equation. As discussed before, a random variable can be continuous or discrete and can be represented using probability density function and probability mass function.

Data can be classified based on data type and data source.

- **Based on the data type:**
  There are two kinds of data types: quantitative and qualitative.
  Quantitative data: Quantitative data is expressed in numbers. Quantitative data can be summarized using statistical analysis for obtaining meaningful information. Examples of quantitative data are daily production from mines (i.e., 50 million tonnes), depth of the deposit (i.e., 100 m), etc.
  Qualitative Data: The data which are not measured numerically is referred to as qualitative data. For example, safety rating in mines (i.e., low, medium, or high), working environment in mines (i.e., poor or good), etc. Safety rating in mines can be referred to as low, medium, or high but not measured with any numeric value. Similarly, the working environment in a mine can be measured with a linguistic value like poor or good.

- **Based on the data source:** There are two kinds of data sources: primary and secondary.
  Primary data: Primary data refers to those collected directly from objects, processes, or individuals. The primary data can be quantitative or qualitative. For example, failure time of mining equipment, roof convergence using extensometer, etc.
  Secondary data: Secondary data refers to those which are obtained from other sources. For example, accident data obtained from federal agencies like Mine Safety and Health Administration (MSHA), USA, Director General of Mines Safety (DGMS), India is secondary data for the user.
- **Based on the continuity of the data:** There are two kinds of data types: discrete and continuous.
  Discrete data: Discrete data refers to the type of quantitative data that relies on counts. Discrete data can take only certain values. For example, the number of persons employed in different mines can be characterized as discrete data (Figure 2.18).
  Continuous data: Continuous data refers to the unlimited number of possible measurements in a specified range. For example, the daily price of dry iron ore in the International market can be characterized as continuous data (Figure 2.19).

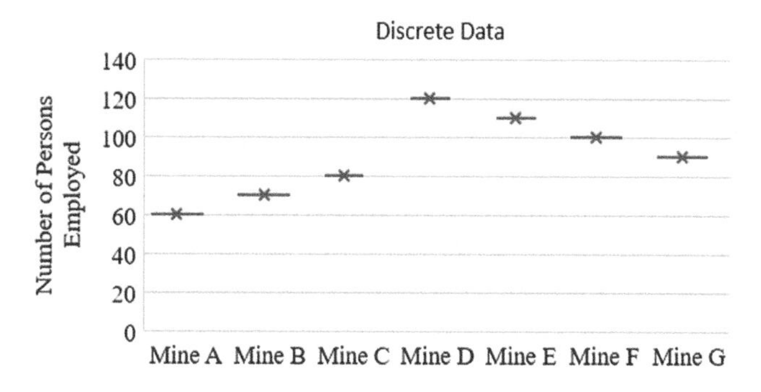

FIGURE 2.18 Graphical representation of a discrete data.

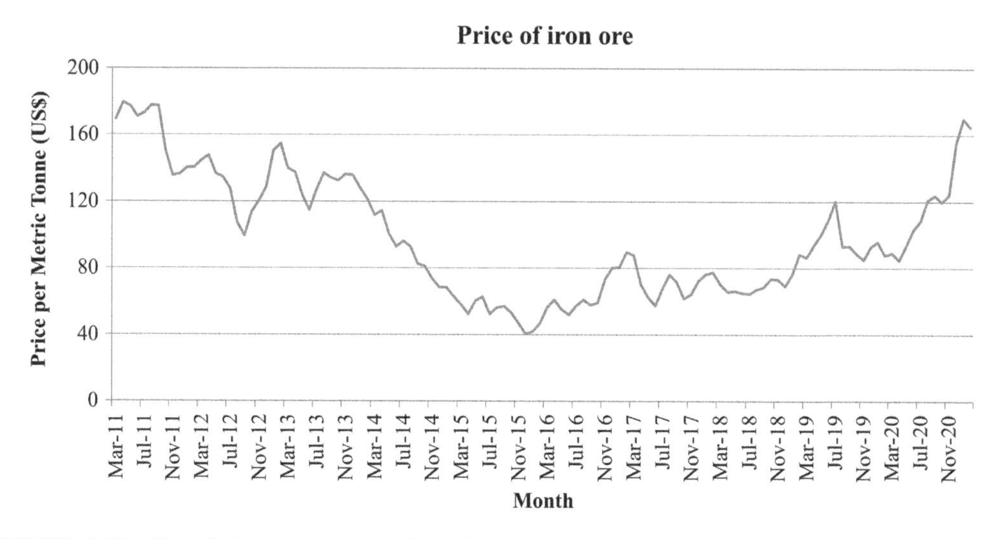

FIGURE 2.19 Graphical representation of continuous data [Source: www.indexmundi.com/commodities/?commodity=iron-ore&months=120].

2.10 STATISTICAL ANALYSES OF DATA

All quantitative data can be statistically analysed. Two main statistical methods are used in data analysis, viz. descriptive statistics and inferential statistics. Descriptive statistics describe the characteristics of data using different indices, including **mean, median, standard deviation, etc.,** whereas inferential statistics allows drawing inferences about the population from that sample data.

2.10.1 Common tools of descriptive statistics

In descriptive statistics, the entire group data, which need to describe, are used to determine the statistics. The descriptive statistics of group data can be represented with complete certainty (outside of measurement error). The statistical measures commonly used in descriptive statistics are explained below.

2.10.1.1 Arithmetic Mean

If a random variable X takes n values $x_1, x_2, \ldots, x_n$, the arithmetic mean can be determined using Eq. 2.21.

$$\text{Arithmetic Mean } (\bar{x}) = \frac{x_1 + x_2 + \cdots + x_n}{n} = \frac{1}{n}\sum_{i=1}^{n} x_i \tag{2.21}$$

If in a distribution, the random variable X has n values $x_1,\ x_2,\ \ldots,\ x_n$ occur with respective frequencies $s_1,\ s_2,\ \ldots,\ s_n$, the arithmetic mean can be determined using Eq. 2.22.

$$\text{Arithmetic Mean} (\bar{x}) = \frac{s_1 x_1 + s_2 x_2 + \cdots + s_n x_n}{s_1 + s_2 + \cdots + s_n} = \frac{\sum_{i=1}^{n} s_i x_i}{\sum_{i=1}^{n} s_i} \tag{2.22}$$

Eq. 2.22 provides weighted arithmetic mean since the value of the variable x_i is weighted by frequency s_i.

If the distribution of the data follows a continuous function, $y = f(x)$ on the interval [a, b], the mean value of the function in the defined range is given by Eq. 2.23.

$$\text{Mean Value} (\bar{x}) = \frac{1}{(\mathrm{b} - \mathrm{a})} \int_{\mathrm{a}}^{\mathrm{b}} f(x)\,dx \tag{2.23}$$

In the above equation, (***b*** - ***a***) represents the width under the curve, and $\bar{x}$ as the mean height of the curve in the range of a to b. This represents the area under the curve in an interval [a, b].

Example 2.17: The daily metal production (in tonne) from a primary process plant of a metal mine for a one-week duration is given as:

160 172.5 161 180 175 165 162.5

Determine the mean production.

Solution

The mean production is given by

$$\text{Mean production} (\bar{x}) = \frac{x_1 + x_2 + x_3 + x_4 + x_5 + x_6 + x_7}{7}$$

$$\Rightarrow \overline{x} = \frac{160+172.5+161+180+175+165+162.5}{7} = 168 \text{ tonnes}$$

Therefore, the mean production level in the week is 168 tonnes.

Example 2.18: The frequency of daily metal production from a primary process plant of a mine for 30 days is given as:

Daily Production (in tonne)	160	172.5	161.	180	175	165	162.5
Frequency	6	5	4	3	2	4	6

Determine the mean production.

Solution

The mean production is given by

$$\text{Mean Production}\left(\overline{x}\right) = \frac{s_1x_1 + s_2x_2 + s_3x_3 + s_4x_4 + s_5x_5 + s_6x_6 + s_7x_7}{s_1 + s_2 + s_3 + s_4 + s_5 + s_6 + s_7}$$

$$\Rightarrow \overline{x} = \frac{6*160+5*172.5+4*161+3*180+2*175+4*165+6*162.5}{6+5+4+3+2+4+6} = \frac{4991.5}{30}$$

$$= 161.38 \text{ tonnes}$$

The mean production level in the primary process plant of a mine in 30 days is 161.38 tonnes.

Example 2.19: In an underground coal mine, the length of the intake airway path is 600 m. The air pressure changes with distance travelled from the pit top to the last point of the intake airway path with the function $P(x) = 600 - \frac{x}{900}$, x is distance, and P is pressure in Pascal. Determine the average air pressure.

Solution

The mean or average air pressure can be determined as

$$\text{Mean or average air pressure }\left(\overline{x}\right) = \frac{1}{(b-a)}\int_a^b f(x)\,dx = \frac{1}{(600-0)}\int_0^{600}(600 - x/900)\,dx$$

$$= \frac{1}{600}\left[600x - x^2/1800\right]_0^{600} = \frac{1}{600}\left[\left(3600 - \frac{3600}{18}\right) - 0\right]$$

$$= \frac{1}{600}(3600 - 200) = 5.67 \text{ Pa}$$

Therefore, the average air pressure $\left(\overline{x}\right)$ is 5.67 Pa.

2.10.1.2 Median

Median is defined as the middle value after sorting the observation either in ascending or decreasing order. If the number of the observation is an odd number, say $2n+1$, where n is an integer, the value of the $(n+1)^{\text{th}}$ observation gives the median value, while if the number is even, say $2n$, then the median is the mean of nth and $(n+1)^{th}$ observation values. The median of a grouped frequency distribution data can be determined using simple interpolation using Eq. 2.24.

$$M_d = L + \frac{\frac{N}{2} - F}{f} * i \tag{2.24}$$

where,

M_d = median
L = lower limit of the median class
f = frequency of the median class
F = total of all the frequencies before the median class
i = width of median class interval
N = number of sample data

Example 2.20: In Example 2.17, determine the median value of metal production.

Solution
The ascending order of the metal production is

160 161 162.5 165 172.5 175 180

Therefore, the median value is 165 tonnes.

Example 2.21: In Example 2.17, determine the median value of metal production based on the first 6 days' production data.

Solution
Given data of metal production:

160 172.5 161 180 175 165

The ascending order of the production for the first 6 days' data is

160 161 162.5 165 172.5 175

Therefore, the median value is given by

$$\text{Median} = \frac{162.5 + 165}{2} = 163.75$$

Example 2.22: The frequency of daily metal production from a mine for 30 days is given as:

Daily Production (tonne)	160–163	163.1–166	166.1–169	169.1–172	172.1–175
Frequency	8	4	6	7	5

Determine the median value of metal production.

Solution
From the problem, the total number of the class is five, and thus the median class is 163.1–169.

The lower limit of the median class (L) = 166.1

Number of sample data (N) = 8 + 4 + 6 + 7 + 5 = 30

Frequency of the median class (f) = 6

Total of all the frequencies before the median class (F) = 8 + 4 = 12

Width of median class interval (i) = 3

The median value of metal production is given by

$$\text{Median}(M_d) = L + \frac{\frac{N}{2} - F}{f} * i$$

$$\therefore M_d = 166.1 + \frac{\frac{30}{2} - 12}{6} * 3 = 166.1 + \frac{3}{6} * 3 = 167.6 \text{ tonne}$$

The median value of the metal production is 167.6 tonnes.

2.10.1.3 Mode

Mode represents the maximum frequency of observations. In other words, the maximum peak of a fitted curve of the actual distribution represents the mode. For a grouped distribution, the mode can be determined using Eq. 2.25.

$$Mode = L + \frac{s_n - s_{n-1}}{2s_n - s_{n+1} - s_{n-1}} * i \tag{2.25}$$

where

L = lower limit of the modal class, which has the highest frequency.
i = width of the modal class
s_n = frequency of the modal class
s_{n+1} and s_{n-1} = frequencies of the following (n+1) and preceding (n−1) of the modal class (n) respectively.

For a moderately symmetrical distribution, the following relationship among mean, median, and mode nearly holds true.

$$\text{Mean} - \text{Mode} = 3 * (\text{Mean} - \text{Median}) \tag{2.26}$$

For a normal distribution, the mean, median, and mode coincide.

Example 2.23: Determine the mode for the problem given in **Example 2.22.**

Solution

The modal class is **160–163**, as this class has the highest frequency.

Therefore, $L = 160$

$$s_n = 8$$

$$s_{n-1} = 0$$

$$s_{n+1} = 4$$

$$i = 3$$

$$\text{Mode} = L + \frac{s_n - s_{n-1}}{2s_n - s_{n+1} - s_{n-1}} * i = 160 + \frac{8-0}{2*8-4-0} * 3$$

$$= 160 + \frac{8}{12} * 3 = 162 \text{ tonnes}$$

The mode value of the given metal production statistics is 162 tonnes.

2.10.1.4 Standard deviation

Standard deviation (σ) is the positive square root of the average of the squares of the deviations of the observations from the mean of the observations. It measures the dispersion of the data from the mean value. The standard deviation of N observations (x_1, x_2, …, x_N) can be determined using Eq.2.27.

$$\text{Standard deviation}(\sigma) = \sqrt{\frac{\sum_{i=1}^{N}(x_i - \bar{x})^2}{N-1}} \tag{2.27}$$

where N is the number of observations, $\bar{x}$ is the mean value of observations, x_i represents the value of ith observation.

If a random variable X has x_1, x_2,…, x_N occurrences with respective frequencies s_1, s_2, …,s_N, the standard deviation can be determined using Eq. 2.28.

$$\text{Standard deviation}(\sigma) = \sqrt{\frac{\sum_{i=1}^{N} s_i (x_i - \bar{x})^2}{\sum_i s_i}} \tag{2.28}$$

where s_i is the frequency of the ith group, and N is the number of groups.

Standard deviation (σ) is the most commonly used measure of dispersion or overall variation in a data set. The low standard deviation value represents observations that are close to their mean value, exhibiting little variation or dispersion. On the other hand, the high standard deviation value represents the observation values are highly dispersed or spread out from their mean, exhibiting more variation. The zero (0) standard deviation means all observations are identical.

The inter-quartile range is another term that can be used to measure the variation of a dataset. In a dataset, the quartiles are the values that divide the data into four equal parts. The inter-quartile range (IQR) represents the difference between the upper quartile (median of the upper half of dataset) and the lower quartile (median of the lower half of the dataset).

Example 2.24: Using Example 2.17 data, determine the standard deviation of the metal production.

Solution

Given metal production data:

160 172.5 161 180 175 165 162.5

The standard deviation of metal production is presented as:

$$\text{Standard deviation}(\sigma) = \sqrt{\frac{\sum_{i=1}^{N}(x_i - \bar{x})^2}{N-1}}$$

Here, N is the number of observations (= 7),
$\bar{x}$ = mean metal production (=168) **[Refer Example 2.17]**

Therefore,

$$\sigma = \sqrt{\frac{\sum_{i=1}^{7}\left(x_i - \bar{x}\right)^2}{7-1}}$$

$$= \sqrt{\frac{\left[(160-168)^2 + (172.5-168)^2 + (161-168)^2 + (180-168)^2 + (175-168)^2 + (165-168)^2 + (162.5-168)^2\right]}{6}}$$

$$= \sqrt{\frac{365.5}{6}} = \sqrt{60.916} = 7.80 \text{ tonnes}$$

Thus, the standard deviation of the metal production is equal to 7.80 tonnes.

Example 2.25: Using Example 2.17 data, determine the interquartile range of the metal production.

Solution
Given metal production data:

160 172.5 161 180 175 165 162.5

The quartiles of the dataset can be determined by arranging the data in ascending order as represented below:

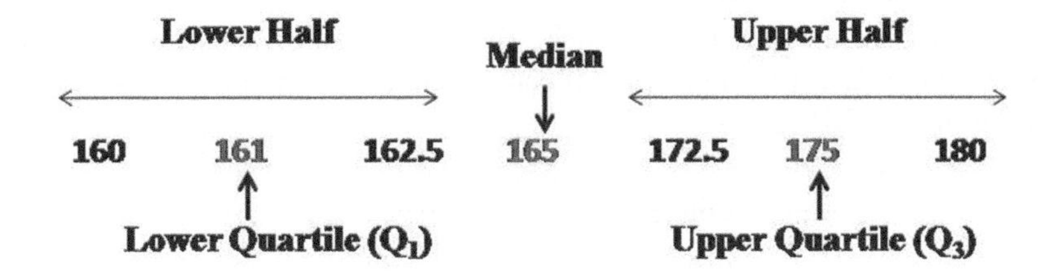

Therefore, the interquartile range (IQR) of the data is given by

$$IQR = Q_3 - Q_1 = 175 - 161 = 14$$

Example 2.26: Using the data from Example 2.22, determine the standard deviation of metal production.

Solution
Given data:

Daily Production	160–163	163.1–166	166.1–169	169.1–172	172.1–175
Frequency	8	4	6	7	5

The mid values of the production range are

Daily Production	160–163	163.1–166	166.1–169	169.1–172	172.1–175
Mid value (x_i)	161.5	164.5	167.5	170.5	173.5
Frequency (s_i)	8	4	6	7	5

The mean value of metal production $(\bar{x})$ is given by

$$\text{Mean Production}(\bar{x}) = \frac{s_1x_1 + s_2x_2 + s_3x_3 + s_4x_4 + s_5x_5}{s_1 + s_2 + s_3 + s_4 + s_5}$$

$$\Rightarrow \bar{x} = \frac{8*161.5 + 4*164.5 + 6*167.5 + 7*170.5 + 5*173.5}{8+4+6+7+5} = \frac{5016}{30} = 167.2\,\text{tonne}$$

The standard deviation of production is given by:

$$\text{Standard deviation}(\sigma) = \sqrt{\frac{\sum_{i=1}^{5} s_i\left(x_i - \bar{x}\right)^2}{\sum_i s_i}}$$

$$= \sqrt{\frac{\left[8(161.5-167.2)^2 + 5(164.5-167.2)^2 + (167.5-167.2)^2 + (170.5-167.2)^2 + (173.5-167.2)^2\right]}{30}}$$

$$= \sqrt{\frac{563.4}{30}} = \sqrt{18.81} = 4.33\,\text{tonne}$$

Thus, the standard deviation of the daily metal production in the given month is equal to 4.33 tonnes.

2.10.1.5 Mean Absolute Deviation

It is the mean of the absolute deviations of a dataset from its mean. This statistic is used to understand the dispersion of data from a measure of central tendency. The mean absolute deviation can be determined using Eq. 2.29.

$$\text{Mean absolute deviation (MAD)} = \frac{1}{N}\Sigma\left|x_i - \bar{x}\right| \tag{2.29}$$

where

x_i = value of ith observation
$\bar{x}$ = mean value of the observations
N = Number of observations

The MAD for a grouped dataset can be determined using Eq. 2.30.

$$\text{Mean absolute deviation (MAD)} = \frac{1}{N}\Sigma\, s_i\left|x_i - x_m\right| \tag{2.30}$$

where s_i = frequency of ith group, N is number of groups, and x_m = Median value of the dataset.

Example 2.27: Determine the mean absolute deviation of the metal production for data given in **Example 2.17**.

Solution

Given data: 160 172.5 161 180 175 165 162.5

The mean of the is

$\bar{x} = 168\,\text{tonne}$ [Refer **Example 2.17**]

The mean absolute deviation is given by

$$\text{MAD} = \frac{1}{N}\Sigma|x_i - \bar{x}| = \frac{1}{7}\left[|160-168|+|172.5-168|+|161-168|+|180-168|\right.$$
$$\left.+|175-168|+|165-168|+|162.5-168|\right] = \frac{47}{7} = 6.71 \text{ tonnes}$$

Therefore, the mean absolute deviation of the metal production is 6.71 tonnes.

2.10.1.6 Skewness

Skewness measures the asymmetry of the probability distribution about its mean. Positive skewness indicates a longer tail of the frequency curve towards the higher value of the data. In other words, the mean is greater than the mode or median, and vice-versa. **For symmetrical normal distribution, the mean, mode, and median coincide.** If the skewness is close to zero, the distribution is approximately symmetric, and the median is close to the mean. The coefficient of skewness can be determined using the following equations.

Pearson's coefficient of skewness uses the mode is given by

$$\text{Coefficient of skewness} = \frac{\text{Mean} - \text{Mode}}{\text{Satandard Deviation}} = \frac{\bar{x} - v}{\sigma} \tag{2.31}$$

If the mode does not exist for the data, the coefficient of skewness can be determined as

$$\text{Coefficient of skewness} = \frac{3(\text{Mean} - \text{Median})}{\text{Satandard Deviation}} = \frac{3(\bar{x} - M_d)}{\sigma} \tag{2.32}$$

where $\bar{x}$, σ, v, and M_d are the mean, standard deviation, mode, and median of the data, respectively.

The coefficient of skewness, represented in Eq. 2.32, uses the median value of the data instead of the mode of the data.

Example 2.28: Determine the skewness of the data given in Example 2.17.

Solution

Given data: 160 172.5 161 180 175 165 162.5

The mean, median, and standard deviation of the given data are as follows:

$\text{Mean}(\bar{x}) = 168 \text{ tonnes}$ [Refer Example 2.17]
$\text{Median}(v) = 165 \text{ tonnes}$ [Refer Example 2.20]
$\text{Standard deviation}(\sigma) = 7.80 \text{ tonnes}$ [Refer Example 2.24]

The mode value of the given data does not exist, and thus the second formula for the coefficient of skewness is used.

$$\text{Coefficient of skewness} = \frac{3(\text{Mean} - \text{Median})}{\text{Satandard Deviation}} = \frac{3(168-165)}{7.8} = 1.15$$

Thus, the coefficient of skewness of the given data is 1.15.

2.10.1.7 Coefficient of variation

The coefficient of variation, *CV*, is used as an alternative to skewness for describing the shape of the distribution. The *CV* can be calculated as:

$$\text{Coefficient of variation}(CV) = \frac{\text{Satandard deviation}(\sigma)}{\text{Mean}(\bar{x})} * 100\ \% \tag{2.33}$$

where σ is standard deviation and $\bar{x}$ is mean of the data. *CV* greater than 100% indicates the presence of erratic high-valued samples in the data set.

Example 2.29: Determine the coefficient of variation of the metal production data given in **Example 2.17**.

Solution

Given data: 160 172.5 161 180 175 165 162.5

The mean, median, and standard deviation of the given data are as follows:

$\text{Mean}(\bar{x}) = 168$ tonne [Refer Example 2.17]
Standard deviation $(\sigma) = 7.80$ tonnes [Refer Example 2.24]

The coefficient of variation (*CV*) is given by

$$CV = \frac{\sigma}{\bar{x}} * 100 = \frac{7.80}{168} * 100 = 0.04\%$$

2.10.1.8 Expectation or expected value

The expectation or expected value of the random variable is the weighted arithmetic mean of all the independent realizations. It is denoted as E(*X*).

The expectation of a discrete random variable, *X*, is given by

$$\text{Expectation of } X = E(X) = p_1.x_1 + p_2.x_2 + p_3.x_3 + \cdots p_n.x_n = \sum_{i=1}^{n} p_i.x_i \tag{2.34}$$

where *X* has only *n* mutually exclusive values, $x_1, x_2, \ldots, x_n$ and their respective probabilities are $p_1, p_2, \ldots, p_n$.

For a continuous random variable, *X*, the expectation of *X* is given by

$$\text{Expectation of } X = E(X) = \int_{-\infty}^{+\infty} x.f(x)dx \tag{2.35}$$

where, *f*(*x*) is the probability density function of the continuous random variable *X*.

The mean (μ) of a continuous random variable, *X*, is equal to the expected value of *X*.

$$\text{i.e., } \mu = E(X) \tag{2.36}$$

Properties of Expectation

- The expectation of a sum of a number of random variables is equal to the sum of their expectations.

$$E(X_1 + X_2 + \cdots X_n) = E(X_1) + E(X_2) + \cdots E(X_n) \tag{2.37}$$

- The expectation of the product of a number of independent random variables is equal to the product of their expectations.

$$E(X_1 * X_2 * \cdots X_n) = E(X_1) * E(X_2) * \cdots E(X_n) \tag{2.38}$$

- The expectation of any discrete or continuous random variable follows the superposition principle.

$$E[a*X+b*Y]=a*E[X]+b*E[Y] \tag{2.39}$$

where X and Y are random variables, and
a and b are constants.

Example 2.30: Let X be a continuous random variable with a probability distribution function

$$f(x)=\begin{cases}2x^{-2}, & \text{for } 1<x<2\\ 0, & \text{Otherwise}\end{cases}$$

Determine $E(X)$.

Solution

We have,

$$E(X)=\int_{-\infty}^{+\infty} xf(x)\,dx=\int_{-\infty}^{+1} xf(x)\,dx+\int_{1}^{2} xf(x)\,dx+\int_{2}^{+\infty} xf(x)\,dx$$

$$\Rightarrow E(X)=\int_{-\infty}^{+1} x*0\,dx+\int_{1}^{2} x*2x^{-2}\,dx+\int_{2}^{+\infty} x*0\,dx$$

$$\Rightarrow E(X)=\int_{1}^{2} 2x^{-1}\,dx=2\log 2=0.602$$

Thus, the expected value of X is 0.602.

Example 2.31: A miner trapped in an underground mine district due to sudden water inrush into the district. The district has three exits paths, as shown in Figure 2.20.

- Exit 1 leads outside the district after 5 minutes.
- Exit 2 leads back to the district after 8 minutes.
- Exit 3 leads back to the district after 10 minutes.

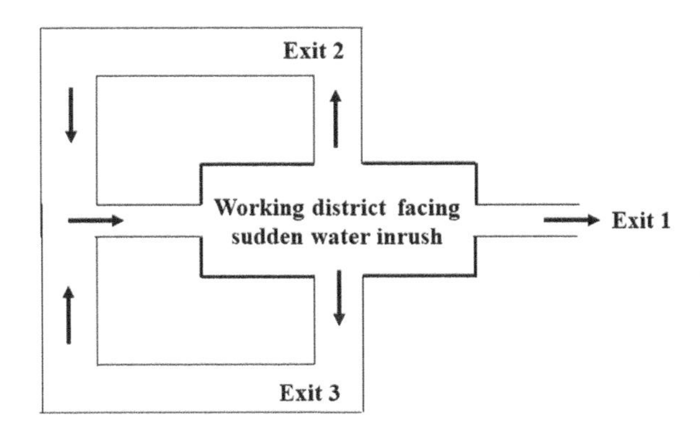

FIGURE 2.20 Escape routes in an underground mine.

Every time the worker makes a choice, it is equally likely to choose any of these three exit paths. What is the expected time taken by the miner to leave the district?

Solution

Let X = time taken for the worker to leave the district.

Y = exit number, the worker chooses first (1, 2, or 3).

$$E(X) = E_Y\left(E(X \mid Y)\right)$$

$$\Rightarrow E(X) = \sum_{y=1}^{3} E(x \mid y = Y) P(y = Y)$$

$$\Rightarrow E(X) = E(X \mid Y = 1) * \frac{1}{3} + E(X \mid Y = 2) * \frac{1}{3} + E(X \mid Y = 3) * \frac{1}{3}$$

We know, $E(X \mid Y = 1) = 5\,\text{minutes}$

$$E(X \mid Y = 2) = 8 + E(X)$$

$$E(X \mid Y = 3) = 10 + E(X)$$

Therefore,

$$E(X) = 5 * \frac{1}{3} + \left[E(X) + 8\right] * \frac{1}{3} + \left[E(X) + 10\right] * \frac{1}{3}$$

$$\Rightarrow \frac{2}{3} E(X) = \frac{23}{3}$$

$$\Rightarrow E(X) = 11.5\ \text{minutes}$$

Therefore, the expected time taken by the worker to leave the district is 11.5 minutes.

2.10.1.9 Variance and covariance

The variance of a random variable, X defined as the mean squared deviation of the same variable from the mean. In other words, for any random variable, it measures the level of spread of data from its mean.

The variance for continuous random variable is equal to the expected value of $(X - \mu)^2$, as represented in Eq. (2.19).

$$\text{Variance of } X = Var(X) = E[(X - \mu)^2] = \int_{-\infty}^{+\infty} (x - \mu)^2 f(x) dx \tag{2.40}$$

Eq. (2.39) can be reduced to

$$\text{Var}(X) = E(X^2 - 2\mu X + \mu^2) = E(X^2) - 2 * E(X) * E(X) + E(X)^2$$

$$\Rightarrow \text{Var}(X) = E(X^2) - \left[E(X)\right]^2 \tag{2.41}$$

The variance for discrete random variable is given by

$$\text{Variance of } X = \text{Var}(X) = E(X-\mu)^2 = \Sigma(X-\mu)^2 P(X) \tag{2.42}$$

A high variance indicates that the observed values of X are highly dispersed from its mean. On the other hand, a lower variance of X indicates that the values are clustered tightly around its mean.

Properties of Variance

- The variance of a constant time random variable is the constant squared times the variance of a random variable and is generally denoted simply as

$$\text{Var}(cX) = c^2\text{Var}(X) \tag{2.43}$$

- Let X and Y be independent random variables. Then

$$\text{Var}(X+Y) = \text{Var}(X) + \text{Var}(Y) \tag{2.44}$$

- If X and Y are dependent random variables, then

$$\text{Var}(X+Y) = \text{Var}(X) + Var(Y) + 2Cov(X,Y) \tag{2.45}$$

In Eq. (2.45), $Cov(X,Y)$ represents the covariance of X and Y. **Covariance** is a measure of the association or dependence between two random variables, X and Y. The covariance between two random variables depends on extend of the data values. Covariance can be either positive or negative.

The covariance between X and Y is given by

$$Cov(X,Y) = E\{(X-\mu_X)(Y-\mu_Y)\} = E(XY) - E(X)E(Y) \tag{2.46}$$

where μ_X = Mean of all $X = E(X)$ and μ_Y = Mean of all $Y = E(Y)$

Example 2.32: Determine the variance of X for the probability density function given in **Example 2.30.**

Solution

The given probability density function is

$$f(x) = \begin{cases} 2x^{-2}, & for\, 1 < x < 2 \\ 0, & \text{Otherwise} \end{cases}$$

We have,

$$\text{Var}(X) = E(X-\mu)^2 = E(X^2) - [E(X)]^2$$

$$E(X) = 2\log 2$$ [Refer Example 2.30]

Again,

$$E(X^2) = \int_{-\infty}^{+\infty} x^2 f(x)\, dx = \int_{-\infty}^{+1} x^2 f(x)\, dx + \int_{1}^{2} x^2 f(x)\, dx + \int_{2}^{+\infty} x^2 f(x)\, dx$$

$$= \int_{-\infty}^{+1} x^2 {*} 0\, dx + \int_{1}^{2} x^2 {*} 2x^{-2}\, dx + \int_{2}^{+\infty} x^2 {*} 0\, dx = 2$$

Therefore, $Var(X) = 2 - (2\log 2)^2 = 2 - 0.36 = 1.64$.

The variance of X for the given probability density function is 1.64.

2.10.1.10 Correlation coefficient

It indicates the relationship between two random variables by measuring the level of change that occurs in one random variable due to change in the other. If the value of one variable is increased, the value of the other variable also increases then correlation is called a positive correlation and vice-versa. A higher value of coefficient indicates a high correlation between the two random variables and vice-versa. The value of the correlation coefficient is always ranges between −1 and +1.

The correlation coefficient can be calculated using the covariance, scaled by the overall variability of two random variables.

If two random variables, X and Y, are measured n number of times and observed as x_i and y_i, where $i = 1$ to n, then the correlation coefficient (Pearson correlation), $r_{X,Y}$ between X and Y can be determined as

$$Corr(X,Y) = r_{X,Y} = \frac{Cov(X,Y)}{\sqrt{Var(X)\ Var(Y)}} = \frac{\sum_{i=1}^{n}(x_i - \bar{x})(y_i - \bar{y})}{\sqrt{\sum_{i=1}^{n}(x_i - \bar{x})^2 \sum_{i=1}^{n}(y_i - \bar{y})^2}} \tag{2.47}$$

where $\bar{x}$ and $\bar{y}$ are the sample means of X and Y.

2.10.2 Standard Analysis Tools of Inferential Statistics

In inferential statistics, the population is defined first, and then a sampling plan is devised to obtain a representative sample. The statistical results involve uncertainty as sample data is used to understand an entire population. The most common types of analyses are hypothesis tests, confidence intervals, and regression analysis. Interestingly, these inferential methods can produce similar summary values as descriptive statistics, such as the mean and standard deviation.

2.10.2.1 Hypothesis Tests

Hypothesis tests use sample data to answer questions like the following:

- Is the population mean greater than or less than a particular value?
- Are the means of two or more populations different from each other?

Example 2.31: A geologist employed in a mine claims that the grades of ore of the deposit are above the cut-off grade (56%). The mean grade of 30 random samples is 60%, and the standard deviation is 5%. Is there sufficient evidence to support the geologist's claim?

Solution

Null hypothesis:

The accepted fact is that the cut-off grade is 56%, and hence the null hypothesis can be defined as:

H_0: $\mu = 56\%$.

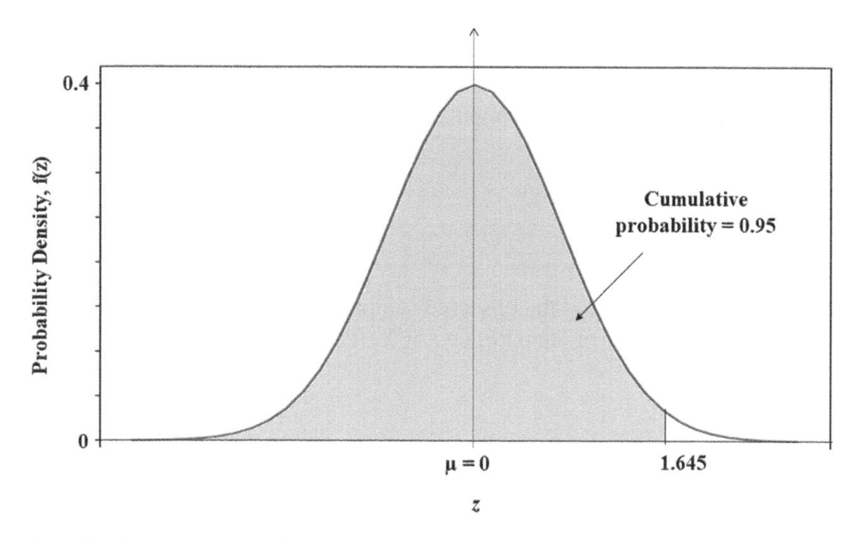

FIGURE 2.21 Shaded area representing the probability for $z \geq 1.645$.

Alternate hypothesis:
It is claimed that the grade of the ore deposit is above the cut-off grade, and hence the alternative hypothesis can be defined as:

$H_1: \mu > 56\%.$

This is a one-tailed test as scores are determined for *greater than* a certain mean value. In a one-tailed test, the critical area of a distribution is one-sided (either greater than or less than a specific value), but not both.

Assuming the data follow a normal distribution, we want to test the hypothesis with a 95% confidence level. Therefore, the value of α is 5% (=0.05). An area of 0.05 is equal to a z-score of 1.645 (Refer to z-table corresponding to row value of 1.6 in z-column and 0.045 across at the top), as shown in Figure 2.21.

In the next step, the rejection region area for the defined α level (=5%) needs to be determined from the z-table. The test statistic can be determined using this formula:

$$z = \frac{(\bar{x} - \mu_0)}{\sigma / \sqrt{n}} = \frac{(60 - 56)}{\frac{5}{\sqrt{30}}} = 2.19$$

If the calculated z-statistics value is less than the z-score obtained for the defined α-level (=0.05), the null hypothesis cannot be rejected. In this case, it is greater (2.19 > 1.645), and hence the null hypothesis can be rejected. Therefore, the claim that the grade of the ore deposit is above the cut-off grade is correct with a 95% confidence level.

Example 2.32: The mean and standard deviation of the of Bauxite ore grades are 40% and 6%, respectively. The estimated grade value, based on the 40 samples, is 42%. Test a hypothesis at $\alpha = 0.05$ that the mean grade in this population differs from 40%.

Solution

Null hypothesis:
In this case, the null hypothesis is that the population mean 40%, and hence the null hypothesis can be written as:

H_0: μ = 40%

Alternative hypothesis:
It needs to be examined that mean of the observed samples (42%) represents a different population from the given population mean of Bauxite, i.e., 40%. Therefore, the alternate hypothesis can be written as:

H_1: μ ≠ 40%

The test statistic can be determined using this formula:

$$z = \frac{(\bar{x} - \mu_0)}{\sigma / \sqrt{n}} = \frac{(42 - 40)}{\frac{6}{\sqrt{40}}} = 2.10$$

The *p*-value corresponding to a *z*-score of 2.10 can be determined from the *z*-statistics table. The value corresponding to a row of 2.1 in *z*-column and 0.00 across at the top represents the *p*-value for a *z*-score of 2.10. This is equal to 0.0178. Thus, there is only a 1.78% chance of getting a mean of 42% or more. At α = 0.05, the null hypothesis, H_0 can be rejected, and it is evident that the observed mean is different from the population mean.

Exercise 2

1. Determine the population mean and standard deviation of the following dataset.

   83 87 93 109 124 126 126 101 102 108

   [**Ans.** 105.9; 14.95]

2. Determine the coefficient of variation of the following dataset:

   83 87 93 109 124 126 126 101 102 108

   [**Ans.** 0.14123]

3. The daily production measured for 50 days in a coal mine follows a normal distribution with a mean 1200 tonnes per day and a standard deviation 100 tonnes per day. Determine the 95% confidence interval of daily production.

   [**Ans.** 1223.26, 1176.73]

4. In an iron ore deposit alumina is distributed with mean of (μ) = 3% and standard deviation of (σ) = 0.5%; whereas silica is distributed with mean of (μ) = 2.5% and standard deviation of (μ) = 0.8%. Determine the combined mean and standard deviation of alumina and silica (as impurities) in percentage.

   [**Ans.** 5.5%; 0.943%]

5. The mean and the standard deviation of the grade of iron ore in a deposit are 62% and 5%, respectively. Determine the coefficient of variation of the grade in %.

   [**Ans.** 8.06%]

6. The correlation coefficient between two variables Y (dependent) and X is 0.5. Given that the coefficient of determination is the ratio of explained variation and total variation of the regression model between the two variables Y and X, determine the percentage of the total variation unexplained by the regression equation.

   [**Ans.** 0.4]

7. A random variable, X is described by a probability density function

$$f(x) = \begin{cases} -3x^2, & 0 \le x \le 1 \\ 0, & \text{otherwise} \end{cases}$$

   Determine the cumulative distribution function.

   [**Ans.** F(x) = 1]

8. A random variable, X is described by a probability density function

$$f(x) = \begin{cases} -3x^2, & 0 \le x \le 1 \\ 0, & \text{otherwise} \end{cases}$$

   Determine the variance of X and expectation of X.

   [**Ans.** –93/80; –3/4]

9. If the variance of random variable X is 0.76, determine the variance of random variable $2X$, and $2X+3$.

   [**Ans.** 3.04; 3.04]

10. If X is a random variable follows a Poisson distribution such that P(X=2) = 0.3, P(X=4) = 0.4, P(X=6) = 0.3, find the variance of X.

    [**Ans.** 2.4]

11. The life of a mining machine (in years) follows an exponential distribution with a mean (λ) value is 1/2. What is the probability that the machine survives at least ten years? It is given that the machine has already been successfully operated for nine years.

    [**Ans.** 0.6]

12. A random variable, X is described by a probability density function

$$f(x) = \begin{cases} -3x^2, & 0 \le x \le 1 \\ 0, & \text{otherwise} \end{cases}$$

    Determine the probability that X is greater and equal to 0.6.

    [**Ans.** 0.784]

13. The grade of copper in a deposit is represented by the following probability density function:

$$f(x) = \begin{cases} x/2, & 0 < x < 2 \\ 0, & \text{otherwise} \end{cases}$$

    Determine the average grade of copper in the deposit.

    [**Ans.** 0.2]

14. Suppose the time (in minutes) that a shovel has to wait at a production bench in an opencast mine for a truck is found to be a random phenomenon, a probability function specified by the distribution function.

$$f(x)=\begin{cases} 0, & x\le 0 \\ 1/2, & 0<x\le 1 \\ 1/3, & 1<x\le 2 \\ 1/4, & 2<x\le 4 \\ 0, & x>4 \end{cases}$$

Determine the probability that the shovel will have to wait (a) more than 3 minutes (b) between 1 and 3 minutes.

[**Ans.** 1/3; 5/6]

15. An iron ore mine recorded an average of 3 accidents per year. The number of accidents is distributed according to Poisson distribution. Determine the probability that there will be exactly two accidents per year.

[**Ans.** 0.07]

16. Pull from an underground tunnel blasting is normally distributed with a mean of 100 tonnes and variance 100 $(\text{tonne})^2$. Determine the probability that the tonnage value from a blast exceeds 110.

[**Ans.** 0.1587]

17. Let X be uniformly distributed over (0, 1); determine the value of $E(X^3)$.

[**Ans.** 1/4]

18. The random variable X has the following probability mass function

$P(4)=1/4,\ P(8)=1/4,\ P(12)=1/4,\ P(16)=1/4$

Determine the expected value of X.

[**Ans.** 10]

19. In a binomial distribution, the probability of success $p\to 0$ and number of trials $n\to\infty$ such that $\lambda=np$ approaches to a finite value. Determine the variance of the distribution.

[**Ans.** np(1−p)]

20. For oil exploration drilling, the chance of striking an oil reservoir is 1 out of 15. If an oil exploration company decides to explore five sites, determine the probability of successful striking of at least one oil reservoir.

[**Ans.** 0.292]

21. The grade values of lead and zinc of 10 samples collected from a mine are as follows:

Sample No.	Zinc (%)	Lead (%)
1	7	2
2	6	1
3	8	3
4	9	3
5	10	4

Sample No.	Zinc (%)	Lead (%)
6	6	2
7	5	1
8	8	2
9	4	1
10	11	5

Determine the value of the correlation coefficient between lead and zinc.

[**Ans.** 0.8557]

22. The cap lamps purchased by a mining company have an average number of two defects in each unit. If the number of defects follows Poisson distribution, determine the probability of finding a product without any defect.

    [**Ans.** 0.135]

23. In an underground mine section, the CH_4 level increases suddenly due to unknown reasons. A miner at return path near station P starts to run out to the main intake of the mine where he will be safe. Refer the Figure 2.22 for the mine section. The probabilities that he will successfully cross the gallery sections P, Q, R, S, T, and U are 0.9, 0.9, 0.8, 0.7, 0.7, and 0.9, respectively. Determine the probability that he will successfully reach the main intake.

    [**Ans.** 0.678]

24. For an explosives company, the probability of producing a defective detonator is 0.02. Determine the probability that a batch of 50 detonators produced by the company contains at most two defective detonators.

    [**Ans.** 0.92]

25. In a mine, the probability that a miner has experience of dump truck operation is 0.4 and that of shovel operation is 0.6, and both shovel and dump truck operation is 0.3. Determine the probability that the miner would have experience of dump truck operation or shovel operation.

    [**Ans.** 0.7]

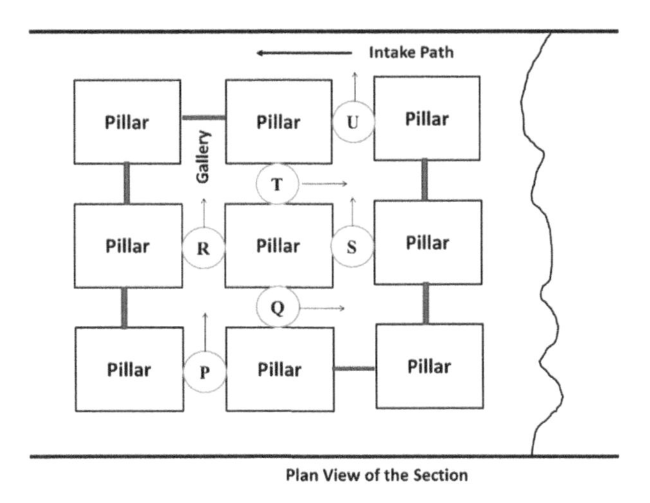

FIGURE 2.22 Plan view of the mine section.

26. The roof bolts produced by a bolts manufacturing company turn out to be defective with an average of 0.20 percentages. The bolts are supplied in a batch of 1000. If the number of defective bolts follows a Poisson distribution, determine the probability that a batch of 1000 contains no defective bolt.

    [**Ans.** 0.135]

27. In an underground mine, the average number of fatal accidents per year is one-half. The number of accidents per year follows the Poisson distribution. Determine the probability that a year will pass without a fatal accident.

    [**Ans.** 0.606]

28. A bauxite deposit has added 5 additional drill holes to increase the confidence in the resource model. The values of alumina (% by weight) and silica (% by weight) in these drill holes are as follows:

Drill hole number	**Alumina (%)–*x***	**Silica (%)–*y***
1	46	1
2	42	5
3	45	2
4	43	4
5	44	3

    Establish the linear relationship between alumina and silica. Also, determine the unbiased estimate of variances of alumina and silica in $(\%)^2$.

    [**Ans.** $y = 47 - x$; 2.5%; 2.5%]

29. The grade of iron ore samples in a mine is normally distributed. The mean (Fe % by weight) and the standard deviation of 30 randomly selected samples are 60% and 4%, respectively. Determine the 95% confidence limits for the population mean.

    [**Ans.** 61.2, 58.8]

3 Linear programming for mining systems

3.1 INTRODUCTION

Linear Programming (LP) is a branch of optimization that minimizes (or maximizes) a linear objective function subject to linear equality and linear inequality constraints (Cococcioni and Fiaschi, 2021). The LP is a widely applied technique both in theoretical and real-world contexts, especially in engineering applications. The simplex algorithm is one of its most used algorithms, which was proposed by Kantorovič and Dantzig in the 1940s. Dantzig independently developed formulation of general linear programming during 1946–1947 to use for the US Air Force planning problems (Dantzig and Thapa, 1997). At the same time, Dantzig (1948) invented the simplex algorithm for solving the linear programming problem efficiently. Subsequently, he planned for solving the industrial and business problems. The simplex method has been applied to solve linear problems in different fields like agriculture, human resources and manufacturing decision making, and so on, to optimize maximum profit and minimize cost. The same method is used in Excel solver to solve linear programming problems. Most researchers in this field (Kurtz, 1992; Taha, 2008; Ezema and Amakon, 2012) applied linear programming in the allocation of scarce resources in the manufacturing industry to boost the output.

3.2 DEFINITION OF LINEAR PROGRAMMING PROBLEM (LPP)

The linear programming problem (LPP) aims to obtain the best solution for the objective of a given problem within the solution space. The objective of the problem is defined by an objective function, and the solution space is defined by a set of constraints. If the objective function and all the constraints of the problem are linear, the problem is said to be an LPP. The objective function of the problem may be of maximizing or minimizing, and the constraints are either equalities or inequalities. In other words, an LPP problem is an optimization model where all the components of the objective function and constraints are linear. In the last few decades, many algorithms have been designed to solve the LPPs. LPP is one of the most popular optimization algorithms applied ever in all systems, including mine systems. Many software packages are available in the market for solving the large-scale LPPs. The most commonly used algorithm to solve the LPPs is the simplex method.

The generalized structure of an LPP can be represented as:

$$\text{Maximize}(\text{Minimize})\,Z = C_1x_1 + C_2x_2 + \cdots + C_nx_n \qquad \text{Objective function}$$

$$\text{Subject to} \quad a_{11}x_1 + a_{12}x_2 + \cdots + a_{1n}x_n \le b_1 \qquad \text{Constraint (1)}$$

$$a_{21}x_1 + a_{22}x_2 + \cdots + a_{2n}x_n \le b_2 \qquad \text{Constraint (2)}$$

$$\cdots$$

$$a_{m1}x_1 + a_{m2}x_2 + \cdots + a_{mn}x_n \le b_m \qquad \text{Constraint (m)}$$

$$x_1, x_2, \ldots, x_n \ge 0 \qquad \text{Non-negativity constraint}$$

DOI: 10.1201/9781003200703-3

In the above LPP, Z is called the objective function that needs to be optimized. The variables $x_1, x_2, \ldots, x_n$ are called decision variables. All the decision variables are subject to non-negative constraints. The problem has m number of constraints, which demarcate the solution space. A solution that falls within the solution space and satisfies all the m number of constraints is called a feasible solution. The solution space is defined as the feasible region.

$C_1, C_2, \ldots, C_n$ are cost coefficients of decision variables in the objective function.

$a_{i1}, a_{i2}, \ldots, a_{in}$ are coefficients of decision variables for i^{th} constraint ($i = 0,1,\ldots,m$)

In any LPP, both the objective function and the constraints are linear functions of decision variables. Any set of values assigned to the decision variables in a given LPP is defined as a solution. It may or may not satisfy any or all of the constraints.

Moreover, if the solution satisfies all the constraints of the problem, the solution is referred to as a feasible solution. On the other hand, any solution that violates at least one of the constraints is referred to as an infeasible solution.

3.3 SOLUTION ALGORITHMS OF LPP

It is important to judge the optimization algorithms before applying them to solve an LPP. To solve an LPP, multiple assumptions are implicit. These assumptions are:

- **Proportionality**: The contribution of each variable, x_i to the objective function, Z is proportional to the coefficient of that variable. Mathematically, this is represented in Eq. (3.1)

$$Z \propto C_i \tag{3.1}$$

  where C_i is coefficient of a decision variable x_i in the objective function, Z.
- **Additivity**: The total contribution of all variables to the objective function is the sum of the individual contribution of the respective variables, as shown in Eq. (3.2).

$$Z = \sum_{i=1}^{n} C_i x_i \tag{3.2}$$

- **Divisibility**: Any decision variable can be fractions if satisfying the functional and non-negativity constraint.
- **Certainty:** All the parameters or coefficients in the objective function and the constraints are known with certainty.

3.3.1 GRAPHICAL METHOD

The graphical method is only applicable to solve an LPP that has a maximum of two decision variables. In the graphical method, the problem can be solved in four major steps, as explained below.

Step 1: **Formulation of the LPP**: The objective function along with all the constraints should be formulated for the defined system.

Step 2: **Plot the graph using all the constraint lines:** In most of the cases, decision variables are ≥ 0, and hence the graph can be drawn only in the first quadrant.

Step 3: Identification of feasible solution space.

Step 4: Identification of the optimal solution from all the feasible solutions. In general, the optimal solution (maximum or minimum) in a feasible region represents by one or more corner points. This is because the maximum/minimum distance from the origin is always represented by any of the corner points in the case of a linear line.

- If the goal is to minimize the objective function, the point closest to the origin gives the minimum value of the objective function.
- If the goal is to maximize the objective function, the point farthest from the origin gives the maximum value of the objective function.

Example 3.1

A mine produces ores, classified into two types based on the grade value. The first one is above the cut-off grade (G1), and the other is below the cut-off grade (G2). Before supplying to consumers, the two types of ores are mixed in order to maintain the cut-off grade. The mixture of two types of ores sends in a maximum of 100-tonne batches. The profit for grade G1 and grade G2 are US$50 per tonne and US$30 per tonne. A typical batch should contain at least 50 tonnes of grade G1 and 20 tonnes of grade G2. Determine the optimal mixture that will maximize profit.

Solution

In the first step, the given problem is formulated as an LPP using mathematical equations.

Let a batch of 100 tonne consists of x_1 tonne of grade G1 and x_2 tonne of grade G2.

The goal of the problem is to maximize the total profit. The objective function of the problem is given by

$$\text{Maximum } Z = 50x_1 + 30x_2 \qquad \text{Objective function}$$

$$\text{Subject to, } x_1 + x_2 \leq 100 \qquad \text{Constraint (1)}$$

$$x_1 \geq 50 \qquad \text{Constraint (2)}$$

$$x_2 \geq 20 \qquad \text{Constraint (3)}$$

$$x_1, x_2 \geq 0 \qquad \text{Non-negativity constraint}$$

Constraint (1) represents that the batch should be at most 100 tonnes of coal. Constraints (2) and (3) indicate the minimum amount of ore of two grades that need to be added in the mixture of 100-tonne batches, respectively. The non-negativity constraints indicate that the decision variables are non-negative.

In the next step, the problem is represented in a graphical diagram, as represented in Figure 3.1(a)–(c).

The next step is to identify the feasible region and feasible solutions. The non-negativity constraint of the decision variables is restricted the solution space only in the first quadrant, where both the decision variables are positive. Constraint (1) further narrowed down the solution space, as shown in Figure 3.1(a). Due to inequality constraint, all the points in the shaded region satisfy the constraint (1), and hence it represents the feasible solution space. In addition to constraint (2) in the problem, the feasible solution space is further narrowed down, as shown in Figure 3.1(b). Since the constraint is greater and equal, the points away from the origin satisfy the constraint. Point C represents the intersection of constraint (1) and constraint (2). The addition of the third constraint further reduces the feasible solution space, as shown in Figure 3.1(c). Constraint (3) intersects the constraint (1) and constraint (2) at A and B, respectively. In Figure 3.1(c), the feasible region is the triangle ABC. Any points within the triangle ABC will satisfy all the constraints of the problem.

The optimal solution needs to be identified from all the feasible solutions to obtain the maximum value of the objective function and hence the profit. The number of solution points in the feasible space ABC is infinite and thus need to identify the optimal solution through a systematic procedure. First, the direction in which the value of the objective function, Z increases (decreases for minimizes

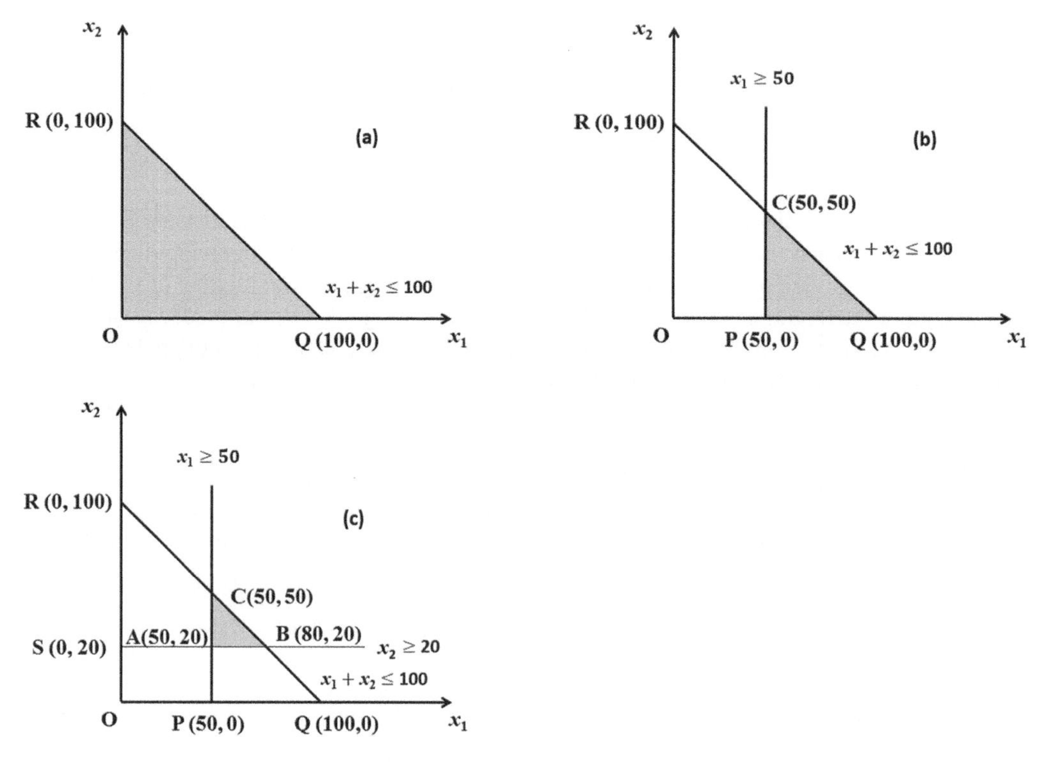

FIGURE 3.1 Graphical representation of the feasible solution space: (a) feasible space satisfying constraint (1), (b) feasible space satisfying constraints (1) and (2), and (c) feasible space satisfying constraints (1), (2), and (3).

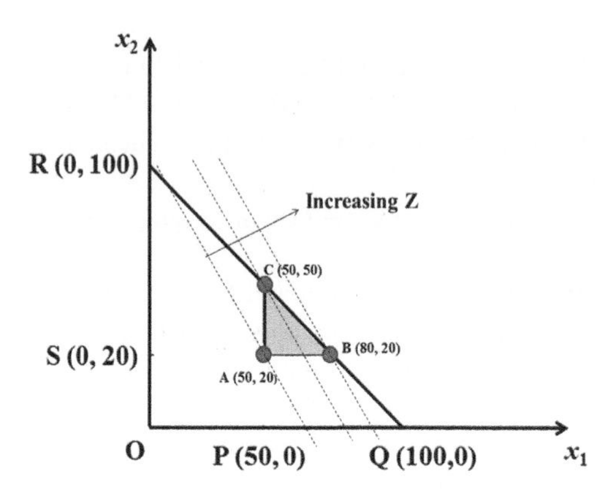

FIGURE 3.2 Searching of the optimal solution in the feasible space for maximizing the objective function.

objective function) need to be identified by arbitrarily choosing the value of the decision variables from the feasible space. The value of Z increases as the solution point is moving away from the origin, as demonstrated in Figure 3.2. The optimal solution is found at point B, which lay on the feasible region beyond which any further increases leads to an infeasible solution. An important characteristic of an LPP is that one or more of the corner point offers the optimal (maximum/minimum) value of the objective function. Thus, the value of the objective function should be examined for all the three corner points (A, B, and C). These are shown in Table 3.1.

TABLE 3.1
Values of the Objective Function at Different Corner Points of the Feasible Region

Point	Values of decision variables at different corner points		Value of objective function (Z)
	x_1	x_2	
A	50	20	310
B	80	20	460
C	50	50	400

The results showed in Table 3.1 clearly indicate that the value of the objective function, Z is maximum at B (80, 20), and this is equal to US$460. Thus, this is the optimum solution out of all the feasible solutions.

It can be inferred that the optimal solution of an LPP is always associated with a corner point, which indicates that the search for the optimum is restricted to a finite number of points rather than infinite. The role of corner points in identifying the optimum solution is the key to the development of the general solution algorithm, called the simplex algorithm.

The above problem has only one optimal solution, and hence the nature of the solution is a **unique solution**. A problem may characterize with other types of solutions like multiple solutions or infinitely many solutions, as explained below.

3.3.1.1 Multiple Solutions

If the LPP has more than one optimal solution, then it is designated as a multiple solutions problem. In the case of multiple solutions, the value of the objective function is optimal for multiple sets of feasible solutions. This is illustrated in **Example 3.2**.

Example 3.2

Maximise $Z = 2x_1 + x_2$ Objective function

Subject to $2x_1 + x_2 \leq 20$ Constraint (1)

$2x_1 + 3x_2 \leq 30$ Constraint (2)

$x_1, x_2 \geq 0$ Non-negativity constraint

The problem is graphically represented in Figure 3.3, and the objective function values for different corner-point solutions are presented in Table 3.2.

It is already seen in Example 3.1 that the optimal solution is associated with one or more of the corner points. But, if the slope of the objective function is parallel to any of the bounded constraint, all the points on the line AB offer the same value of the objective function. Thus, an infinite number of optimal solutions exist for the problem.

The results shown in Table 3.2 indicate that the value of objective function Z is maximum at two points, A (10, 0) and B (15/2, 5), and this is equal to 20. Thus, there is more than one solution at which the value of the objective function is optimum. All the points lying on the line AB offer the same maximum value of the objective function. The problem has infinitely many solutions.

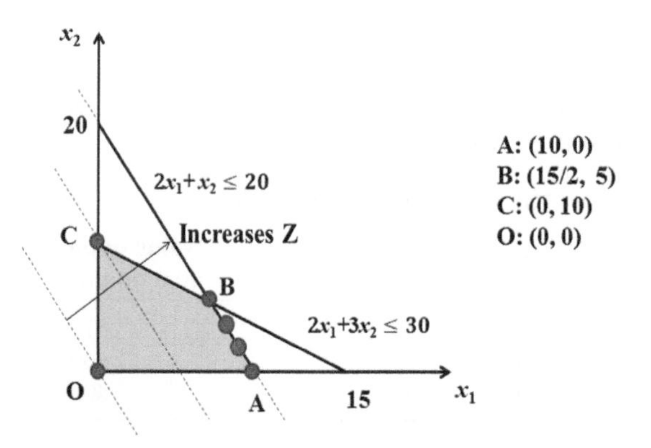

FIGURE 3.3 Graphical representation of the feasible solution space for the problem defined in Example 3.2.

TABLE 3.2
Values of the Objective Function at Different Corner Points of the Feasible Region

Point	Values of decision variables at different corner points		Value of objective function, Z
	x_1	x_2	
A	10	0	20
B	15/2	5	20
C	0	10	10
O	0	0	0

3.3.1.2 Unbounded solution

An LPP has an unbounded solution if the value of the objective function can be reached up to infinity without violating any of the constraints. This is illustrated in **Example 3.3**.

Example 3.3

Maximise $Z = x_1 + x_2$	Objective function
Subject to $x_1 \geq 5$	Constraint (1)
$x_2 \leq 10$	Constraint (2)
$x_1, x_2 \geq 0$	Non-negativity constraint

The problem is graphically represented in **Figure 3.4**.

Figure 3.4 clearly tells that the value of the objective function can reach up to ∞ (Z = ∞) at $x_1 = \infty$, $x_2 = 10$. This indicates that the problem has an unbounded solution in one direction. In the above diagram, there is no boundary or intersection points beyond the point B to give a bounded feasible solution. Thus, the above problem has an unbounded solution.

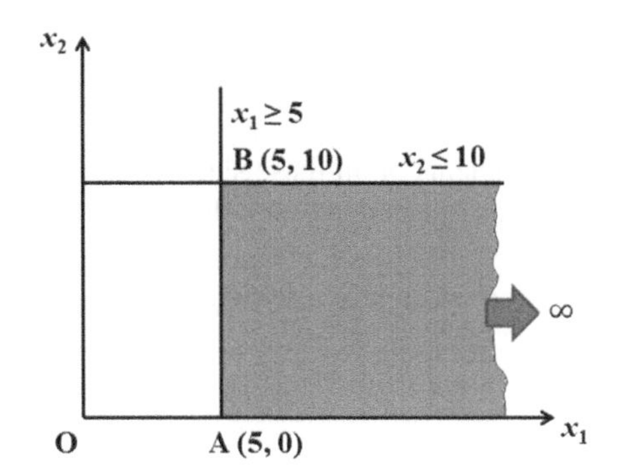

FIGURE 3.4 Graphical representation of the problem.

3.3.2 Simplex Method

The simplex method is a linear-programming solution algorithm that can solve problems with any number of decision variables in contrast to a graphical method, which can handle only two decision variables. Any linear programming problem can be defined in two forms, viz. primal and dual. The simplex algorithm on a primal problem starts with a feasible but non-optimal solution. It was observed that the optimal solution of a linear problem is represented by one or more corner points. The primal simplex algorithm starts with a feasible solution and reaches optimality through the iterative method with satisfying the feasibility conditions. The initial feasible solution corresponds to the origin, and subsequent iteration is developed by shifting to an adjacent corner point in the direction that yields the better value.

On the other hand, the dual problems are those for which it is easy to obtain an initial basic solution, which is infeasible but satisfies the optimality criterion.

Simplex algorithm

The algorithm is demonstrated with a generalized example of a linear programming problem. In the first case, the algorithm is explained for an LPP with only slack variables; that is, the problem has only less and equal type (inequality) constraints. The algorithm is explained with a generalized form of LPP.

Generalised form of an LPP

Maximize $Z = C_1x_1 + C_2x_2 + \cdots + C_nx_n$ Objective function

Subject to $a_{11}x_1 + a_{12}x_2 + \cdots + a_{1n}x_n \leq b_1$ Constraint (1)

$a_{21}x_2 + a_{22}x_3 + \cdots + a_{2n}x_n \leq b_2$ Constraint (2)

...

$a_{m1}x_1 + a_{m2}x_2 + \cdots + a_{mn}x_n \leq b_m$ Constraint (m)

$x_1, x_2, \ldots, x_n \geq 0$ Non-negativity constraint

where

Z is the objective function,

$x_1, x_2, \ldots x_n$ are decision variables,

$C_1, C_2, \ldots C_n$ are coefficients of $x_1, x_2, \ldots x_n$, respectively, in the objective function.

$a_{i1}, a_{i2} \ldots a_{in}$ are coefficients of $x_1, x_2, \ldots x_n$, respectively, in the ith constraints (i =1 to m),

$b_1, b_2, \ldots, b_n$ are non-negative right hand side values of the m constraints, respectively.

The steps involved in solving an LPP using the simplex method are explained below.

Step 1: Formulation of the standard form of the problem

Each inequality constraint should be converted into an equality constraint by introducing a slack variable to the left-hand side. The coefficients of all these slack variables in the objective function will be zero. The nature of variables needs to be added for a different form of constraints is summarized in Table 3.3.

In the simplex method, the LPP is written in a standard form with a maximization problem, as demonstrated below.

A standard form of the LPP:

$$Z - C_1x_1 - C_2x_2 - \cdots C_nx_n - 0y_1 - 0y_2 - \cdots 0y_m = 0 \qquad \text{Objective function}$$

Subject to

$$a_{11}x_1 + a_{12}x_2 + \cdots a_{1n}x_n + y_1 = b_1 \qquad \text{Constraint (1)}$$

$$a_{21}x_1 + a_{22}x_2 + \cdots a_{2n}X_n + y_2 = b_2 \qquad \text{Constraint (2)}$$

$$\cdots$$

$$a_{m1}x_1 + a_{m2}x_2 + \cdots a_{mn}x_n + y_m = b_m \qquad \text{Constraint (m)}$$

$$x_1, x_2, \ldots, x_n, \; y_1, y_2, \ldots, y_m \geq 0 \qquad \text{Non-negativity constraint}$$

In the standard form of LPP, $S_1, S_2, \ldots, S_m$ are called slack variables. The coefficients of all the slack variables in the objective function are equal to zero.

Step 2: Finding an initial basic feasible solution or starting solution

The starting solution of an LPP, which is **basic** and **feasible**, is termed as an initial basic feasible solution. A basic solution, which is feasible, is called a **basic feasible solution**.

In an LPP, the number of basic variables is equal to the number of equality constraints. The set of basic variables in a solution is defined as a basis. All the variables under the basis are non-negative in nature. All variables outside the basis are designated as non-basic variables. All the non-basic variables are zeros.

If an LPP has n decision variables and m slack variables, then the basic solution has m number of basic variables and n number of non-basic variables. The basic solution should have at most m non-zero variables, which is equal to the number of equality constraint or slack variables and at least n zero-valued variables, number of decision variables. Therefore, the initial solution of the above generalized problem is a basic solution in which $x_1, x_2, x_3, \ldots, x_n$ are non-basic variables and $y_1, y_2, \ldots, y_m$ are basic variables.

TABLE 3.3
Transformation of Inequality Constraints into Equality Constraints

Constraint type	Variable to be added
$\leq$	+ Slack (s)
$\geq$	– Surplus (s) + artificial (A)
=	+ Artificial (A)

In the standard form of the generalized LPP, the total number of variables is $(n + m)$. Out of $(n + m)$ variables, n numbers are decision variables, and m numbers are slack variables. A unique solution of m number of slack variables in the problem can be determined by setting the remaining n decision variables equal to zero.

$$\text{That is, } \left. \begin{aligned} x_1 = x_2 = x_3 = \cdots x_n = 0 \\ y_1 = b_1, y_2 = b_2, \ldots, y_m = b_m \end{aligned} \right\} \text{Starting or Initial Solution}$$

Step 3: Formulation of the initial simplex tableau

The initial simplex tableau is formed with the structure as shown in Table 3.4(a).

Step 4: Method of iteration

Step 4.1: Selection of entering variable

A new basis will be formed by replacing one of the current basic variables (y_1/ y_2/…y_m) with a non-basic variable (x_1/ x_2/…x_n). The non-basic variable selected to **enter into the basis** depends on the extent of the improvement in the objective function value. The non-basic variable, which decreases (increases for minimization problem) the Z-value maximum, is chosen as a basic variable for the next iteration. The variable is also called **entering variable**. In case of a tie, either of the variables can be chosen. The highest negative coefficient increases the value of the objective function most. The column of entering variable is called the pivotal column.

A non-basic variable with the most negative value (in case of maximization) or with the most positive (in case of minimization) in the Z-row should be selected as the entering variable. The entering variable column is called the 'pivot column', as shown in Table 3.4(b). It is assumed that C_r is the highest negative value in the Z-row under the non-basic variable, and thus x_r is selected as entering variable.

Step 4.2: Selection of departing variable

- Identify strictly positive (>0) coefficients (a_{1r}, a_{2r},…,) in the pivot column for all constraints.

TABLE 3.4(a)
Structure of Initial Simplex Tableau

		Coefficients of								
Row	**Basis**	x_1	x_2	…	x_n	y_1	y_2	…	y_m	**R.H.S**
Constraint 1	y_1	a_{11}	a_{12}		a_{1n}	1	0		0	b_1
Constraint 2	y_2	a_{21}	a_{22}		a_{2n}	0	1		0	b_2
⋮				⋮	⋮		⋮		⋮	
Constraint m	y_m	a_{m1}	a_{m2}		a_{mn}	0	0	…	1	b_m
Z-row	Z	$-C_1$	$-C_2$	…	$-C_n$	0	0		0	0

TABLE 3.4(b)
Simplex tableau showing the pivot column

		Coefficients of										
Row	**Basis**	x_1	x_2		x_r		x_n	y_1	y_2	…	y_m	**R.H.S**
Constraint 1	y_1	a_{11}	a_{12}		a_{1r}		a_{1n}	1	0		0	b_1
Constraint 2	y_2	a_{21}	a_{22}		a_{2r}		a_{2n}	0	1		0	b_2
⋮				⋮		⋮	⋮					
Constraint m	y_m	a_{m1}	a_{m2}		a_{mr}		a_{mn}	0	0		1	b_m
Z-row	Z	$-C_1$	$-C_2$		$-C_r$		$-C_n$	0	0	…	0	0

Pivot Column

TABLE 3.4(c)
Simplex tableau showing pivot row and pivot column

		Coefficients of										
	Basis	x_1	x_2		x_r		x_n	y_1	y_2	…	y_m	**R.H.S**
Constraint 1	y_1	a_{11}	a_{12}		a_{1r}		a_{1n}	1	0		0	b_1
Constraint 2	y_2	a_{21}	a_{22}	⋮	a_{2r}	⋮	a_{2n}	0	1		0	2
⋮							⋮					
Constraint i	y_i	a_{i1}	a_{i2}		a_{ir}		a_{il}	0	0	…	1	b_i
⋮							⋮					
Constraint m	y_m	a_{m1}	a_{m2}		a_{mr}		a_{mn}	0	0		1	b_m
Z-row	Z	$-C_1$	$-C_2$		$-C_r$		$-C_n$	0	0	…	0	0

Pivot Row (Constraint i)

Pivot Column

- Divide each of these positive coefficients into the respective R.H.S (right-hand side) values.
- Identify the row, which exhibits the smallest ratios (b_i/a_{ir}).

$$(\text{Ratio})_i = b_i / a_{ir}, \quad \text{for } a_{ir} > 0 \tag{3.3}$$

In Eq. (3.3), 'i' represents the row, and 'r' represents the pivot column (corresponding to the entering variable, x_r). If all the coefficients (a_{ir}) in the pivot column, r, are negative in nature, then the problem is ill-defined, and the problem leads to an **unbounded solution**.

The basic variable (y_i) correspond to the smallest ratio is identified as the departing variable. The departing variable row is called the 'pivot row', as indicated in Table 3.4(c). In case of a tie, either of the variables can be chosen.

- Replace the departing variable by the entering variable in the basic variable column of the next simplex tableau.

Step 4.3: Solve for the new solution by using elementary row operations using Gauss-Jordan transformation as:

- The new pivot row can be updated by dividing the old pivot row by the 'pivot number' (the number at the intersection of the pivot row and pivot column). Thus, the new pivot row is

$$\text{New coefficients for pivot row} = \frac{(\text{Old pivot row})_i}{\text{Pivot coefficients}(a_{ir})} \tag{3.4}$$

- All other rows can be updated as:

$$\begin{aligned}(\text{New coefficients for all other row})_j &= (\text{Old row})_j \\ &- (\text{Pivot column coefficients})_j * \text{ New pivot row}\end{aligned} \tag{3.5}$$

Step 5: Checking the Optimality

The optimality of the problem should be examined after obtaining a new solution. If all the coefficients in the Z-row are non-negative for the maximization problem (non-positive for minimization problem), then the solution is optimal, and no further iteration is needed.

If there is a zero coefficient under any of the non-basic variables in the Z-row, then an alternate solution of the problem exists. This can be determined using a similar procedure with further iteration.

The above simplex algorithm is applied to solve the LPP, as demonstrated in **Example 3.4**.

Example 3.4

An open-pit mine produces ores of two designated grades (G1 and G2) from two different working benches in two stages, viz. ore breakage through drilling-blasting and transportation through haul-truck. Each stage has an operating time of 8 hours in a shift. It is observed that 1000 units of G1 type ore require 2 hours of operation on Stage 1 and 3 hours on Stage 2. For G2 type of ore, 1000 units require 3 hours of operation at Stage 1 and 2 hours at Stage 2. The revenues per unit of G1 and G2 types of ores are US$180 and US$150, respectively. Determine the optimal production plan.

Solution

Mathematical Formulation of the LPP

Assuming x_1 and x_2 represent the number of thousand units ore produced of Grade 1 and Grade 2, respectively, in 8 hours. The LPP of the defined problem can be formulated as

$$\begin{array}{lll} \text{Maximise} & Z = 180x_1 + 150x_2 & \text{Objective function} \\ \text{Subject to,} & 2x_1 + 3x_2 \le 8 \ (\text{Stage 1}) & \text{Constraint (1)} \\ & 3x_1 + 2x_2 \le 8 \ (\text{Stage 2}) & \text{Constraint (2)} \\ & x_1, x_2 \ge 0 & \text{Non-negativity constraint} \end{array}$$

The problem has two constraints, and thus two slack variables y_1 and y_2, are introduced. Both the slack variables are non-negative.

Standardized form of the LPP

$$Z - 180\,x_1 - 150\,x_2 - 0y_1 - 0y_2 = 0$$

$$\text{Subject to } 2x_1 + 3x_2 + y_1 + 0y_2 = 8$$

$$3x_1 + 2x_2 + 0y_1 + y_2 = 8$$

$$x_1, x_2, y_1, y_2 \geq 0$$

In the next step of the simplex method, an initial basic feasible solution needs to be identified.

The initial basic feasible solution is identified by setting all the decision variables (in this case, x_1 and x_2) equal to zero. That is, all the decision variables are non-basic variables, and all the slack variables are basic variables in the initial basic feasible solution. The values of slack variables at the initial basic feasible solution can be determined from the constraint functions.

From the first constraint, $2*0+3*0+y_1+0*y_2=8$

$$\Rightarrow y_1 = 8$$

Similarly, from the second constraint, $3x_1 + 2x_2 + 0y_1 + y_2 = 8$

$$\Rightarrow y_2 = 8$$

The initial basic feasible solution of the given problem is

$$\left(x_1 = 0,\ x_2 = 0,\ y_1 = 8, y_2 = 8\right)$$

At the starting solution, the value of the objective function is given by

$$180*0+150*0+0*8+0*8=0$$

The basic feasible solution at the initial tableau is ($x_1 = 0$, $x_2 = 0$, $y_1 = 8$, $y_2 = 8$). The initial basic feasible solution represents one of the corner points (origin: 0, 0). Thus, y_1 and y_2 are basic variables, and x_1 and x_2 are non-basic variables in the initial basic feasible solution. The initial basic feasible solution is represented in Table 3.5(a).

The next step is to find a better solution through an iterative process. In a subsequent iteration, the solution is shifted to a new corner point adjacent to the previous solution in the feasible region, which offers a better solution.

Selection of entering variable: In this case, the most positive coefficient in Z-row is 180, which is under x_1. Thus, x_1 should be chosen as an entering variable. The highest positive coefficient

TABLE 3.5(a)
Initial simplex tableau

		Coefficient of				
	Basis	x_1	x_2	y_1	y_2	RHS
Row 1	y_1	2	3	1	0	8
Row 2	y_2	3	2	0	1	8
Row Z	Z	−180	−150	0	0	0

TABLE 3.5(b)
Simplex Tableau Showing the Pivot Row and Pivot Column

	Basis	Coefficient of				RHS	Ratio	
		x_1	x_2	y_1	y_2			
Row 1	y_1	2	3	1	0	8	8/2 = 4	
Row 2	y_2	3	2	0	1	8	8/3	⇦ Departing Variable
Row Z	Z	–180	–150	0	0	0		

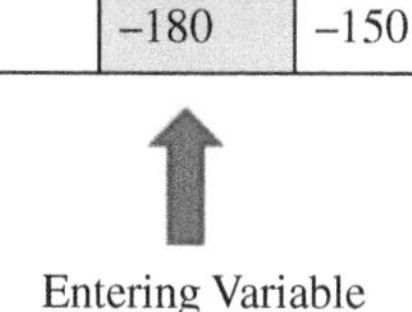

increases the value of the objective function most. Thus, column x_1 is chosen as the pivotal column.

Departing Variable: In this case, out of the two basic variables (y_1 or y_2), one should depart the basis to allow entry of x_1 into the new basis. The ratio, as demonstrated in Eq.(3.1), is used to select the departing variable.

In this case, the minimum ratio is obtained for basic variable y_2, as shown in Table 3.5(b). Therefore, y_2 is chosen as a departing variable.

Thus, x_1 replaces y_2 in the new solution, where y_1 and x_1 are the member of the basis. The new pivot row (Row 2) is obtained using the Gauss-Jordan transformation, as explained in Eq. (3.2) and Eq. (3.3).

Thus, the coefficient of the new pivot row is $\left(\frac{3}{3},\frac{2}{3},\frac{0}{3},\frac{1}{3},\frac{8}{3}\right)=\left(1,\frac{2}{3},0,\frac{1}{3},\frac{8}{3}\right)$.

Thus, the coefficients of all other row are given by

$$\text{New Row 1}=\left[2,3,1,0,8\right]-(2)*\left[1,\frac{2}{3},0,\frac{1}{3},\frac{8}{3}\right]=\left[2,3,1,0,8\right]-\left[2,\frac{4}{3},0,\frac{2}{3},\frac{16}{3}\right]=\left[0,\frac{5}{3},1,-\frac{2}{3},\frac{8}{3}\right]$$

In the same way, the new z row can be determined as

$$\text{New Z Row}=\left[-180,-150,0,0,0\right]-(-180)*\left[1,\frac{2}{3},0,\frac{1}{3},\frac{8}{3}\right]$$

$$=\left[-180,-150,0,0,0\right]+\left[180,120,0,60,480\right]$$

$$=\left[0,-30,0,60,480\right]$$

The revised simplex table, after the first iteration, is shown in Table 3.5(c).

After completion of the first iteration, the value of the objective function is increased to 480 from 0.

Checking the optimality

The optimality of the problem should be examined after obtaining a new solution. The solution is not optimal as there is still one negative coefficient (–30) in the Z-row under non-basic variable (x_2). Thus, a further iteration is required to obtain the optimal solution. This time, the entering variable is x_2, as it has the maximum negative coefficient in the Z-row. On the other hand, the departing variable is y_1, as it has the minimum ratio. The revised value after the second iteration is shown in Table 3.5(d).

TABLE 3.5(c)
Simplex Tableau after the First Iteration

	Basis	Coefficient of x_1	x_2	y_1	y_2	RHS	Ratio	
Row 1	y_1	0	5/3	1	−2/3	8/3	8/5	← Departing Variable
Row 2	x_1	1	2/3	0	1/3	8/3	4	
Row z	Z	0	−30	0	60	480		
			↑ Entering Variable					

TABLE 3.5(d)
Revised Simplex Table after the Second Iteration

	Basis	Coefficient of x_1	x_2	y_1	y_2	RHS
Row 1	x_2	0	1	3/5	−2/5	8/5
Row 2	x_1	1	0	−2/5	3/5	8/5
Row Z	Z	0	0	18	48	528

It is clear from the simplex table that no negative coefficient exists in the Z-row under the basic variable. Thus, the solution reached the optimal stage, and no further iteration was required.

However, in this case, no zero exists in the Z-row under the non-basic variable, and hence the problem has a unique solution.

Thus, the optimal solution is

$$x_1 = 8/5, x_2 = 8/5, Z = 180*8/5 + 150*8/5 = 528$$

Therefore, the maximum revenue can be obtained as US$528.

Special cases of LPP solutions

- **Multiple solutions or infinitely many solutions**: If zero is found under one or more non-basic variables in the optimal solution tableau, the problem has multiple optimal solutions.
- **Unbounded solution**: If there is no positive ratio found while calculating the leaving variable in any simplex tableau (all the entries in the pivot column are negative and zeroes), the solution is unbounded.
- **Degenerated solution**: If the basic variable is assigned to zero value in the solution, the solution is termed as degenerate solution.

3.3.3 Big-M method

Although the simplex algorithm is the most powerful to solve the LPP, it has an initialization problem like the starting solution must be a basic feasible to solve the problem. If the LPP has any equality or greater and equality constraints, it has to incorporate an artificial or surplus variable to convert into a standardized form. In that case, the simplex algorithm cannot be directly applied due to the initialization issue. To overcome the initialization problem, two approaches (the Big-M method and two-phase method) may be used to solve the problem. These methods can accommodate both positive and negative aspects of the initial basic feasible solution. The Big-M method is explained with **Example 3.5**.

Example 3.5

A company has two iron ore mines and a processing plant, located 20 km away from the first mine and 5 km from the second mine. Both the mines produce ores and are transported to the processing plant through conveyor belts. The hourly production of ore and waste in the first mine is 300 tonnes and 100 tonnes, respectively, and the corresponding operational cost is US$4000. Similarly, the hourly production of ore and waste in the second mine is 100 tonnes and 100 tonnes, respectively, and the corresponding operational cost is US$1000. To meet customer demands, at least 6000 tonnes of ores must be processed daily. Also, in order to comply with government regulations, at most, 5000 tonnes of waste can be produced daily from both mines. Determine the optimum working hours in two mines for minimizing the operational cost.

Solution

Assuming that x_1 is the number of operating hours per day of process 1 and x_2 is the number of operating hours per day of process 2, the mathematical formulation of the linear program becomes

Minimize	$Z = 4000x_1 + 1000x_2$	Objective function
Subject to	$300x_1 + 100x_2 \geq 6000$	Constraint (1)
	$100x_1 + 100x_2 \leq 5000$	Constraint (2)
	$x_1; x_2 \geq 0$	Non-negativity constraint

The objective of the problem is to minimize the operational cost subject to satisfying the constraints of productions of mineral and waste. The constraint with ≥ sign needs to incorporate a surplus in addition to an artificial variable to make it an equality constraint, as explained in Table 3.3. The standardized objective function is derived by adding (subtracting for maximization problem) the right-hand side with the multiplication of artificial variables and a big number (M), as shown below in the standard form of the objective function. The reason for choosing a big number, M, is that the values of artificial variables $(A_1, A_2, \ldots, A_n)$ becomes zero in the final (optimal) tableau. The rules for adding artificial variables are as follows:

- For Minimization objective function, $A_1, A_2, \ldots, A_n$ must be added to the RHS of the objective function multiplied by a big number (M).
- For Maximization objective function, $A_1, A_2, \ldots, A_n$ must be subtracted from the RHS of the objective function multiplied by a very large number (M)

Standardized form of the problem

$$Z - 4000x_1 + 1000x_2 - 0.y_1 - MA_1 = 0$$

$$\text{Subject to } 300x_1 + 100x_2 - V_1 + A_1 = 6000$$

$$100x_1 + 100x_2 + y_1 = 500$$

$$x_1,\ x_2,\ S_1,\ A_1,\ y_1 \geq 0$$

TABLE 3.6(a)
Initial Simplex Tableau

	Basis	Coefficient of					RHS
		x_1	x_2	V_1	y_1	A_1	
Row 1	A_1	300	100	–1	0	1	6000
Row 2	y_1	100	100	0	1	0	5000
Row Z	Z	–4000	–1000	0	0	–M	0

TABLE 3.6(b)
Revised Initial Simplex Tableau

	Basis	Coefficient of					RHS	Ratio	
		x_1	x_2	V_1	y_1	A_1			
Row 1	A_1	300	100	–1	0	1	6000	20	← Departing Variable
Row 2	y_1	100	100	0	1	0	5000	50	
Row Z	Z	300M–4000	100M–1000	–M	0	0	6000M		
		↑ Entering Variable							

The initial solution of the problem is obtained by equating the decision and surplus variable zero. Thus, in the initial solution, y_1 and A_1 are basic variables. The initial simplex tableau is shown in Table 3.6(a).

To apply the simplex method, the coefficients in the Z-row under all basic variables should be equal to zero except on surplus variables, which have no restrictions. This is not followed in the initial simplex tableau, as demonstrated in Table 3.2(a). This can be done through elementary row operations, as shown in Eq. 3.6.

$$\text{New (Z row)} = \text{Old (Z row)} \pm \text{M } (\text{A}_1 \text{ row}) \pm \text{M } (\text{A}_2 \text{ row}) \cdots \text{M } (\text{A}_n \text{ row}) \tag{3.6}$$

In this case, the new Z-row is obtained as follows:

Old (Z row):	–4000	–1000	0	0	–M	0
M (A_1 row)	300M	100M	–M	0	M	6000M
New (Z row):	300M–4000	100M–1000	–M	0	0	6000M
					↑ Becomes zero	

The revised simplex tableau is shown in Table 3.6(b).

Now the same simplex algorithm can be applied to solve the problem. Since there is a positive value in the Z-row under non-basic variables excluding surplus variable in Table 3.6(b), the initial solution is not optimal. The entering variable is the most positive value in the Z-row under non-basic variables. It is clear from Table 3.6(b) that x_1 has the maximum positive value (300M-4000) and is thus selected as an entering variable. The departing variable is chosen based on the smallest ratio. The ratios showed in Table 3.6(b) indicate that A_1 has the smallest ratio and is thus chosen as a departing variable. The new rows are determined using Gauss Jordan transformation as explained in the simplex method. The revised simplex tableau is shown in Table 3.6(c).

TABLE 3.6(c)
Revised Simplex Tableau after First Iteration

	Basis	x_1	x_2	V_1	y_1	A_1	RHS	Ratio	
Row 1	x_1	1	1/3	–1/300	0	1/300	20	60	
Row 2	y_1	0	200/3	1/3	1	–1/3	3000	45	← Departing Variable
Row Z	Z	0	1000/3	–40/3	0	–M + 40/3	80000		
			↑ Entering Variable						

(Coefficient of: x_1, x_2, V_1, y_1, A_1)

TABLE 3.6(d)
Revised Simplex Tableau after the Second Iteration

	Basis	x_1	x_2	V_1	y_1	A_1	RHS
Row 1	x_1	1	0	1/200	–1/200	1/200	25/2
Row 2	x_2	0	1	1/200	3/200	–1/200	45/2
Row Z	Z	0	0	–15	–5	–M + 15	72500

(Coefficient of: x_1, x_2, V_1, y_1, A_1)

In Table 3.6(c), it can be observed that still there is a positive value in the Z- row under non-basic variable excluding the surplus variable, and thus the solution is not optimal. The entering variable is the most positive value in the Z-row. It is clear from Table 3.6(c) that the entering variable is x_2 (maximum positive value =1000/3), and the leaving variable is y_1 (smallest ratio =45).

The current solution is optimal no positive coefficients exist in the Z-row under the basic variable, excluding the surplus variable.

Thus, the optimal solution is

$$x_1 = 25/2 = 12.5, x_2 = 45/2 = 22.5, Z = 72500$$

Thus, the total operational cost is US$72500 for 12.5 working hours in the first mine and 22.5 hours in the second mine.

Special cases of solution of Big-M method

- If one or more artificial variables (A_1, A_2, ..., A_n) are found to be in the basis and have a non-zero value in the RHS of the optimal tableau, then the problem has an infeasible solution.
- The other conditions explained for the simplex method are also applicable for this method.

The readers, who are interested to learn the algorithm of the two-phase simplex method, can refer to Taha (2016). In this method, the problem is solved in two phases. In phase I, a starting basic feasible solution is identified; whereas, in phase 2, the original problem is solved.

3.4 SENSITIVITY ANALYSIS

Sensitivity analysis of an LPP also referred to as post-optimality analysis, is an exercise to understand the impacts of the coefficients of the objective function and resource limits of the constraints on the final solution. It finds a new solution without solving afresh. The sensitivity analysis of an LPP is done to examine the effect of change in the data of the original problem on the optimality. This type of analysis is important for a dynamic system, where the coefficients of the objective function and availability of resources of the constraints are not estimated with certainty.

3.4.1 Graphical Method of Sensitivity Analysis

The graphical method of sensitivity analysis can be applied to an LPP with a maximum of two decision variables. The graphical method of sensitivity analysis is explained using a decision-making problem, as defined in **Example 3.4**.

Mathematical form of the problem

$$\text{Maximise} \quad Z = 180x_1 + 150x_2$$

$$\text{Subject to} \quad 2x_1 + 3x_2 \le 8 \ (\text{Stage 1})$$

$$3x_1 + 2x_2 \le 8 \ (\text{Stage 2})$$

$$x_1, x_2 \ge 0$$

The feasible solution space of the given LPP is shown in Figure 3.5. The values of the objective function at different corner points are shown in Table 3.7.

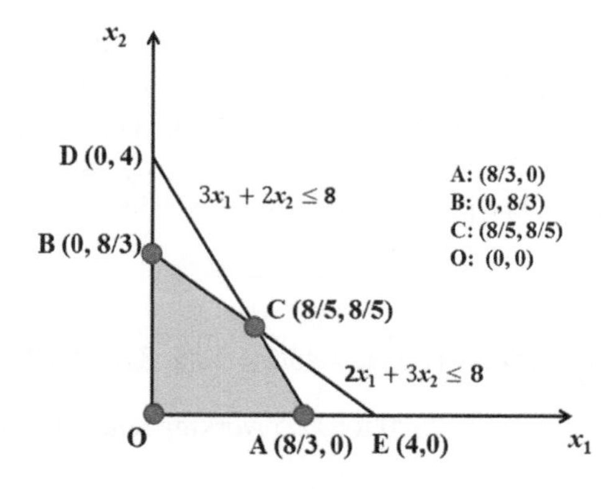

FIGURE 3.5 Graphical representation of feasible solution space.

TABLE 3.7
Values of the Objective Function at Different Corner Points of the Feasible Region

	Feasible solution of variables		
Point	x_1	x_2	**Value of objective function Z**
O	0	0	0
A	8/3	0	480
B	0	8/3	400
C	8/5	8/5	528

The optimal value of the objective function, $Z = 528$ for $x_1 = 8/5$ and $x_2 = 8/5$

The sensitivity analysis of an LPP is conducted in two ways, viz. change in the input resources or right-hand side of the constraint and change in the cost coefficients of the objective functions. The following types of post-optimal analyses can be done in sensitivity analysis of the problem.

Case I: Changes in the available resources or RHS of the constraints

The sensitivity of the model is analyzed for change in operating hours of Stage 1/Stage 2. It is clear that increasing or decreasing the available resources or RHS of the constraints, which represents total working hours of Stage 1 or Stage 2, moves the corresponding constraint parallel to itself provided the coefficients in the constraint are not changed. The current solution is optimal until the first and second constraints intersecting with each other.

Figure 3.6(a) illustrates the change in the optimum solution when changes are made in the total working hours of Stage 1. If the working hour of Stage 1 is increased from 8 hours to 9 hours, the new optimum will occur at point C1. The rate of change in the objective function (Maximise Z) resulting from changing working hour of Stage 1 from 8 hours to 9 hours can be determined as follows:

$$\text{Rate of change in the objective function value} = \frac{Z_{C1} - Z_C}{(\text{Change in available resources})} = \frac{546-528}{9-8}$$

$$= \text{US\$18 / hour} \; [\text{Optimal point shifted from C to C1}]$$

The computed revenue change rate directly links the available resources (input) and the total revenue (output). That is, the rate represents the change in the optimal objective function value per unit change in the input resource availability is defined as the dual price. The result indicates that unit increase (decrease) in the total working hours of stage 1 will increase (decrease) the revenue by US$18. That is, the dual price of Stage 1 is US$18. The dual price represents the change in the price for a unit change in the conditions of service for the same product or service.

The dual price (prices in different conditions for the same product or service) of US$18/hour remains valid for changes (increases or decreases) in the total working hours in Stage 1 that move the constraint parallel to itself to any point on the line segment *AD*, as shown in Figure 3.6(a). That

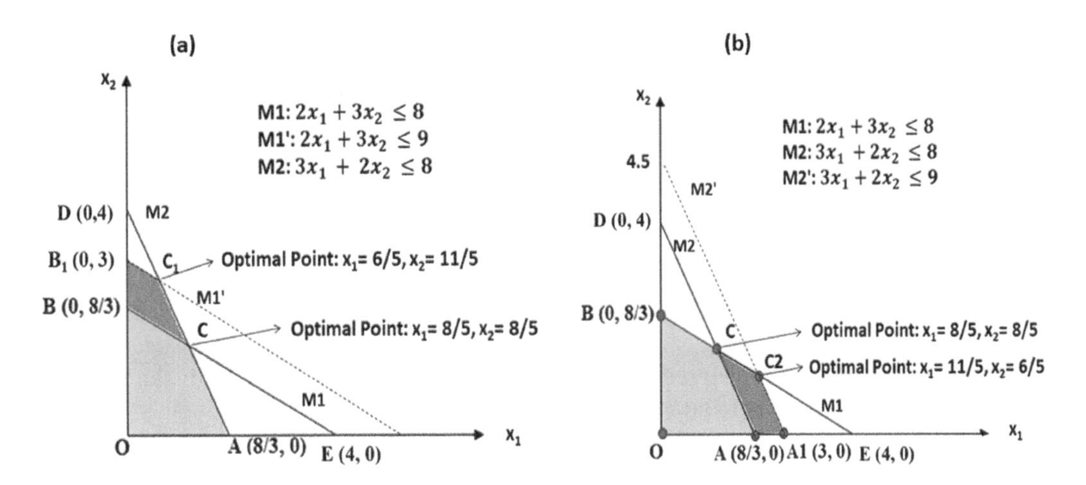

FIGURE 3.6 Graphical representations of sensitivity of optimal solution with change in the availability of resources (right-hand side of the constraints): (a) Unit increase in the resource of Stage 1; and (b) Unit increase in the resource of Stage 2.

is, the range of working hours for Stage 1 for which the dual price is valid should lay at points A and D. This means that the range of applicability of the given dual price can be computed as follows:

$$\text{Working hours of Stage 1 at point A} = 2x_1 + 3x_2 = \left(2*\frac{8}{3}\right) + 3*0 = 5.34 \text{ hours}$$

$$\text{Working hours of Stage1 at point D} = 2x_1 + 3x_2 = 2*0+3*4=12\,\text{hours}$$

That is, the dual price of US$18/hour will remain valid for the range

$$5.34\,\text{hours} \leq \text{Working hours of Stage1} \leq 12\,\text{hours}$$

Changes outside this range will produce a different dual price, which needs different calculations with additional information.

In a similar way, Figure 3.6(b) indicates that the dual price of US$48/hour for Stage 2, which remains valid for changes (increases or decreases) in the range of B to E. In this case, the dual price remains valid for changes (increases or decreases) that move its constraint parallel to itself to any point on the line segment *BE*, as shown in Figure 3.6(b).

$$\text{Rate of change in the objective function value} = \frac{Z_{C2} - Z_C}{(\text{Capacity change})} = \frac{576-528}{9-8}$$

$$= \text{US\$48 / hour [Optimal point shifted from C to C2]}$$

$$\text{Working hours of Stage 2 at point B} = 3*0+2*8/3 = 5.34\,\text{hours}$$

$$\text{Working hours of Stage 2 at point E} = 3*4+2*0 = 12\,\text{hours}$$

That is, the dual price of US$48/hour for Stage 2 will remain applicable in the following range.

$$5.34\,\text{hours} \leq \text{Operating hours of stage 2} \leq 12\,\text{hours}$$

The computed range shown above for Stage 1 and 2 is designated as the range of feasibility. The economic decision is made based on the value of dual prices. The following questions of decision-making analysis are answered through sensitivity analysis.

Q1. If the mining authority needs to increase the working hours for more revenue generation, which stage should be given higher priority?

The dual prices for Stage 1 and Stage 2 are US$18/hour, and US$48/hour, respectively. This indicates that each additional hour of Stage 2 will increase revenue by US$48/hour, as opposed to only US$18/hour for Stage 1. Thus, priority should be given to Stage 2.

On the other hand, if the problem has a minimization objective function, the constraint which has the lowest dual price should be given first priority.

Q2. If the additional cost incurred in increasing the working hours for both the stages is US$30/hour, is it economical to adopt the same?

For Stage 2, the resultant net revenue per hour is US$(48 – 30) = US$18; and for Stage 1, the resultant net revenue is US$(18 – 30) = US$–12. Therefore, it is advisable to increase the working hours of Stage 2 only.

Q3. If the working hour of Stage 1 is increased from 8 hours to 12 hours, how will the optimum revenue change?

The dual price for Stage 1 is US$18/hour in the time range of 5.34 to 12 hours. The planned increase to 12 hours lies within the feasible range. The revenue gain is US$18*(12–8) = US$72. This result indicates that revenue will be increased to US$600 [= current revenue (US$528) + revenue gain (US$72)].

Q4. If the working hour of Stage 1 is increased from 8 hours to 16 hours, estimate the impact on optimal revenue.

The planned increase to 16 hours lies outside the feasible range of 5.34 hours to 12 hours, for which the dual price of US$48 is applicable. Based on the current information, the decision can be taken to increase the working hours of Stage 1 up to 12 hours. Beyond that, more information is needed to make the optimal decision. It should be noted that falling outside the feasible range does not indicate that the problem has infeasible or no solution.

Case 2: Changes in the coefficients of the objective functions

The feasible solution space for the problem, which is defined in Example 3.4, is shown in Figure 3.7. The problem has an optimum solution at point C $\left(x_1 = \frac{8}{5}, x_2 = \frac{8}{5}, Z = 528\right)$.

The generalized form of the given objective function can be written as:

$$\text{Maximize } Z = C_1 x_1 + C_2 x_2$$

Here, $C_1 = 180$ and $C_2 = 150$.

The optimum solution will remain at point C as long as the slope of the objective function $\left(= -C_1/C_2\right)$ lies between the slope of the two lines $\mathbf{2x_1 + 3x_2 \le 8}$ **and** $\mathbf{3x_1 + 2x_2 \le 8}$. It means that the ratio C_1/C_2 can vary between 2/3 and 3/2, which yields the following condition:

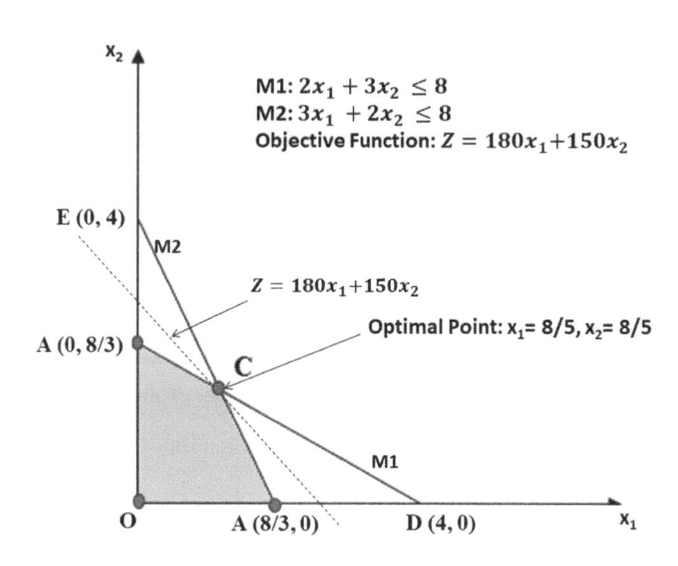

FIGURE 3.7 Graphical method of sensitivity analysis for changing the coefficients of the objective function.

$$\frac{2}{3} \le \frac{C_1}{C_2} \le \frac{3}{2}$$

$$\Rightarrow 0.666 \le \frac{C_1}{C_2} \le 1.5$$

Any change in the objective function's coefficients, the slope of the objective function, Z will change. But the optimum solution will remain at point C as long as the slope of the objective function lies between slopes of constraint lines AE (= –2/3) and BD (= –3/2). This indicates the optimum solution of the problem will remain at C as long as the slope of the objective function lies in the range of –2/3 to –3/2.

This information offers answers to the questions related to the optimal solution, as demonstrated below:

Q1. If the unit profits for Grade 1 and Grade 2 ores are changed to US\$200 and US\$140, respectively, will the optimum solution remain the same (at point C)?

The new objective function is given by

$$\text{Maximize } Z = 200\,x_1 + 140\,x_2$$

The optimal solution will remain the same at C as the ratio $\frac{C_1}{C_2}\left(=\frac{20}{14}=1.42\right)$ remains within the optimality range $(0.666, 1.5)$. If the ratio falls outside the optimality range, the optimal point will shift to a new point. It should be noted that though the optimum point, C remains unchanged but the optimum value of the objective function, Z changes to $200\times(8/5)+140\times(8/5) = \text{US\$544}$

Q2. If the unit profit of Grade 2 ore is fixed at its current value of C_2 = US\$150, determine the associated range of C_1 for which the optimum point is unchanged.

Substituting $C_2 = 150$ in the condition $\frac{2}{3} \le \frac{C_1}{C_2} \le \frac{3}{2}$,

$$\frac{2}{3}\times 150 \le C_1 \le \frac{3}{2}\times 150$$

$$\Rightarrow 100 \le C_1 \le 150$$

That is, the optimal point remains unchanged in the range of $100 \le C_1 \le 150$ and $C_2 = 150$.

3.4.2 Sensitivity analysis of the model using simplex method

The sensitivity analysis of an LPP using the simplex method is explained using the same problem, as given in Example 3.4.

Mathematical form of the problem

Maximise $Z = 180\,x_1 + 150x_2$ Objective function

Subject to $2x_1 + 3x_2 \le 8$ (Stage 1) Constraint (1)

$$3x_1 + 2x_2 \le 8 \text{ (Stage 2)} \qquad \text{Constraint (2)}$$

$$x_1, x_2 \ge 0 \qquad \text{Non-negativity constraint}$$

Case 1: Change in the available resources: In the first case, the sensitivity analysis is done to understand the change in the objective function value with per unit change in the available resources of a particular item keeping all other conditions remain constant. The amount of change in the objective function value is termed as **Shadow price**.

Assuming T_1 and T_2 are the changes made in the working hours of Stage 1 and Stage 2, respectively. Then the original problem can be modified as

$$\text{Maximise} \quad Z = 180x_1 + 150\,x_2$$

$$\text{Subject to} \quad 2x_1 + 3x_2 \le 8 + T_1 \text{ (Stage 1)} \quad \text{Constraint (1)}$$

$$3x_1 + 2x_2 \le 8 + T_2 \text{ (Stage 2)} \quad \text{Constraint (2)}$$

$$x_1, x_2 \ge 0$$

Let y_1 and y_2 be the slack variables incorporated in Stage 1 and Stage 2 constraints to make them equal. The starting simplex tableau of the revised problem is shown in Table 3.8(a).

In the above simplex tableau, two shaded regions have similar characteristics, and hence all the simplex iterations offer similar values in both portions. The optimum simplex tableau, derived in Table 3.5(d) for the same problem, can be written with similar values in both the shaded portion. The optimal simplex tableau, in this case, is shown in Table 3.8(b).

In the optimal simplex tableau [Table 3.8(b)], the optimal value of the objective function can be obtained by adding the RHS of the Z-row with the coefficients of T_1 and T_2 in the Z-row. Similarly, the decision variables at an optimal solution are the respective row's RHS plus the coefficients of T_1 and T_2. Thus, the new optimal tableau offers the following relationships:

TABLE 3.8(a)
Initial Simplex Tableau

		Coefficient of				Solution		
	Basis	x_1	x_2	y_1	y_2	RHS	T_1	T_2
Row 1	y_1	2	3	1	0	8	1	0
Row 2	y_2	3	2	0	1	8	0	1
Row Z	Z	−180	−150	0	0	0	0	0

TABLE 3.8(b)
Optimal Simplex Tableau

		Coefficient of				Solution		
	Basis	x_1	x_2	y_1	y_2	RHS	T_1	T_2
Row 1	x_2	0	1	3/5	−2/5	8/5	3/5	−2/5
Row 2	x_1	1	0	−2/5	3/5	8/5	−2/5	3/5
Row Z	Z	0	0	18	48	528	18	48

$$Z = 528 + 18\,T_1 + 48\,T_2 \tag{3.7}$$

$$x_1 = 8/5 - 2/5\,T_1 + 3/5\,T_2 \tag{3.8}$$

$$x_2 = 8/5 + 3/5\,T_1 - 2/5\,T_2 \tag{3.9}$$

Eq. (3.7) gives the value of the optimal solution of the revised LPP at x_1 and x_2, which can be determined from Eq. (3.8) and Eq. (3.9), respectively. These three equations can be used to determine the shadow price and feasibility range.

A unit change in working hours in Stage 1 ($T_1 = \pm$ 1 hour) leads to a change in the Z value by US$18.

Similarly, a unit change in working hours in Stage 2 ($T_2 = \pm$ 1 hour) leads to a change in the Z value by US$48.

This indicates that the corresponding shadow prices are US$18/hour and US$48/hour, respectively, for Stage 1 and Stage 2. The same results were obtained in the graphical method of sensitivity analyses.

In the optimal simplex tableau, the basic variables are x_1 and x_2. The current optimal solution remains feasible until all the basic variables (x_1 and x_2) are non-negative in the optimal simplex tableau.

Therefore, from Eq. (3.8) and Eq. (3.9), we can write

$$x_1 = \frac{8}{5} - \frac{2}{5}\,T_1 + \frac{3}{5}\,T_2 \geq 0 \tag{3.10}$$

$$x_2 = \frac{8}{5} + \frac{3}{5}T_1 - \frac{2}{5}T_2 \geq 0 \tag{3.11}$$

The above conditions can estimate the individual constraint's feasibility range associated with changing the resources one at a time. In the first case, if the working hour of Stage 1 changes keeping Stage 2 remain constant, then Eq. (3.10) and Eq. (3.11) become,

$$\frac{8}{5} - \frac{2}{5}\,T_1 + \frac{3}{5} * 0 \geq 0 \tag{3.12}$$

$$\frac{8}{5} + \frac{3}{5}T_1 - \frac{2}{5} * 0 \geq 0 \tag{3.13}$$

From Eq. (3.12) and Eq. (3.13), we can derive the following relationship:

$$\Rightarrow -2.6 \leq T_1 \leq 4$$

In the same way, we can determine the range for T_2 as

$$-2.6 \leq T_2 \leq 4$$

The feasibility ranges of Stage 1 and Stage 2 can be determined from the constraint (1) and Constraint (2) by putting the minimum and maximum values of T_1 and T_2, respectively.

For Stage 1

Minimum value = 8+T_1 = 8–2.6 = 5.4
Maximum value = 8+T_1 = 8 +4 = 12

TABLE 3.9
Summary of Shadow Prices and their Feasibility Ranges

Stage	Shadow price	Feasibility range	Working hours		
			Minimum	Current	Maximum
1	18	$-2.6 \le T_1 \le 4$	5.4	8	12
2	48	$-2.6 \le T_1 \le 4$	5.4	8	12

For Stage 2

Minimum value = $8+T_2 = 8–2.6 = 5.4$
Maximum value = $8+T_2 = 8 +4 = 12$

The shadow prices and their feasibility range for the LPP (Example 3.4) are listed in Table 3.9.

For any simultaneous changes in the right-hand side, the shadow prices may remain applicable that keep the solution feasible even if the changes violate the individual ranges.

For example: $T_1 = 5$, $T_2 = 1$

$$x_1 = \frac{8}{5} - \frac{2}{5}T_1 + \frac{3}{5}T_2 = \frac{8}{5} - \frac{2}{5}*5 + \frac{3}{5}*1 = 0.2 \ge 0 \quad \text{(Feasible)}$$

$$x_2 = \frac{8}{5} + \frac{3}{5}T_1 - \frac{2}{5}T_2 = \frac{8}{5} + \frac{3}{5}*5 - \frac{2}{5}*1 = 4.4 \ge 0 \quad \text{(Feasible)}$$

That is, $T_1 = 5$ violates the feasibility range, but the solutions are feasible. This indicates that the shadow prices will remain applicable, and the new optimum value of the objective function can be determined as:

$$Z = 180x_1 + 150\,x_2 = 180*0.2 + 150*4.4 = 36 + 660 = \text{US\$696}$$

Case 2: Changes in the coefficients in the objective function: In the second case, the sensitivity analysis is done to understand the change in the objective function value per unit change in the value of decision variables. The reduced cost for a decision variable is nonzero only when the variable's value is equal to its upper or lower bound at the optimal solution.

Let c_1 and c_2 are the change in revenue (coefficients of the objective function) for two types of ore, respectively. The revised objective function for the LPP (Example 3.4) is given by

$$\text{Maximize} \quad Z' = (180 + c_1)x_1 + (150 + c_2)x_2$$

$$\text{Subject to} \quad 2x_1 + 3x_2 \le 8 \text{ (Stage 1)}$$

$$3x_1 + 2x_2 \le 8 \text{ (Stage 2)}$$

$$x_1, x_2 \ge 0$$

Therefore, the initial simplex tableau of the revised LPP is shown in Table 3.10(a).

The optimal simplex tableau can be derived directly without a fresh calculation of the entire steps using the following method. A new row, R, and a new column, C, are added in the optimal simplex tableau, which was derived in Table 3.5(d) of the original problem (Example 3.4). The

TABLE 3.10(a)
Initial Simplex Tableau of the Revised LPP

	Basis	Coefficient of x_1	x_2	y_1	y_2	RHS
Row 1	y_1	2	3	1	0	8
Row 2	y_2	3	2	0	1	8
Row Z	Z'	$-180-c_1$	$-150-c_2$	0	0	0

TABLE 3.10(b)
Revised Simplex Tableau Derived from the Optimal Tableau of the Original LPP

			Coefficient of				
		R	c_1	c_2	0	0	
	C	Basis	x_1	x_2	y_1	y_2	RHS
Row 1	c_2	x_2	0	1	3/5	−2/5	8/5
Row 2	c_1	x_1	1	0	−2/5	3/5	8/5
Row Z	1	Z'	0	0	18	48	528

revised simplex tableau after adding a new row and column is shown in Table 3.10(b). Table 3.10(b) is used to derive the revised optimal simplex tableau. The entries in the shaded row are the change c_i associated with variable x_j. For the shaded column, the Z-row element is 1 followed by a change in c_i for the basic variable x_i. The c_i value corresponds to all the non-basic variables being zero.

The method of deriving the elements of the optimal simplex tableau of the revised problem is demonstrated below.

The coefficients of the Z-row and RHS value can be determined as:

Coefficient of x_1 = x_1-column * C-column- $c_1 = 0 * 1 + 1 * c_1 + 0 * c_2 - c_1 = 0$
Coefficient of x_2 = x_2-column * C-column- $c_2 = 0 * 1 + 0 * c_1 + 1 * c_2 - c_2 = 0$
Coefficient of y_1 = y_1-column * C-column- 0

$$= 18 * 1 - 2/5 * c_1 + 3/5 * c_2 - 0 = 18 - 2/5\, c_1 + 3/5\, c_2$$

Coefficient of y_2 = y_2-column*C-column- 0

$$= 48 * 1 + 3/5 * c_1 - 2/5 * c_2 - 0 = 48 + 3/5\, c_1 - 2/5\, c_2$$

Revised RHS = RHS-column*C-column

$$= 528 * 1 + 8/5 * c_1 + 8/5 * c_2 = 528 + 8/5\, c_1 + 8/5\, c_2$$

The revised optimal simplex tableau is shown in Table 3.10(c).

The current solution remains optimal so long as the new coefficients in Row Z are non-negative for the maximization problem (non-positive for minimization problem). The mathematical representations of the non-negativity are shown in Eq. (3.14) and Eq. (3.15).

TABLE 3.10(c)
Optimal Simplex Tableau of the Revised LPP

	Basis	Coefficient of				RHS
		x_1	x_2	y_1	y_2	
Row 1	x_2	0	1	3/5	−2/5	8/5
Row 2	x_1	1	0	−2/5	3/5	8/5
Row Z	Z'	0	0	$18\text{-}2/5c_1\text{+}3/5\ c_2$	$48\text{+}3/5c_1\text{-}2/5\ c_2$	$528 + 8/5\ c_1 + 8/5\ c_2$

Thus,

$$18 - 2/5\ c_1 + 3/5\ c_2 \geq 0 \tag{3.14}$$

$$48 + 3/5\ c_1 - 2/5\ c_2 \geq 0 \tag{3.15}$$

To determine the optimality range of individual variables, the cost coefficient of one variable is changed, keeping the rest of the coefficient unchanged.

For changing the cost coefficient of x_1 variable by c_1 and no change in x_2 ($c_2 = 0$), the Eq. (3.14) can be reduced to

$$18 - \frac{2}{5}c_1 \geq 0$$

$$\Rightarrow c_1 \leq 45$$

Similarly, the Eq. (3.15) can be reduced to

$$48 + 3/5\ c_1 \geq 0$$

$$\Rightarrow c_1 \geq -80$$

Therefore, the range of optimality for x_1 is

$$-80 \leq c_1 \leq 45$$

Similarly, for changing the cost coefficient of x_2 variable by c_2 and no change in x_1 ($c_1 = 0$), Eq. (3.14) can be reduced to

$$18 + 3/5\ c_2 \geq 0$$

$$\Rightarrow c_2 \geq -30$$

Eq. (3.15) can be reduced to

$$48\ - 2/5\ c_2 \geq 0$$

$$\Rightarrow c_2 \leq 120$$

Therefore, the range of optimality for x_2 is

$$-30 \leq c_2 \leq 120$$

Suppose the value of c_1 and c_2 are –60 and 10, the given objective function becomes

$$Z' = (180 - 60)x_1 + (150 + 10)x_2$$

$$\Rightarrow Z' = 120x_1 + 160x_2$$

Substituting the values of c_1 and c_2 in Eq. (3.14) and Eq. (3.15), we have

$$18 - \frac{2}{5}*c_1 + \frac{3}{5}c_2 = 18 - \frac{2}{5}*(-60) + \frac{3}{5}*10 = 48 \geq 0\,(\textbf{Non-nragtivity condition satisfied})$$

$$48 + \frac{3}{5}c_1 - \frac{2}{5}c_2 = 48 + \frac{3}{5}*(-60) - \frac{2}{5}*10 = 8 \geq 0\ (\textbf{Non-nragtivity condition satisfied})$$

The results indicate that the proposed changes will keep the current solution optimal ($x_1 = 8/5, x_2 = 8/5$) with a new value of $Z' = 528 + \frac{8}{5}c_1 + \frac{8}{5}c_2 = 528 + \frac{8}{5}*(-60) + \frac{8}{5}*10 = 448.$

If any of the conditions is not satisfied, a new solution needs to be determined.

It should be noted that the changes in c_1 and c_2 may be within their respective allowable range, but that may not satisfy the simultaneous conditions and vice-versa.

For example, if the value of c_1 and c_2 are –70 and 20, the objective function becomes

$$Z' = 110x_1 + 170x_2$$

Substituting the values of c_1 and c_2 in Eq. (3.14) and Eq. (3.15), we have

$$18 - \frac{2}{5}*c_1 + \frac{3}{5}c_2 = 18 - \frac{2}{5}*(-70) + \frac{3}{5}*20 = 58 \geq 0\ (\textbf{Non-negativity condition satisfied})$$

$$48 + \frac{3}{5}c_1 - \frac{2}{5}c_2 = 48 + \frac{3}{5}*(-70) - \frac{2}{5}*20 = -2 \leq 0\ (\textbf{Non-negativity condition satisfied})$$

That is, though the individual changes are within the optimality range, the simultaneous change may not satisfy the conditions, and thus the optimal point for the new objective function will shift.

3.5 THE DUAL PROBLEM

For every primal LPP, there is a dual LPP, which describes the original problem. The primal to dual conversion states that every maximization (or minimization) problem in primal form takes the form of minimization (or maximization) in the dual form. It is observed that solving a dual problem is easier than solving the primal problem if the primal problem has fewer decision variables and more constraints.

Furthermore, a primal problem with artificial variables is relatively cumbersome to solve, and this can be eliminated in the dual simplex method.

3.5.1 Formulation of dual problem for a given primal LPP

The method is explained with a generalized primal LPP as defined below:

Maximize $Z = C_1x_1 + C_2x_2 + \cdots + C_nx_n$ Objective function

Subject to $a_{11}x_1 + a_{12}x_2 + \cdots + a_{1n}x_n \leq b_1$ Constraint (1)

$$a_{21}x_2 + a_{22}x_3 + \cdots + a_{2n}x_n \le b_2 \quad \text{Constraint (2)}$$

$$\ldots$$

$$a_{m1}x_1 + a_{m2}x_2 + \cdots + a_{mn}x_n \le b_m \quad \text{Constraint (m)}$$

$$x_1, x_2, \ldots, x_n \ge 0 \quad \text{Non-negativity constraint}$$

Followings are the rules to convert a primal LPP into a dual problem:

- The problem should be written in a standard form. If the problem is maximization (minimization) then all the constraints should be converted into ≤ (≥) sign. The sign of the constraints is reversed in the dual problem. In the case of equality constraint, the corresponding dual variable will be unrestricted in nature. Similarly, if any variable in the primal problem is unrestricted in nature, then the corresponding dual constraint will be an equality constraint.
- For every constraint in the primal problem, there is a dual variable and vice-versa. If in a primal problem, there are m constraints, then the dual problem will have *n* variables (assuming $x_1^d, x_2^d, \ldots, x_m^d$). On the other hand, if the number of decision variables in the primal problem is n ($x_1, x_2, \ldots, x_n$), the number of constraints in the dual problem is n.
- The nature of the objective function is reversed in the dual problem. If the primal problem is maximization, the nature of the dual problem will be minimization and vice-versa.
- The right-hand sides of the primal problem ($b_1, b_2, \ldots, b_m$) form the coefficients of the dual variable in the objective function of the dual problem.
- The coefficients of the objective function of the primal problem ($C_1, C_2, \ldots, C_n$) form the right-hand side of the constraints in the dual problem.
- All the decision variables in the dual problem should be greater or equal to zero.

The summarised form of the conversion of primal LPP into dual form is demonstrated in Table 3.11.

TABLE 3.11
Summarised Rules for Conversion from Primal to Dual

Primal	Dual
Objective function is a maximization	Objective function is a minimization
Objective function is a minimization	Objective function is a maximization
i^{th} number of variable	i^{th} number of constraints
j^{th} number of constraints	j^{th} number of variable
variable $x_i \ge 0$	Inequality sign of ith constraint ≥ type, if dual is minimization ≤ type, if dual is maximization
i^{th} variable is unrestricted	i^{th} constraint should be = sign
j^{th} constraint with = sign	j^{th} variable should be unrestricted
j^{th} constraint with inquality sign	j^{th} variable is ≥ type if primal constraint is ≤ type j^{th} variable is ≤ type if primal constraint is ≥ type
Coefficients of the ith decision variable in the objective function are C_i	Right-hand side of ith constraint is C_i
Right-hand side of ith constraint is b_i	Coefficients of the ith dual variable in the objective function are b_i

Dual form of the generalized primal LPP

$$\text{Minimize } Z' = b_1 x_1^d + b_2 x_2^d + \cdots + b_m x_m^d \qquad \text{Objective function}$$

$$\text{Subject to} \quad a_{11} x_1^d + a_{21} x_2^d + \cdots + a_{m1} x_m^d \geq C_1 \qquad \text{Constraint (1)}$$

$$a_{12} x_1^d + a_{22} x_2^d + \cdots + a_{m2} x_m^d \geq C_2 \qquad \text{Constraint (2)}$$

$$\ldots$$

$$a_{1n} x_1^d + a_{2n} x_2^d + \cdots + a_{mn} x_m^d \geq C_m \qquad \text{Constraint (m)}$$

$$x_1^d,\ x_2^d, \ldots,\ x_m^d \geq 0 \qquad \text{Non-negativity constraint}$$

The optimal value of any primal problem and its corresponding dual problem should always be the same. This is demonstrated with the problem, as demonstrated in **Example 3.4.**

Primal Problem	**Dual Problem**
Maximize $Z = 180x_1 + 150x_2$	Minimize $Z' = 8x_1^d + 8x_2^d$
Subject to, $2x_1 + 3x_2 \leq 8(\text{Stage } 1)$	Subject to, $2x_1^d + 3x_2^d \geq 180$
$3x_1 + 2x_2 \leq 8(\text{Stage2})$	$3x_1^d + 2x_2^d \geq 150$
$x_1, x_2 \geq 0$	$x_1^d, x_2^d \geq 0$
Solution:	**Solution:**
The feasible solution space of the primal problem is shown in Figure 3.8(a).	The feasible solution space of the dual problem is shown in Figure 3.8(b).

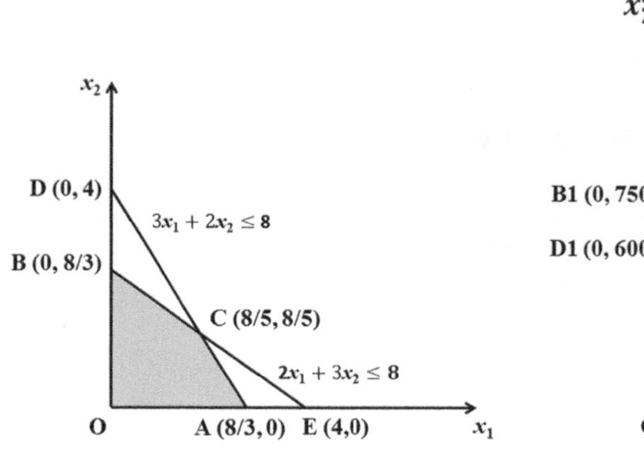

FIGURE 3.8(a) Feasible solution space of primal LPP.

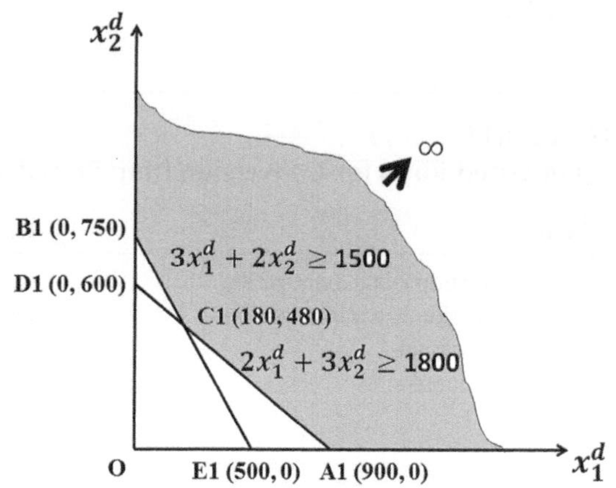

FIGURE 3.8(b) Feasible solution space of dual LPP.

Primal Problem	Dual Problem
The value of the objective function $Z\ at\ O = 0$ $Z\ at\ A = 180*\frac{8}{3}+150*0 = 480$ $Z\ at\ B = 180*0+150*\frac{8}{3} = 400$ $Z\ at\ C = 180*\frac{8}{5}+150*\frac{8}{5} = 528$ Therefore, Maximum $Z = 528$	The value of the objective function $Z'\ at\ A1 = 8*90+8*0 = 720$ $Z'\ at\ B1 = 8*0+8*75 = 600$ $Z'\ at\ C1 = 8*18+8*48 = 528$ Therefore, Minimum Z'= 528

Thus, Maximum Z of the Primal LPP = Minimum Z′of the Dual LPP = 528

Example 3.6: Convert the primal problem into dual form.

Minimize $Z = 40000x_1 + 10000x_2$ Objective function

Subject to $300x_1 + 100x_2 \geq 1200$ Constraint (1)

$100x_1 + 100x_2 \leq 500$ Constraint (2)

$x_1, x_2 \geq 0$ Non-negativity constraint

This is a minimization problem, and thus all the constraints should be of ≥ types. Thus, multiplying by −1 on both the sides of constraint (2), we get

$$-100x_1 - 100x_2 \geq -500$$

Now the standardized form of the primal problem is

Minimize $Z = 40000x_1 + 10000x_2$

Subject to $300x_1 + 100x_2 \geq 1200$

$-100x_1 - 100x_2 \geq -500$

$x_1, x_2 \geq 0$

Dual form of the problem

Since there are two constraints, the number of dual variables is two. Let x_1^d and x_2^d are two dual variables corresponding to the first and second constraints.

Maximize $Z' = 1200x_1^d - 500x_2^d$

Subject to $300x_1^d - 100x_2^d \leq 40000$

$100x_1^d - 100x_2^d \leq 10000$

$x_1^d, x_2^d \geq 0$

Example 3.7: Convert the following primal problem into dual form.

Minimize $Z = 2x_1 - 4x_2 - 3x_3$

Subject to $4x_1 - 2x_2 + 3x_3 \leq 8$

$3x_1 - 5x_2 \geq 13$

$$-5x_1 + 4x_2 + 9x_3 = 11$$

$$x_1, x_2 \geq 0 \text{ and } x_3 \text{ is unrestricted in sign}$$

This is a minimization problem, and thus all the ≤ type constraints are converted into ≥ type by multiplying both sides by -1.

$$\text{Minimize} \quad Z = 2x_1 - 4x_2 - 3x_3$$

$$\text{Subject to} \quad -4x_1 + 2x_2 - 3x_3 \geq -8$$

$$3x_1 - 5x_2 \geq 13$$

$$-5x_1 + 4x_2 + 9x_3 = 11$$

$$x_1, x_2 \geq 0 \; and \; x_3 \text{ unrestricted in sign}$$

Let x_1^d, x_2^d, and x_3^d are the dual variables corresponding to three constraints.

Dual form of the problem

$$\text{Maximize} \quad Z' = -8x_1^d + 13x_2^d + 11x_3^d$$

$$\text{Subject to} \quad -4x_1^d + 3x_2^d - 5x_3^d \leq 2$$

$$2x_1^d - 5x_2^d + 4x_3^d \leq -4$$

$$-3x_1^d + 9x_3^d = -3$$

$$x_1^d, x_2^d \geq 0 \text{ and } x_3^d \text{ unrestricted}$$

3.5.2 Dual simplex algorithm

In a dual simplex method, the LP starts with an optimum value of the objective function, which is infeasible. In every iteration, the solution moves toward feasibility without violating optimality. The algorithm ends when feasibility is restored.

The method will be demonstrated with three variables generalized dual problem, as given below.

$$\text{Minimize } Z = C_1x_1^d + C_2x_2^d + C_3x_3^d \qquad \text{Objective function}$$

$$\text{Subject to} \quad a_{11}x_1^d + a_{12}x_2^d + a_{13}x_3^d \geq b_1 \qquad \text{Constraint (1)}$$

$$a_{21}x_1^d + a_{22}x_2^d + a_{23}x_3^d \leq b_2 \qquad \text{Constraint (2)}$$

$$a_{31}x_1^d + a_{32}x_2^d + a_{33}x_3^d = b_3 \qquad \text{Constraint (3)}$$

$$x_1^d, x_2^d, x_3^d \geq 0 \qquad \text{Non-negativity constraint}$$

Steps of dual simplex algorithm

Step 1: Write the problem in standard form. Convert the problem into a minimization problem if it is maximization by multiplying with −1. All the inequalities with a ≤ sign should be converted into ≥ sign constraints by multiplying with −1 in both the side of the inequality. All equality constraints should be replaced with two constraints, one with a ≥ sign and another with a ≤ sign.

In the given problem, Constraint (1) is ≥ type and needs to convert ≤ type by multiplying with -1 on both sides. Therefore, constraint (1) becomes

$$-a_{11}x_1^d - a_{12}x_2^d - a_{13}x_3^d \le -b_1 \qquad \text{Constraint (1a)}$$

Similarly, Constraint (3) is = type, and thus two new constraints need to be derived. One with ≥ type and another with ≤ type, as shown below:

$$a_{31}x_1^d + a_{32}x_2^d + a_{33}x_3^d \le b_3 \qquad \text{Constraint (3a)}$$

$$a_{31}x_1^d + a_{32}x_2^d + a_{33}x_3^d \ge b_3 \qquad \text{Constraint (3b)}$$

Constraint (3b) further needs to convert into ≤ type by multiplying with −1, as given below:

$$-a_{31}x_1^d - a_{32}x_2^d - a_{33}x_3^d \le -b_3 \qquad \text{Constraint (3b')}$$

Thus, the revised problem is

$$\text{Minimize } Z = C_1x_1^d + C_2x_2^d + C_3x_3^d \qquad \text{Objective function}$$

$$\text{Subject to} \quad -a_{11}x_1^d - a_{12}x_2^d - a_{13}x_3^d \le -b_1 \quad \text{Constraint (1a)}$$

$$a_{21}x_1^d + a_{22}x_2^d + a_{23}x_3^d \le b_2 \quad \text{Constraint (2)}$$

$$a_{31}x_1^d + a_{32}x_2^d + a_{33}x_3^d \le b_3 \quad \text{Constraint (3a)}$$

$$-a_{31}x_1^d - a_{32}x_2^d - a_{33}x_3^d \le -b_3 \quad \text{Constraint (3b')}$$

$$x_1^d, x_2^d, x_3^d \ge 0 \qquad \text{Non-negativity constraint}$$

The standard form of a dual problem is shown below.

Standardised form of a dual LPP

$$Z - C_1x_1^d - C_2x_2^d - C_3x_3^d = 0$$

$$\text{Subject to} \quad -a_{11}x_1^d - a_{12}x_2^d - a_{13}x_3^d + y_1 = -b_1$$

$$a_{21}x_1^d + a_{22}x_2^d + a_{23}x_3^d + y_2 = b_2$$

$$a_{31}x_1^d + a_{32}x_2^d + a_{33}x_3^d + y_3 = b_3$$

$$-a_{31}x_1^d - a_{32}x_2^d - a_{33}x_3^d + y_4 = -b_3$$

$$x_1^d, x_2^d, x_3^d, y_1, y_2, y_3, y_4 \ge 0$$

Let y_1, y_2, y_3, and y_4 are the slack variables added to constraints (1a), (2), (3a), and (3b') respectively to make equality constraints.

The structure of the initial dual simplex tableau is shown in Table 3.12(a).

TABLE 3.12(a)
Initial Dual Simplex Tableau

	Basis	Coefficient of x_1^d	x_2^d	x_3^d	y_1	y_2	y_3	y_4	RHS
Row 1	y_1	$-a_{11}$	y_2	$-a_{13}$	1	0	0	0	$-b_1$
Row 2	y_2	a_{21}	y_3	a_{23}	0	1	0	0	b_2
Row 2	y_3	a_{31}	y_4	a_{33}	0	0	1	0	b_3
Row 2	y_4	$-a_{31}$	$-a_{32}$	$-a_{33}$	0	0	0	1	$-b_3$
Row Z	Z	$-C_1$	$-C_2$	$-C_3$	0	0	0	0	0

TABLE 3.12(b)
Initial Dual Simplex Tableau Showing the Departing Variable

	Basis	Coefficient of x_1^d	x_2^d	x_3^d	y_1	y_2	y_3	y_4	RHS	
Row 1	y_1	$-a_{11}$	$-a_{12}$	$-a_{13}$	1	0	0	0	$-b_1$	← Departing Variable
Row 2	y_2	a_{21}	a_{22}	a_{23}	0	1	0	0	b_2	
Row 2	y_3	a_{31}	a_{32}	a_{33}	0	0	1	0	b_3	
Row 2	y_4	$-a_{31}$	$-a_{32}$	$-a_{33}$	0	0	0	1	$-b_3$	
Row Z	Z	$-C_1$	$-C_2$	$-C_3$	0	0	0	0	0	

Step 2: Selection of departing variable: In the dual simplex method, the departing variable is chosen first. The departing variable is the basic variable having the most negative value in the RHS column. If there is a tie, it can be broken arbitrarily. If all the basic variables are non-negative, the algorithm ends.

Let b_1 is the most negative value in the RHS column, and thus y_1 is chosen as the departing variable, as indicated in Table 3.12(b). Therefore, Row 1 is the pivot-row.

Step 3: Selection of entering variable: The entering variable is selected from the non-basic variable, which has negative coefficients using the following ratio criteria, represented in Eq. (3.16).

$$\text{Ratio} = \frac{\text{Coefficients of Z-row}}{\text{Coefficients of pivot row}\left(\text{only forve coefficients under non-basic variables}\right)} \tag{3.16}$$

The above ratio is determined only for non-basic variables, which have negative coefficients in the pivot row. The variable offers the minimum ratio should be selected as an entering variable.

In Table 3.12(b), only three variables are non-basic ($x_1^d, x_2^d, \text{and } x_3^d$). Since all the coefficients in the pivot row under non-basic variables are negative and thus the ratio should be calculated for all. Let the ratio under non-basic variable x_2^d (= C_2/a_{12}) is minimum, and thus x_2^d is chosen as entering variable, as indicated in Table 3.12(c). The x_2^d column is the pivot column.

Step 4: The row transformation should be carried out using a similar method, as explained in the Primal Simplex algorithm.

TABLE 3.12(c)
Initial Dual Simplex Tableau Showing the Departing and Entering Variables

	Basis	Coefficient of x_1^d	x_2^d	x_3^d	y_1	y_2	y_3	y_4	RHS	
Row 1	y_1	$-a_{11}$	$-a_{12}$	$-a_{13}$	1	0	0	0	$-b_1$	← Departing Variable
Row 2	y_2	a_{21}	a_{22}	a_{23}	0	1	0	0	b_2	
Row 2	y_3	a_{31}	a_{32}	a_{33}	0	0	1	0	b_3	
Row 2	y_4	$-a_{31}$	$-a_{32}$	$-a_{33}$	0	0	0	1	$-b_3$	
Row Z	Z	$-C_1$	$-C_2$	$-C_3$	0	0	0	0	0	
Ratio		C_1/a_{11}	C_2/a_{12}	C_3/a_{13}			--	--		
			↑ Entering Variable							

The algorithm is applied to solve the dual problem derived for Example 3.4.

Minimize $z' = 8x_1^d + 8x_2^d$ Objective function

Subject to $2x_1^d + 3x_2^d \geq 180$ Constraint (1)

$3x_1^d + 2x_2^d \geq 150$ Constraint (2)

$x_1^d, x_2^d \geq 0$ Non-negativity constraint

Both the constraints with a ≥ sign are converted into ≤ constraints by multiplying both sides by -1. Thus, the revised forms of the constraints are as follows:

$-2x_1^d - 3x_2^d \leq -180$ Constraint (1a)

$-3x_1^d - 2x_2^d \leq -150$ Constraint (2a)

Let y_1 and y_2 be the slack variables added to inequality constraints (1a) and (2a), respectively, to make them equality constraints.

The standardized form of the problem is given as:

$$Z' - 8x_1^d - 8x_2^d = 0$$

Subject to $-2x_1^d - 3x_2^d + y_1 = -180$

$$-3x_1^d - 2x_2^d + y_2 = -150$$

$$x_1^d, x_2^d, y_1, y_2 \geq 0$$

The initial dual simplex tableau is shown in Table 3.13(a).

The coefficients in the initial dual simplex tableau are negative, and thus the solution is not feasible. The departing variable is y_1, as the coefficient in the RHS corresponding to y_1 is most negative. Furthermore, the ratio is minimum corresponding to column y_2 and thus chosen as entering variable. The revised table after the first iteration is shown in Table 3.13(b).

TABLE 3.13(a)
Initial Dual Simplex Tableau

	Basis	Coefficient of x_1^d	x_2^d	y_1	y_2	RHS	
Row 1	y_1	–2	–3	1	0	–180	← Departing Variable
Row 2	y_2	–3	–2	0	1	–150	
Row Z′	Z′	–8	–8	0	0	0	
Ratio		4 (=–8/–2)	8/3 (=–8/–3)	—	—		
			↑ Entering Variable				

TABLE 3.13(b)
Revised Dual Simplex Tableau after the First Iteration

	Basis	Coefficient of x_1^d	x_2^d	y_1	y_2	RHS	
Row 1	x_2^d	2/3	1	–1/3	0	60	
Row 2	y_2	–5/3	0	–2/3	1	–30	← Departing Variable
Row Z′	Z′	–8/3	0	–8/3	0	480	
Ratio		8/5	—	4	—		
		↑ Entering Variable					

TABLE 3.13(c)
Revised Dual Simplex Tableau after the Second Iteration

	Basis	Coefficient of x_1^d	x_2^d	y_1	y_2	RHS
Row 1	x_2^d	2/3	1	–1/3	0	48
Row 2	x_1^d	1	0	2/5	–3/5	18
Row Z’	Z	0	0	–8/5	–8/5	528

Still, one of the coefficients in the RHS column in Table 3.13(b) is negative, and thus the solution is not feasible. Thus, further iteration is required to obtain the optimal solution. The revised table after the first iteration is shown in Table 3.13(c).

In the above simplex tableau, feasibility is reached, and thus the solution is optimal. The optimal solution is

$$Z' = 528$$
$$\text{at } x_1^d = 18 \text{ and } x_2^d = 48$$

The solution obtained is the same as that of the primal problem.

3.6 CASE STUDY OF THE APPLICATION OF LPP IN OPTIMIZATION OF COAL TRANSPORTATION FROM MINE TO POWER PLANTS

The case study was carried out in a coal mining area in Indonesia. The area has a total of 17 active open-pit coal mines. A morphological condition of the study area is undulated, with the highest point at 325 m above mean sea level (AMSL) and the lowest point at 20 m at AMSL. In the mining lease area, several rivers divide the mining area from South to North. The coal deposits of the study area were formed during the Tertiary period. Different geological events, i.e., folds and faults, caused the undulating topography around the study area, which makes the area sensitive to landslides. The major geology structure of the area is dominated by folds that form solid anticlines and have a north-east strike. The main overburden types are siltstone, mudstone, and sandstone. The inter-burden thickness is relatively constant. According to ASTM classification (American Society for Testing Material), the coal of the study area is categorized into *High Volatile C Bituminous Coal* with a calorific value of 6,424.58–7,262.57 kcal/kg and weight of contents between 1.3–1.5 gr/cm^3.

This study used data from one of the active open-pits from the study area. Within the current pit, it is 12-mineable coal seams (Seam 1 to Seam 12). The average dip of the coal seams is 13.3°–14.9°. The quality of the coal in the study area is represented by four quality parameters: total moisture (TM), ash (ASH), total sulphur (TS), and calorific value (CV). Average quality parameters of all the 12 seams for a specific production period (day) are presented in Table 3.14.

The mine is supplying coal to three neighbouring power plants. Each power plant has its capacity and coal quality requirements. Also, the mining company has long-term price contracts with each of the power plants, based on the quality requirements. The mine has the flexibility to send coal to the power plants from any coal seam as long as quality and quantity requirements are satisfied. However, there is limited production from each coal seam and different transportation costs involved for transporting coal from the coal seam to the power plant. Table 3.15 shows the daily coal production from 12 coal seams and the coal prices from each coal seam based on their quality parameters.

Similarly, the transportation costs from the coal seam locations to the power plant locations are presented in Table 3.16. The quality and quantity limits for the three power plants are shown in Table 3.17.

Given these data in hand, the goal of the mining company is to allocate the specific amount of coal from different coal seams to different power plants to maximize the profit and minimize the transportation cost. The problem is a linear programming problem and can be solved using

TABLE 3.14
Coal Quality Parameters of All 12 Coal Seams

Seam	TM (%)	ASH (%)	TS (%)	CV (kcal/kg)
1	19.56	21.71	0.28	6217.70
2	16.10	14.32	0.60	5651.30
3	25.65	3.34	0.37	5198.98
4	12.75	26.52	0.23	5485.42
5	26.97	5.47	0.23	5870.43
6	17.51	17.88	0.35	5653.94
7	20.25	2.52	0.53	5565.96
8	17.76	10.50	0.28	5669.70
9	17.89	2.21	0.57	6366.48
10	18.18	25.66	0.69	6367.12
11	17.13	3.62	0.33	6300.41
12	16.51	8.61	0.89	6628.91

TABLE 3.15
Coal Production and Coal Price of All 12 Coal Seams

Seam	Daily production (ton/day)	Price of coal (US $/ton)
1	900	52.3
2	2000	29.1
3	600	22.6
4	1000	25.2
5	1600	42.3
6	700	29.1
7	2000	29.1
8	1400	29.1
9	1000	52.3
10	1200	52.3
11	800	52.3
12	900	58.4

TABLE 3.16
Transportation Cost (US $/ton) from Coal Seam Locations to Power Plants

Seam	Plant 1	Plant 2	Plant 3
1	2.28	3.22	3.82
2	1.96	2.8	3.42
3	1.56	2.48	3.12
4	1.06	2	1.8
5	1.12	2.04	2.72
6	0.82	1.6	2
7	0.86	1.8	2.42
8	0.82	1.46	2.02
9	0.94	1.2	1.8
10	1.3	0.8	1.62
11	1.34	0.84	1.72
12	2.02	1.74	0.9

TABLE 3.17
Quality and Quantity Limits of Coals in Three Power Plants

Plant	TM (%)	ASH (%)	TS (%)	CV (GAR)	Total coal (tonne/day)
1	18.2	10.5	0.3	5980	2300
2	18.6	11.2	0.4	5880	4200
3	18.9	11.5	0.5	5800	5900

a simplex algorithm. The symbols and formulation of the objective function and constraints are presented below:

Symbols description and definition of different terms used in the coal optimization model

Superscripts and subscripts

p– Power plant index
TS^p, TM^p, Ash^p, CV^p– Target parameters for power plan p, TS for the total sulfur (%), TM for total moisture (%), Ash for ash content (%), and CV^p for calorific value
s– Coal seam index

Variables used in the model

x_s^p – Real variable for amount of material transported from coal seam $s$to power plant$p$

Known constants in the model

S, P– Total number of seams and power plants considered for optimization
r_s^p– Revenue generated by sending one ton of coal transporting from coal seam $s$to power plant$p$
c_s^p– Transportation cost of sending one ton of coal from coal seam $s$to power plant$p$
TS_s, TM_s, Ash_s, CV_s– Total sulfur (%), total moisture (%), ash content (%), and calorific value of coal seam s
X_s – Total daily production from coal seam s
TS^p– Total daily capacity of power plant$p$

Objective function

$$\text{Maximize} \sum_{s=1,p=1}^{S,P} r_s^p * x_s^p - \sum_{s=1,p=1}^{S,P} c_s^p * x_s^p$$

The objective function for the proposed optimization model consists of two parts. Part I in the objective function represents the revenue generated by sending coals from seam s to power plant t. The goal here is to maximize Part I. Part II consists of total transportation costs from coal seams to power plants. So, this transportation cost needs to minimize. The overall objective function is a maximization problem, where the transportation cost (Part II) is subtracted from the total revenue (Part I) of the problem.

Constraints

Several constraints need to satisfy for this optimization problem. The following are the set of constraints for this coal mine optimization problem.

$$\sum_{p=1}^{P} x_s^p \le X_s \qquad \forall s \in 1, S \quad \text{Constraint(1)}$$

$$\sum_{s=1}^{S} x_s^p \le X^p \qquad \forall p \in 1, P \quad \text{Constraint(2)}$$

$$\sum_{s=1}^{S} x_s^p * \left(TS_s - TS^p\right) \le 0 \qquad \forall p \in 1, P \quad \text{Constraint(3)}$$

$$\sum_{s=1}^{S} x_s^p * \left(TM_s - TM^p\right) \le 0 \qquad \forall p \in 1, P \quad \text{Constraint(4)}$$

$$\sum_{s=1}^{S} x_s^p * \left(Ash_s - Ash^p\right) \le 0 \quad \forall p \in 1, P \quad \text{Constraint}(5)$$

$$\sum_{s=1}^{S} x_s^p * \left(CV_s - CV^p\right) \ge 0 \quad \forall p \in 1, P \quad \text{Constraint}(6)$$

Constraint (1) signifies that the coal transported from a coal seam to power plants should be less than or equal to total daily production from the seam. Constraint (2) ensures total coal transported from coal seams to a power plant should be less than the maximum target coal requirement. It is noted here that the mining company and the power plan had a long-term contract where the plant is not willing to accept more than the total coal tonnage mentioned in Table 3.17. However, there is no lower limit restriction imposed in that contract. It means the mining company can decide not to send any coal to a specific power plan on a specific date. The reason for this flexibility is that the power plants also purchased coal from the other two companies. Although there is the flexibility of sending total daily coal to the power plants; however, power plants are very particular about the quality of the coal they are receiving. Constraints (3)–(5) ensure that average total sulfur, total moisture, and ash content, respectively, in coal transported to a power plant should be less than or equal to the target total sulfur, total moisture, and ash content of the power plant. Constraint (6) ensures that the average calorific value in coal transported to a power plant should be greater than or equal to the target calorific value of the power plant. All three power plants want their coal quality should have an upper limit value for total moisture, total sulfur, and ash content and a lower limit for calorific value. If a mining company fails to send the coal with that desired quality, the power plant will not accept that coal. So, the real challenge of the mining company is to optimize the coal transported from a specific seam to a specific plant such that the quality requirement is made and at the same time maximize the profit.

Given this problem set and the data, the optimization can optimally be solved using a simplex algorithm. The optimization problem can be solved using any available optimizer; however, CPLEX software (www.ibm.com/analytics/cplex-optimizer) is used for solving the problem. The linear programming formulation was created using ZIMPL software (https://zimpl.zib.de/). Using ZIMPL and CPLEX to solve optimization problems is demonstrated in the later chapter in this book. It is noted here that the quality and quantity parameters from a coal seam for a specific day were obtained from long-range and subsequent short-range mine planning. The target coal quality and quantity parameters for the power plant were obtained from the annual contract agreement between the mining and power plants companies. An assumption was made here that coal quality and quantity parameters from the seam are known with full certainty. However, due to the geological uncertainty, these parameters can significantly be different from the obtained data from the long- and short-range planning. We understand the limitation of ignoring the uncertainty; however, the purpose of this case study is to show how linear programming can be used for solving real-case mining problems. For optimization under uncertainty, the reader can follow these literature (Paithankar et al., 2021; Lamghari and Dimitrakopoulos, 2012).

The optimum solution was achieved after creating the linear programming formulation of this problem using ZIMPL and solving it in CPLEX using dual simplex method. The objective function value is $4.35\text{x}10^5$, which is total revenue for selling coals to power plants minus total transportation cost (i.e., total profit) in the US dollar. The amount of materials sent from each coal seam to each power plant, the total amount of coal sent from coal seams, and unused coals in each coal seam are presented in Table 3.18. It was observed from the results that all produced coals from eight coal seams were transported to power plants. In four coal seams (Seam 2, 3, 7, and 10), there are some excess coals that were not transported to any power plants. Out of these four coal seams, no coal is being used from coal seam 3. It was also observed that in Plant 2 and 3, the full required coal quantity was supplied; however, from plan 1, the amount sent was less than the upper limit of the required

TABLE 3.18
Amount of Coal Sent (tonne/day) from Coal Seams to Plants and Unused Coals (tonne/day) from Coal Seams

	Plant				
Seam	**1**	**2**	**3**	**Total sent from seam to plant**	**Unused Coal**
1	329.76	570.24	0.00	900.00	0.00
2	0.00	694.27	822.24	1516.51	483.49
3	0.00	0.00	0.00	0.00	600.00
4	0.00	2.50	997.50	1000.00	0.00
5	43.94	429.15	1126.91	1600.00	0.00
6	0.00	700.00	0.00	700.00	0.00
7	0.00	0.00	1455.52	1455.52	544.48
8	758.78	641.22	0.00	1400.00	0.00
9	41.44	817.85	140.71	1000.00	0.00
10	0.00	0.00	457.12	457.12	742.88
11	455.24	344.76	0.00	800.00	0.00
12	0.00	0.00	900.00	900.00	0.00
Total	**1629.15**	**4200.00**	**5900.00**		

capacity. As mentioned in the constraint, the plant only has the upper limit tonnage requirement; therefore, the solution doesn't violate the constraints.

Exercise 3

Q1: An explosive company produces two types of commercial explosives, including E1 and E2. The unit revenues of E1 and E2 are US$7.5 per kg and US$8 per kg, respectively. Two types of raw materials, R1 and R2, are used in the manufacture of the two explosives. The daily availabilities of the two types of raw materials are 1500 kg and 50 kg, respectively. One kg of E1 uses 0.8 kg of R1 and 0.1 kg of R2, and 1 kg of E2 uses 0.9 kg of Rl and 0.05 kg of R2.

a. Determine the optimal production plan.
b. Calculate the dual prices and feasibility ranges of R1 and R2.
c. The company is planning to purchase 400 kg of additional raw material of R1 with the cost of US$8 per kg. Is the additional purchase increase the revenue?
d. Determine the maximum recommended purchasing cost for R2.
e. If the availability of R2 is increased by 10 kg, calculate the associated optimum revenue.

[**Ans.** a. E1 = 300, E2 = 1400), Revenue = 13450; b. Dual Price of R1 =8.5 in the range of 800 and 1800, Dual Price of R2 =7 in the range of 83.34 and 187.5; c. Yes; d. US$7; e. US$13520]

Q2. A mining company produces ores of two grades (G1 and G2). G1 grade ore is directly saleable in the market, and G2 requires processing before sale. The production of 1 unit of G2 type ore requires twice as much resources as to produce 1 unit of G1 type of ore. If all the available resources are dedicated to producing G2 type ore alone, the mine can produce 4000 tonnes per day. The demands for the two types are 1500 and 2000 tonnes per day respectively. The profit is US$30 per tonne and US$10 per tonne.

a. Calculate the optimal production of two types of ore that maximize the profit.

b. Calculate the dual price of the production capacity of the G2 type ore and the feasibility range.
c. If the demand of G1 type ore is decreased to 1200 tonne/day, calculate the associated effect on the optimal profit based on the dual price.

[**Ans**. a. G1 = 4000, G2 = 2000, Profit = 140000; b. Dual Price of G2 = −50 in the range of 0 to 3250; c. US$ 140000]

Q3. The ore produces in a Limestone mine are classified into two viz. G1 and G2. The CaO content of G1 types is > 45%, and that of G2 types is > 40%. A cement industry requires the limestone ore for cement manufacturing with CaO % to be more than the cut-off grade. The production of G1 type ore should not exceed 1200 tons per day. The production cost for G1 type ore is US$25 per tonne and for G2 types is US$20 per tonne. The total production of the day has to be at least 2600 tonnes to meet the demand of the cement industry, and the production of G2 types should not exceed G1 type by more than 1000 tonnes. Determine the level of the daily production of G1 and G2 type ore for minimizing the production cost and the minimum supply grade percentage.

[**Ans.** G1 = 800, G2 = 1800), Production cost = US$56000; Minimum supply grade = 41.53%]

Q4. In an open-pit mine, two types of dump trucks (40 tonnes capacity and 60 tonnes capacity) are deployed for the transportation of materials at the rate of 1200 tonnes per hour. The average cycle time is assumed to be 15 minutes for 40-tonne dump truck and 18 minutes for 60-tonne dump truck. The average diesel consumptions are 70 L/ hour and 90 L/hour, respectively, for a 40-tonne dump truck and 60-tonne dump truck. The overall diesel consumption level is a maximum of 560 L/hour. Determine the minimum number of trucks needed for each type for transportation of the materials with minimum diesel consumption.

[**Ans.** Number of 40 te trucks = 46, Number of 90 te trucks = 42]

Q5. An XYZ company has four diamond mines in four locations in South Africa. The mines differ in terms of production capacities, treatment costs, grades, and number of stones per tonne of production, as follows:

Mine	**Capacity (maximum production in tonne per month)**	**Production cost ($per tonne)**	**Treatment cost ($per tonne)**	**Grade (carats per tonne)**	**Count (stones per tonne)**
Mine 1	60,000	5	2.00	0.40	0.60
Mine 2	2,00,000	3	2.00	0.20	0.30
Mine 3	1,00,000	2	2.50	0.30	0.20
Mine 4	1,50,000	3	1.75	0.25	0.40

The monthly production of at least 1,50,000 stones is desired to fulfill the demand. The mines should produce at least 1,40,000 carats. Determine optimum production plan for minimizing the total production cost.

[**Ans.** Mine 1: 60000, Mine 2: 242500, Mine 3: 100000, Mine 4: 150000, Objective function value = US$2795000]

Q6. An iron ore mine supply ore of two grades, G1, and G2, by blending four types of iron ores, O1, O2, O3, and O4. Steel, G1 type uses ores O1, O2, O3, and O4 in the ratio of 1:1:2:4, and Steel, G2 uses ores in the ratio of 2:2:1:3. The production limits for O1, O2, O3, and O4 are 600, 800, 800, and 1000 tonnes/day, respectively. The selling price of G1 and G2 are US$120 and US$150 per tonne, respectively. The minimum supply for G1 and G2 are 1500 and 1600 tonnes/day, respectively. Determine the optimal supply plan for G1 and G2.

[**Ans.** Supply of G1 grade = 1600, Supply of G2 grade = 1600, Objective function value = US$43200]

Q7. A company uses two metals, M1 and M2 to produce two products, P1 and P2, with the following specifications.

Product	Specifications	Selling price (US$/kg)
P1	At most, 70% of M1 between 30% and 50% of M2	2000
P2	Between 20% and 50% of M1 At least 20% of M2	3000

These metals are extracted from three types of ores according to the following data:

Ore	Maximum availability of ore (tonnes)	Constituents (%)			Price/tonne in US$
		MI	M2	Waste	
O1	1000	40	10	50.0	500
O2	2000	30	15	45.0	400
O3	3000	25	20	45.0	600

How much of each type of product should produce for maximizing the revenue? It is given that entire ores are processed and the extracted metals are allocated to produce products.

[**Ans.** P1 = 1666.667 kg, P2 = 1000 kg, Revenue = 3233333]

Q8. A thermal power station use 300 tonnes per hour pulverized coal for generating electricity. The pulverized coals are produced by mixing three types of coal, C1, C2, and C3. The burning of coal emits sulfur dioxide, which must meet the Environmental Protection Agency (EPA)'s specifications of at most 30 tonnes per day. The company installed an air pollution control device, which reduces SO_2 emissions by 95%. The following table summarizes the data of the situation:

	C_1	C_2	C_3
Sulphur content (%)	0.50	0.40	0.40
Pulverizer capacity (tonne/hour)	100	140	120
Costs per tonne	2000	2500	2200

It is known that 1 gm of sulfur burned produces 2 gm of SO_2. Determine the optimal mix of the coal for minimizing the cost.

[**Ans.** C1 = 50 tonne/hour, C2 = 140 tonne/hour, C3 = 110 tonne/hour, Cost = US$608000]

Q9. A mining company procures detonators and explosive cartridges for blasting operations. The daily maximum availability of the detonator and explosive cartridges are 600 and 800 units. Ten tonnes of ore production require two units of detonators and three units of explosive cartridges, and 10 tonnes of overburden removal requires two units of detonators and two units of explosive cartridges. The unit production costs of ore and overburden are US$50 per tonne and US$20 per tonne, respectively. The daily production of ore and overburden should be at least 1000 tonne and 1500 tonne, respectively.

a. Determine the dual prices of detonator and explosive cartridges and their feasibility ranges.
b. The company is planning to purchase four additional units of detonator with the cost of US$2 per unit. Is the additional purchase increase the revenue?

c. Determine the maximum recommended purchasing cost for explosive cartridges.
d. If the availability of explosive cartridges is increased by five units, calculate the associated optimum cost.

Q10. An XYZ company owns two iron ore mines. The ore produced by two mines are graded into three classes including High, Medium, and Low. The company need to supply ores to a plant with 1200 tonnes of high-grade, 1000 tonnes of medium-grade, and 800 tonnes of low-grade ores per day. The operating conditions two mines are given below.

Mine	Operating cost per hour (US$)	Production (tonnes/hour)		
		High	**Medium**	**Low**
M1	6000	300	200	100
M2	4000	100	300	200

Calculate the minimum operating hours of each mine in order to make the supply the ores to plant with minimum production cost.

[**Ans.** M1 = 4 hours, M2 = 4 hours, Production cost = US$40000]

4 Transportation and assignment problems in mines

4.1 DEFINITION OF A TRANSPORTATION PROBLEM

The transportation problem applied to the mine system in which mineral is transported from multiple production points (referred to as source) to multiple consumers (referred to as destinations). In general, let there be n mines $M_1, M_2, \ldots, M_n$ have $S_1, S_2, \ldots, S_n$ unit of production capacity to be transported among m consumers, $P_1, P_2, \ldots, P_m$ with $D_1, D_2, \ldots, D_m$ unit of demand.

Let C_{ij} be the costs of transportation of one unit of the material from mine i to destination j for each route. Suppose x_{ij} represents the units transported from mine i to destination j. The objective is to make an optimal transportation planning to minimize the total transportation cost satisfying production capacity and demand requirements. The transportation model can be mathematically represented as

$$\text{Minimize total cost of transportation, } Z = \sum_{i=1}^{n}\sum_{j=1}^{m} C_{ij}x_{ij}$$

Subject to

$$\sum_{j=1}^{m} x_{ij} = S_i, \qquad i = 1, 2, \ldots, n \left(\textit{supply} \text{ constraint}\right)$$

$$\sum_{i=1}^{n} x_{ij} = D_j, \qquad j = 1, 2, \ldots, m \left(\textit{demand constraint}\right)$$

And, $\quad x_{ij} \geq 0 \textit{ for all } i \textit{ and } j$

4.2 TYPES OF TRANSPORTATION PROBLEM

The transportation problem can be broadly classified into two types:

Balanced Transportation Problem: The problem is called a balanced transportation problem if the supply capacity by the mines is equal to the demand of consumers.

Thus, if the equation $\sum_{i=1}^{n} S_i = \sum_{j=1}^{m} D_j$ holds true; the problem is a balanced transportation problem.

Example 4.1

The following example is used to demonstrate the formulation of the transportation model. The produced coals from three mines M_1, M_2, and M_3, are transported to three power plants P_1, P_2, and P_3. The distances from mines to the plants, along with the normal supply and demand capacities of mines and plants, are indicated below in Figure 4.1.

DOI: 10.1201/9781003200703-4

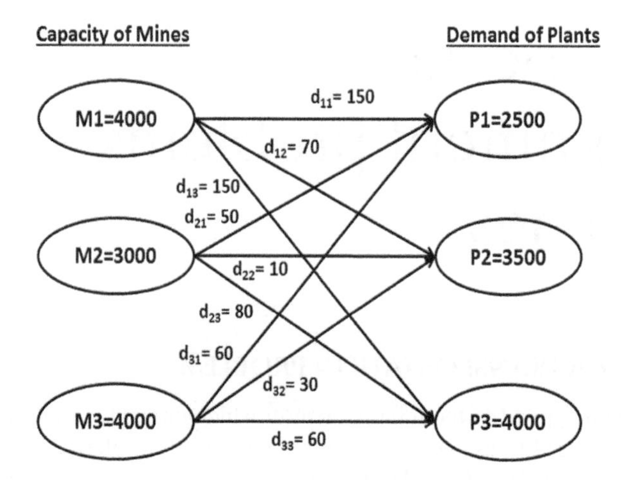

FIGURE 4.1 Supply and demand sources in a mining system.

In Figure 4.1, the sum of total supply from three mines ($=\sum_{i=1}^{3} S_i = 4000+3000+4000$) is equal to the sum of the total demand of three plants ($=\sum_{j=1}^{3} D_j = 2500+3500+4000$), i.e., 11000. Therefore, the problem is a balanced transportation model.

Unbalanced Transportation Problem: The problem is called an unbalanced transportation problem if the supply capacity by the mines is not equal to the capacity consumed by the plants.

If the equation $\sum_{i=1}^{n} S_i \neq \sum_{j=1}^{m} D_j$ holds true; then, the problem is called an unbalanced transportation problem.

4.3 SOLUTION ALGORITHMS OF A TRANSPORTATION MODEL

The transportation model is perhaps the earliest application of linear programming in real-life problems. The solution algorithm is explained with the same problem, as explained in **Example 4.1.** The data given in Figure 4.1 are represented in the standard transportation tableau, as shown in Table 4.1.

The above problem is a balanced transportation problem as the total supply capacity of mines is equal to the total demand of the plants.

Let x_{ij} is the amount of ore transported from mine i to processing plant j in a tonne.

d_{ij} represents the distance between mine i and power plant j in km.

Assuming the cost of transportation (including loading and unloading) of coal is US\$1 per ton per km in each route. Therefore, the cost of transportation from mine i to power plant j = US\$1*d_{ij} per ton = US\$ d_{ij} per tonne.

The objective is to make optimal allocation of ore from each mine to each plant in order to minimize the transportation cost. The linear programming model for this transportation problem is formulated as follows.

TABLE 4.1
Transportation tableau showing the distances of transportation from each mine to each plant along with the demand of plants and production capacity of mines

Mines	Power Plants			Production capacity of mines in tonnes (S_i)
	P_1	P_2	P3	
M1	150	70	150	4000
M2	50	10	80	3000
M3	60	30	60	4000
Demand of plants in tonnes (D_j)	2500	3500	5000	**11000**

$$\text{Minimum } Z = 150(x_{11}) + 70(x_{12}) + 150(x_{13}) + 50(x_{21}) + 10(x_{22}) + 80(x_{23}) + 60(x_{31}) + 30(x_{32}) + 60(x_{33}) \quad \text{Objective Function}$$

$$\begin{aligned}
\text{Subject to } & x_{11} + x_{12} + x_{13} \le 4000 && \text{Constraint (1)}\\
& x_{21} + x_{22} + x_{23} \le 3000 && \text{Constraint (2)}\\
& x_{31} + x_{32} + x_{33} \le 4000 && \text{Constraint (3)}\\
& x_{11} + x_{21} + x_{31} = 2500 && \text{Constraint (4)}\\
& x_{12} + x_{22} + x_{32} = 3500 && \text{Constraint (5)}\\
& x_{13} + x_{23} + x_{33} = 5000 && \text{Constraint (6)}\\
& x_{ij} \ge 0, \quad i, j = 1, 2, 3 && \text{Non-negativity constraints}
\end{aligned}$$

In this model, the decision variables, x_{ij}, represent the amount of coal to be transported from mine, i (i =1, 2, 3), to power plant, j (j = 1, 2, 3). The objective of the problem is to minimize the total transportation cost of the system. The cost of the transportation for each individual path is represented by one term in the objective function. For example, the cost of transportation from M1 to P1 is given by the multiplication of tonnes of coal transported in the path (= x_{11}) and the unit transportation cost (= US$150).

Constraints (1)–(3) represent the production constraints of three mines, and Constraints (4)–(6) represent the demand at three power plants, respectively. The production constraint indicates that the supply from individual mines to three processing plants (P_1, P_2, and P_3) should not exceed the production capacity of the respective mine. The inequality (≤) sign indicates that the maximum amount transported from M_1, M_2, and M_3 should be at most 4000 tonnes, 3000 tonnes, and 4000 tonnes, respectively. Similarly, the demand constraint indicates that the coal receipt in the three power plants should be equal to the demand of the respective power plant.

The amount of coal transported from each mine to each plant is shown in Table 4.2. Each cell in the transportation table (Table 4.2) is assigned by a variable x_{ij} (i, j =1, 2, 3), representing the amount of coal transported from i^{th} mine to j^{th} power plant. Each cell also contains the unit (1 tonne) transportation cost for that route. For example, in cell 11, the value 150 is the cost of transportation of 1 tonne of coal from mine 1(M1) to plant 1 (P1). In Table 4.2, the last column and last row represent the production capacity of mines and demand value of plants, respectively.

The initial feasible solution of the transportation model can be found by three methods: the **north-west corner method**, **the minimum cell cost method**, and **Vogel's approximation method**. There are two methods for obtaining an optimal solution of a transportation model. These are the

TABLE 4.2
Transportation table showing the amount of coal transported from each mine to each plant

Mines	Plants			Production capacity of mines
	P_1	P_2	P3	
M_1	150 (x_{11})	70 (x_{12})	150 (x_{13})	4000 (S_1)
M_2	50 (x_{21})	10 (x_{22})	80 (x_{23})	3000 (S_2)
M_3	60 (x_{31})	30 (x_{32})	60 (x_{33})	4000 (S_3)
Demand of plants	2500 (D_1)	3500 (D_2)	5000 (D_3)	11000

modified distribution method (also known as MODI) and the **stepping stone method**. Both the optimal solution methods offer the same solution of a typical transportation model irrespective using of either of the initial solution methods.

4.3.1 Initial basic feasible solution

4.3.1.1 The north-west corner method

This rule is a simple method for generating a basic feasible solution to the transportation problem. The solution starts from the north-west corner and approaches towards the south-east corner. The only variables required to obtain this type of solution are supply and demand and not the cost of transportation.

The steps of the north-west corner method are as follows:

Step 1: Allocate as much as possible to the cell in the upper left-hand corner, subject to the supply and demand constraints.
Step 2: Allocate as much as possible to the next adjacent feasible cell.
Step 3: Repeat step 2 until all demands are fulfilled, or supplies are exhausted.

In the north-west corner method, the first allocation is made to the cell in the upper left-hand corner of the tableau (i.e., the 'north-west corner' or cell 11). The allocation is based on the minimum supply and demand value corresponding to that cell. In **Example 4.1**, the amount of allocation made to cell 11 is 2500 tonnes, since this is the maximum demand of plant P_1, even though the supply capacity of M1 is 4000 tonnes. We next allocate to a cell adjacent to cell 11, in this case, either cell 21 or cell 12. However, cell 21 no longer represents a feasible allocation because the demand for coal at P_1 (i.e., 2500 tons) has already been allocated. Thus, cell 12 represents the only feasible alternative, which needs to allocate next. The amount allocated to cell 12 is the minimum of remaining supply available at M1 (=4000–2500=1500 tons) and demand at P2 (=3500 tonnes). Similarly, the allocation process continues till all the supply and demand constraints are fulfilled. The steps involved in allocating all the cells using north-west corner method are demonstrated below.

Step I: $x_{11} = \min (S_1, D_1) = D_1 = 2500$; i.e. if $D_1 \le S_1$; $x_{i1} = 0$ $[i > 1]$
Step II: $x_{12} = \min (S_1 - D_1, D_2) = (1500, 3500) = S_1 - D_1 = 1500$; thus $x_{1j} = 0$ $[j > 2]$
Step III: $x_{22} = \min [D_2 - (S_1 - D_1), S_2)] = (2000, 3000) = 200$; $x_{i2} = 0$ $[i > 2]$
Step IV: $x_{23} = \min [S_2 - \{D_2 - (S_1 - D_1)\}, D_3] = (1000, 5000) = 1000$
Step V: $x_{33} = \min [[D_3 - [S_2 - \{D_2 - (S_1 - D_1)], S_3] = 4000$

TABLE 4.3
Initial allocation transportation tableau using north-west corner method

Mines	Plants			Production capacity of mines
	P_1	P_2	P3	
M_1	150 (x_{11}=2500)	70 (x_{12}=1500)	150 (x_{13}=0)	**4000 (S_1)**
M_2	50 (x_{21}=0)	10 (x_{22}=2000)	80 (x_{23}=1000)	**3000 (S_2)**
M_3	60 (x_{31}=0)	30 (x_{32}=0)	60 (x_{33}=4000)	**4000 (S_3)**
Demand of plants	**2500 (D_1)**	**3500 (D_2)**	**5000 (D_3)**	**11000**

The allocation made to each cell is shown in Table 4.3.

Therefore, the cost of transportation = 150*2500 + 70*1500 + 10*2000 + 80*1000 + 60*4000
= $8, 20,000

4.3.1.2 Matrix minimum method

This is an efficient method for finding the initial basic feasible solution to a transportation problem. The steps involved in the matrix minimum method is demonstrated below:

Step 1: This method solution starts with searching the cell of minimum cost and allocating the units as maximum as possible. If there are multiple minima, say C_{mn} (cost of cell mn) = C_{pq} (cost of cell pq), the allocation is made to lower-order cells. That is, if there are two or more minimum costs, then we should have to select the cell corresponding to the lower row number. If they appear in the same row; we should select the cell corresponding to the lower column number.

Step 2: Search the next minimum cost cell among all feasible cells and allocate as maximum as possible, which satisfies the demand and supply constraints.

Step 3: The process of Step 2 continues until all the supply and demand constraints are fulfilled.

The steps involved in allocating the materials in Example 4.1 is demonstrated below:

Step I: Minimum cost cell is 22 (C_{22}=10). Thus, the first allocation made to cell 22, which is equal to $x_{22} = \min(S_2, D_2) = 3000$; $x_{2j} = 0$ [$j \neq 2$].

Step II: The next minimum cost cell is 32 ($C_{32} = 30$). Therefore, the next allocation is made to cell 32, which is equal to $x_{32} = \min(D_2 - S_2, S_3) = 500$; $x_{i2} = 0$ [$i \neq 2, 3$].

Step III: The next minimum cost cell is 31 and 33 both ($C_{31} = C_{33} = 30$) and thus the lower order cell (31) is chosen to allocate. Therefore, the next allocation is made to cell 31, which is equal to $x_{31} = \min[S_3 - (D_2 - S_2), D_1) = \min(3500, 2500) = 2500$; $x_{i1} = 0$ [$i \neq 3$].

Step IV: The next minimum cost cell is 33 ($C_{33} = 60$). Therefore, the next allocation is made to cell 33, which is equal to $x_{33} = \min[D_3, S_3 - D_1 - (D_2 - S_2)] = 1000$.

Step V: The next minimum cost cell is 13 ($C_{13} = 150$). Therefore, the next allocation is made to cell 13, which is equal to $x_{13} = \min(D_3 - \{S_3 - D_1 - (D_2 - S_2)\}, S_1) = 4000$.

TABLE 4.4
Initial allocation transportation tableau using matrix minimum method

Mines	Plants			Production capacity of mines
	P_1	P_2	P_3	
M_1	150 (x_{11}=0)	70 (x_{12}=0)	150 (x_{13}=4000)	**4000 (S_1)**
M_2	50 (x_{21}=0)	10 (x_{22}=3000)	80 (x_{23}=0)	**3000 (S_2)**
M_3	60 (x_{31}=2500)	30 (x_{32}=500)	60 (x_{33}=1000)	**4000 (S_3)**
Demand of plants	**2500 (D_1)**	**3500 (D_2)**	**5000 (D_3)**	**11000**

The solution will proceed in the same way. The initial solution obtained using the matrix minimum method is shown in Table 4.4.

Therefore, cost of transportation = 60*2500 + 10*3000 + 30*500 + 150*4000 + 60*1000
= $8,55,000

4.3.1.3 Vogel Approximation Method (VAM)

The third method for determining an initial solution, Vogel's approximation model (also called VAM), is based on the concept of penalty cost or regret. This method requires many calculations for finding the initial basic feasible solution to the transportation problem. This method, however, usually yields a starting solution that is very close, if not equal, to the optimum solution.

The steps of Vogel's approximation method are as follows:

Step 1: In the first step, the penalty cost for each row and column should be determined by subtracting the lowest cell cost in the row or column from the next lowest cell cost in the same row or column. Mathematically, it can be represented as:

$$\text{Penalty} = \text{Minimum cost of any row} - \text{Next minimum cost of any row}$$
$$= \text{Minimum cost of any column} - \text{Next minimum cost of any column}$$

Step 2: Consider the row or column, which has the highest penalty cost. If there is two highest penalty cost found, break the ties by arbitrarily choosing any of them.

Step 3: Allocate as maximum as possible to the feasible cell with the lowest transportation cost in the row or column with the highest penalty cost.

Step 4: Repeat steps 1, 2, and 3 until all demand and supply requirements have been met.

The allocation made in using the above-described algorithm for Example 4.1 is demonstrated below:

Step I: In the first step, the maximum penalty found in M1 (=80), i.e., $C_{11} - C_{12,}$ and thus the first allocation made to cell C_{12}. The maximum possible allocation to cell 12 is Min (S_1, D_2), which is equal to 3500 tonnes. This is indicated in Table 4.5. Since the demand of plant P2 is fulfilled,

TABLE 4.5
Transportation table showing the first allocation using Vogel approximation method

Mines	Plants			Production capacity of mines	Penalty
	P_1	P_2	P3		
M_1	150 (x_{11})	70 (x_{12} = 3500)	150 (x_{13})	**4000 (S_1)**	80
M_2	50 (x_{21})	10 (x_{22} = 0)	80 (x_{23})	**3000 (S_2)**	40
M_3	60 (x_{31})	30 (x_{32}=0)	60 (x_{33})	**4000 (S_3)**	30
Demand of plants	**2500 (D_1)**	**3500 (D_2)**	**5000 (D_3)**	**11000**	
Penalty	10	20	20		

TABLE 4.6
Revised transportation tableau after first allocation

Mines	Plants		Production capacity of mines	Penalty
	P_1	P3		
M_1	150 (x_{11} = 0)	150 (x_{13})	**4000–3500 = 500 ($S_1 - D_2$)**	0
M_2	50 (x_{21} = 2500)	80 (x_{23})	**3000 (S_2)**	30
M_3	60 (x_{31} = 0)	60 (x_{33})	**4000 (S_3)**	0
Demand of plants	**2500 (D_1)**	**5000 (D_3)**	**11000**	
Penalty	10	20		

no further allocation is possible in the P2 column. Thus, the allocations in all the other cells in the P2 column are zero. In the next step, the P2 column should be eliminated before penalty calculation.

Step II: The revised table after the first allocation is shown in Table 4.6. The penalty values for each row and column of the revised table are determined. This time the maximum penalty is also observed in the M2 row, and thus the allocation should be made to cell 12. The maximum possible allocation to cell 12 is Min (S_2, D_2), which is equal to 2500 tonne. The allocation made in Step II is shown in Table 4.6. After allocating cell 12, the demand of plant P1 is fulfilled, and thus no further allocation is possible in the P1 column. Thus, the allocations in all the

TABLE 4.7
Revised transportation tableau after the second allocation

Mines	Plants	Production capacity of mines
	P3	
M_1	150 (x_{13}=500)	**4000–3500=500 ($S_1 - D_2$)**
M_2	80 (x_{23}=500)	**3000–2500=500 ($S_2 - D_1$)**
M_3	60 (x_{33}=4000)	**4000 (S_3)**
Demand of plants	**5000 (D_3)**	**11000**
Penalty	**20**	

TABLE 4.8
Revised transportation tableau after the third allocation

Mines	Plants	Production capacity of mines
	P3	
M_1	150 (x_{13}=500)	**4000–3500=500 ($S_1 - D_2$)**
M_2	80 (x_{23}=500)	**3000–2500=500 ($S_2 - D_1$)**
Demand of plants	**5000–4000 =1000 (D_3–S_3)**	**11000**
Penalty	**20**	

other cells in the P1 column are zero. In the next step, P1 should be eliminated before penalty calculation.

Step III: The revised table after the second allocation is shown in Table 4.7. In Table 4.7, only one column (P3) is available. Thus, the allocation can be directly made from low-cost to high-cost cells, as the same sequence will offer the highest penalty values. In P3 Column, the lowest cost cell is 33, and thus allocation should be made to cell 33. The maximum possible allocation to cell 33 is Min(S_3, D_3), which is equal to 4000 tonnes. The allocation made in Step III is shown in Table 4.7. After allocating cell 33, the supply of mine M3 is exhausted, and thus no further allocation is possible in the M3 row. In the next step, the M3 row should be eliminated before penalty calculation.

Step IV: The revised table after the second allocation is shown in Table 4.8. In Table 4.8, only one column (P3) and two rows are available. Thus, the allocation can be directly made from

TABLE 4.9
Initial allocation transportation tableau using VAM

Mines	Plants			Production capacity of mines
	P_1	P_2	P3	
M_1	150 (x_{11}=0)	70 (x_{12}=3500)	150 (x_{13}=500)	**4000 (S_1)**
M_2	50 (x_{21}=2500)	10 (x_{22}=0)	80 (x_{23}=500)	**3000 (S_2)**
M_3	60 (x_{31}=0)	30 (x_{32}=0)	60 (x_{33}=4000)	**4000 (S_3)**
Demand of plants	**2500 (D_1)**	**3500 (D_2)**	**5000 (D_3)**	**11000**

low-cost to high-cost cells. In Table 4.8, the lowest cost cell is 23, and thus allocation should be made to cell 23. The maximum possible allocation to cell 23 is Min (D_3–S_3, $\mathbf{S_2 - D_1}$), which is equal to 500 tonnes.

Step V: After allocating cell 23, the remaining cell (13) should be allocated with the balance amount.

The solution obtained using the VAM method is shown in Table 4.9.

Therefore, cost of transportation = 70*3500 + 150*500 +50*2500 + 80*500 + 60*4000
= \$7, 25,000/-

4.3.2 Determination of optimal solution

Once any of the previous three methods determine an initial basic feasible solution, the next step is to solve the optimal solution model to minimize the transportation cost. There are two basic solution methods: the modified distribution method (MODI) and the stepping stone solution method.

4.3.2.1 The Modified Distribution method

The initial solution obtained by the north-west corner method is chosen as the starting solution for explaining the MODI method. The solution derived from the north-west corner method is shown below in transportation Table 4.10.

The corresponding cost of transportation obtained for the above allocation is US\$8,20,000.

It is noted here that non-basic cells/variables are assigned a zero value, and all other variables/cells with a non-zero assignment are basic cells. In the above initial solution, x_{11}, x_{12}, x_{22}, x_{23}, and x_{33} are basic variables and x_{13}, x_{21}, x_{31}, and x_{32} are non-basic variables.

The iterative steps of the MODI method are as follows:

Step I: Identify the entering variable from the non-basic variables using the optimality condition, as used in the simplex method. If the current allocation satisfies the optimal condition, stop the iteration; otherwise, go to step II.

Step II: Identify the leaving variable using the feasibility condition, as used in the simplex method. Change the basis, and Go to Step I.

TABLE 4.10
Initial solution using north-west corner method

Mines	Plants			Production capacity of mines
	P_1	P_2	P3	
M1	150 (x_{11}=2500)	70 (x_{12}=1500)	150 (x_{13}=0)	**4000 (S_1)**
M2	50 (x_{21}=0)	10 (x_{22}=2000)	80 (x_{23}=1000)	**3000 (S_2)**
M3	60 (x_{31}=0)	30 (x_{32}=0)	60 (x_{33}=4000)	**4000 (S_3)**
Demand of plants	**2500 (D_1)**	**3500 (D_2)**	**5000 (D_3)**	**11000**

The optimal allocation in a transportation problem indicates that any other shifting of the allocation towards empty cells leads to a higher transportation cost. To start the iterative algorithm, the following condition should be fulfilled in the initial solution.

$$m\ (= \text{number of rows}) + n\ (=\text{number of columns}) - 1 = \text{Number of cells with allocations}$$

In this case, the number of allocations in the north-west corner method is 5, which is equal to the desired number of allocations (= 3 + 3 – 1). Thus, an iterative process can start.

If the number of allocations in the initial basic solution is less than $(\boldsymbol{m + n - 1})$, the problem has a degenerate solution**.** In that case, a non-basic cell (empty cell) with the lowest cost should be considered a basic cell with zero allocation to start the iterative algorithm.

The selection of entering variables from the current non-basic cells (which have zero allocations) is computed using the method of multipliers. In this method, the multipliers U_i and V_j are considered respective to the ith row and j^{th} column, as shown in Table 4.11.

For each current basic variable x_{ij}, these multipliers satisfy the following equations:

$$U_i + V_j = C_{ij}$$

In the above equation, C_{ij} indicates the transportation cost from i^{th} source to j^{th} destination.

The number of cells that have positive allocations in the initial basic solution is 5 (x_{11}, x_{12}, x_{22}, x_{23}, x_{33}), and thus the number of basic variables is 5. Now there are five equations with six unknowns (U_1 to U_3, and V_1 to V_3). To solve these equations, it is necessary to assign a zero value to any one of the unknowns. In this case, U_1 is assigned a zero value. After assigning U_1 to 0, the remaining variables' values can be calculated, as shown below:

Basic Variables	(U,V) Equation	Solution	Summarized solution
x_{11}	$U_1 + V_1 = 150$	Set $U_1 = 0, V_1 = 150$	$U_1 = 0, V_1 = 150$
x_{12}	$U_1 + V_2 = 70$	$U_1 = 0, V_2 = 70$	$U_2 = -60, V_2 = 70$
x_{22}	$U_2 + V_2 = 10$	$V_2 = 70, U_2 = -60$	$U_3 = -80, V_3 = 140$
x_{23}	$U_2 + V_3 = 80$	$U_2 = -60, V_3 = 140$	
x_{33}	$U_3 + V_3 = 60$	$V_3 = 140, U_3 = -80$	

TABLE 4.11
Transportation tableau showing the U_i and V_j variables corresponding to each cell

Mines	Plants			Production capacity of mines	
	P_1	P_2	P3		
M1	150 (x_{11}=2500)	70 (x_{12}=1500)	150 (x_{13}=0)	**4000 (S_1)**	U1
M2	50 (x_{21}=0)	10 (x_{22}=2000)	80 (x_{23}=1000)	**3000 (S_2)**	U2
M3	60 (x_{31}=0)	30 (x_{32}=0)	60 (x_{33}=4000)	**4000 (S_3)**	U3
Demand of plants	**2500 (D_1)**	**3500 (D_2)**	**5000 (D_3)**	**11000**	
	V1	V2	V3		

In the next step, the optimal allocation is examined using the K_{ij} value for each non-basic variable x_{ij}. The value of K_{ij} is determined from U_i and V_j using the following equation

$$K_{ij} = U_i + V_j - C_{ij}$$

The K_{ij} value of all four non-basic cells is calculated, as presented here:

Non-basic variables	$K_{ij} = U_i + V_j - C_{ij}$
x_{13}	$K_{13} = U_1 + V_3 - C_{13} = 0 + 140 - 150 = -10$
$\mathbf{x_{21}}$	$\mathbf{K_{21} = U_2 + V_1 - C_{21} = -60 + 150 - 50 = 40}$
x_{31}	$K_{31} = U_3 + V_1 - C_{31} = -80 + 150 - 60 = 10$
x_{32}	$K_{32} = U_3 + V_2 - C_{32} = -90 + 70 - 30 = -50$

For optimality, the value of K_{ij} for all the non-basic cells should be negative or zero. If not, identify the entering cell based on the highest positive value and follow the same process. In this case, all K_{ij} values are not negative and zero, and the highest positive value is found for cell 21 (=40); thus cell 21 is the entering variable. Once the entering variable is identified, the leaving variable needs to be identified to replace in the basis. To maintain the supply and demand balance, a cycling shifting is done, as shown in Table 4.12. The cycle should form by considering only basic variables, including the basic entering variable in such a way that it does not violate any of the demand or supply constraints.

Let the number of units allocated to cell $x_{21} = \theta$ (non-negative value). Note, variable x_{21} is the entering variable into the basis.

It should be noted that the value of θ should be chosen in such a way that no decision variable becomes negative. Thus, the maximum possible value of θ can be determined from the non-negativity constraints, derived from Table 4.12, as

$$2500 - \theta \geq 0, \text{ and } 2000 - \theta \geq 0$$

From the above two equations, the maximum possible value of θ is 2000. That is, the maximum allocation can be done to new basic cell x_{21} is 2000. After cycling shifting of allocation, the cell

TABLE 4.12
Transportation table showing the selection of entering and departing variable

Mines	Plants				
	P1	**P2**	**P3**		
M_1	x_{11}=2500-θ (C_{11}= 150)	x_{12}=1500+θ (C_{12}= 70)	x_{13}=0 (C_{13} =150)	**4000 (S_1)**	U_1
M_2	x_{21}=0+θ (C_{21} = 50)	x_{22}=2000-θ (C_{22} =10)	x_{23}=1000 (C_{23} = 80)	**3000 (S_2)**	U_2
M_3	x_{31}=0 (C_{31}= 60)	x_{32}=0 (C_{32}= 30)	x_{33}=4000 (C_{33} = 60)	**4000 (S_3)**	U_3
	2500 (D_1)	**3500 (D_2)**	**5000 (D_3)**		
	V_1	V_2	V_3		

TABLE 4.13
Revised allocation after Iteration 1

Mines	Plants				
	P1	**P2**	**P3**		
M_1	x_{11}=500 (C_{11}= 150)	x_{12}=3500 (C_{12}= 70)	x_{13}=0 (C_{13} =150)	**4000 (S_1)**	U_1
M_2	x_{21}=2000 (C_{21} = 50)	x_{22}=0 (C_{22} =10)	x_{23}=1000 (C_{23} = 80)	**3000 (S_2)**	U_2
M_3	x_{31}=0 (C_{31}= 60)	x_{32}=0 (C_{32}= 30)	x_{33}=4000 (C_{33} = 60)	**4000 (S_3)**	U_3
	2500 (D_1)	**3500 (D_2)**	**5000 (D_3)**		
	V_1	V_2	V_3		

x_{22} becomes zero, and thus the same is chosen as departing variable. The variable x_{21} is now a non-basic variable, and x_{22} is a basic variable The revised transportation tableau after the first iteration is represented in Table 4.13.

Therefore, the revised cost of transportation after first iteration = 150*500 + 70*3500 + 50*2000 + 80*1000 + 60*4000 = US$7, 40,000

The basic cells are now $x_{22,}$ x_{13}, x_{31}, and x_{32}. This is the end of the first iteration. The optimality condition of the current solution is calculated by computing U_1 to U_3, V_1 to V_3, values for basic variables, and K_{ij} values for non-basic variables. First, the U_1 to U_3, V_1 to V_3 values are calculated and presented here:

Checking for optimality

Basic Variables	(U,V) Equation	Solution	Summarized solution
x_{11}	$U_1 + V_1 = 150$	Set $U_1 = 0, V_1 = 150$	$U_1 = 0, V_1 = 150$
x_{12}	$U_1 + V_2 = 70$	$U_1 = 0, V_2 = 70$	$U_2 = -100, V_2 = 70$
x_{21}	$U_2 + V_1 = 50$	$V_1 = 150, U_2 = -100$	$U_3 = -120, V_3 = 180$
x_{23}	$U_2 + V_3 = 80$	$U_2 = -100, V_3 = 180$	
x_{33}	$U_3 + V_3 = 60$	$V_3 = 180, U_3 = -120$	

Next, we use U_i and V_j to evaluate the non-basic variables by computing $U_i + V_j - C_{ij}$ for each non-basic x_{ij}

Non-basic variables	$K_{ij} = U_i + V_j - C_{ij}$
x_{13}	$\mathbf{K_{13} = U_1 + V_3 - C_{13} = 0 + 180 - 150 = 30}$
x_{22}	$K_{22} = U_2 + V_2 - C_{22} = -100 + 70 - 10 = -40$
x_{31}	$K_{31} = U_3 + V_1 - C_{31} = -120 + 150 - 60 = -30$
x_{32}	$K_{32} = U_3 + V_2 - C_{32} = -120 + 70 - 30 = -80$

The K_{ij} value is still positive for cell 13, and thus optimality is not reached. In this case, the highest positive value is found for $x_{13,}$ and thus x_{13} is the entering variable. Let the allocation made to entering variable (x_{13}) is θ. To maintain the supply and demand balance, a cycling shifting is done, as shown in Table 4.14.

The value of θ can be determined from the non-negativity constraints, derived from Table 4.14, as

$$500 - \theta \geq 0, \text{ and } 1000 - \theta \geq 0$$

From the above two equations, the maximum value of θ is 500. The revised allocation is shown in transportation Table 4.15.

TABLE 4.14
Entering and departing cells for iteration 2

Mines	**Plants**				
	P1	**P2**	**P3**		
M1	x_{11}=500-θ (C_{11}= 150)	x_{12}=3500 (C_{12}= 70)	x_{13}=0+θ (C_{13} = 150)	**4000 (S_1)**	U1
M2	x_{21}=2000+θ (C_{21} = 50)	x_{22}=0 (C_{22} =100)	x_{23}=1000-θ (C_{23} =80)	**3000 (S_2)**	U2
M3	x_{31}=0 (C_{31}= 60)	x_{32}=0 (C_{32}= 30)	x_{33}=4000 (C_{33} = 60)	**4000 (S_3)**	U3
	2500 (D_1)	**3500 (D_2)**	**5000 (D_3)**		

TABLE 4.15
Revised allocations after iteration 2

Mines	**Plants**				
	P1	**P2**	**P3**		
M1	x_{11}=0 (C_{11}= 150)	x_{12}=3500 (C_{12}= 70)	x_{13}=500 (C_{13} = 150)	**4000 (S_1)**	U1
M2	x_{21}=2500 (C_{21} = 50)	x_{22}=0 (C_{22} =100)	x_{23}=500 (C_{23} =80)	**3000 (S_2)**	U2
M3	x_{31}=0 (C_{31}= 60)	x_{32}=0 (C_{32}= 30)	x_{33}=4000 (C_{33} = 60)	**4000 (S_3)**	U3
	2500 (D_1)	**3500 (D_2)**	**5000 (D_3)**		
	V1	V2	V3		

Therefore, the revised cost of transportation = 70*3500 + 150*500 + 50*2500 + 80*500 + 60*4000 = US$ 7, 25,000

That is, the transportation cost is decreased in the subsequent iteration.

After cycling shifting of allocation, x_{11} becomes 0, and thus it is now a non-basic variable along with the cells x_{22}, x_{31}, and x_{32}. After identifying basic and non-basic variables, the optimality criterion is tested, and the results are presented here:

Checking for optimality

Basic Variables	**(U,V) Equation**	**Solution**	**Summarized solution**
x_{12}	$U_1 + V_2 = 70$	Set $U_1 = 0, V_2 = 70$	$U_1 = 0, V_1 = 120$
x_{13}	$U_1 + V_3 = 150$	$U_1 = 0, V_3 = 150$	$U_2 = -70, V_2 = 70$
x_{21}	$U_2 + V_1 = 50$	$U_2 = -70, V_1 = 120$	$U_3 = -90, V_3 = 150$
x_{23}	$U_2 + V_3 = 80$	$V_3 = 150, U_2 = -70$	
x_{33}	$U_3 + V_3 = 60$	$V_3 = 150, U_3 = -90$	

Non-basic variables	$K_{ij} = U_i + V_j - C_{ij}$
x_{11}	$K_{11} = U_1 + V_1 - C_{11} = 0 + 120 - 150 = -30$
x_{22}	$K_{22} = U_2 + V_2 - C_{22} = -70 + 70 - 10 = -10$
x_{31}	$K_{31} = U_3 + V_1 - C_{31} = -90 + 120 - 60 = -30$
x_{32}	$K_{32} = U_3 + V_2 - C_{32} = -90 + 70 - 30 = -50$

In this case, all the K_{ij} are non-positive, and thus optimality has been reached. No further iteration is required. Therefore, the allocation shown in transportation Table 4.15 represents an optimal one. The optimal cost of transportation is US$ 7,25,000.

4.3.2.2 Stepping Stone Method

The **Stepping Stone Method** is used to obtain the optimal solution of a transportation problem from the initial feasible solution, which is determined by using any of the three methods like a north-west corner, matrix minimum method, or Vogel's approximation method. Thus, the stepping stone method is a procedure for finding the potential of any non-basic variables (empty cells) in terms of the objective function. In this method, the effect of one-unit allocation to the empty or non-basic cells on the transportation cost is examined to identify the optimal solution.

Contrary to the MODI method, where only one closed path for the non-basic cell with the highest opportunity cost is drawn, the stepping stone method requires drawing as many closed paths as equal to the non-basic cells for their evaluation.

The steps involved in the stepping stone method are demonstrated with the same problem, as given in **Example 4.1**.

The initial basic feasible solution derived using the north-west corner method for the problem is considered for obtaining the optimal solution using the stepping stone method. The allocations made using the north-west corner method are shown in Table 4.16.

The first step in the stepping stone method is to evaluate these non-basic cells by examining whether their allocation would reduce transportation costs. If we find such a route, then we will allocate as much as possible to it. In this case, the basic cells are 11, 12, 22, 23, and 33, and non-basic cells are 13, 21, 31, and 32.

The change in transportation cost with unit allocations to each non-basic cell should be examined one by one. The cell, which offers maximum cost reduction, should be selected as an entering variable. If the cost is not reduced with allocation made to either of the non-basic cells, the existing allocation is optimal. First, let us consider allocating one tonne of material to cell 13. If one tonne is allocated to cell 13, the cost will be increased by Rs.150, which is equal to the transportation cost for cell 13. However, by allocating one ton to cell 13, a similar amount should be deducted from any allocated or basic cell in the same row and in the same column to maintain the supply and demand constraints. Thus, a cyclic shifting of one-unit coal should be done, as shown in Table 4.16. The same should be examined for all other non-basic cells (21, 31, and 32).

TABLE 4.16
Initial solution derived using north-west corner method

Mines	Plants			Production capacity of mines
	P1	P2	P3	
M1	150 (x_{11}= 2500)	70 (x_{12}= 1500)	150 (x_{13} =0)	4000 (S_1)
M2	50 (x_{21} = 0)	10 (x_{22} =2000)	80 (x_{23} = 1000)	3000 (S_2)
M3	60 (x_{31}= 0)	30 (x_{32}= 0)	60 (x_{33} = 4000)	4000 (S_3)
Demand of plants	2500 (D_1)	3500 (D_2)	5000 (D_3)	

First, we always start with a non-basic cell and form a closed path using basic cells. In developing the path, it is possible to skip over both basic and non-basic cells. In any row or column, there can be only one addition and one subtraction.

Path 1 for cell 13: 13 → 23→22→12 = 150 – 80 + 10 – 70 = 10
Path 2 for cell 21: 21→ 11 → 12 → 22 = 50 – 150 + 70 – 10 = –40
Path 3 for cell 31: 31→ 11→12→22→23→33 = 60 – 150 + 70 – 10 + 80 – 60 = –10
Path 4 for cell 32: 32→22→23→33 = 30 – 10 + 80 – 60 = 40

In the above optimality check, the allocation is optimal provided all the paths show non-negative values. A negative value in any of the paths indicates that the initial solution is not optimal because a lower cost can be achieved by allocating to a cell with a negative value. Our goal is to determine the cell or entering 'variable' that will reduce transportation costs the most. Thus, the cell which is having the maximum negative value will be the entering cell or new basic cell. In this case, path two is showing negative value solely, and thus cell 21 is the entering variable, as shown in Table 4.17.

Let the amount of allocation made to cell 21 is θ. The value of θ can be determined in a similar method as used in the MODI method. The non-negativity constraints can be derived from Table 4.17, as

$$2500 - \theta \geq 0, \text{ and } 2000 - \theta \geq 0$$

From the above two constraints, the maximum possible value of θ is 2000. The revised allocation is shown in transportation Table 4.18.

After the first iteration, the allocation in cell 22 is 0, and thus cell 22 becomes a non-basic cell along with cells 13, 31, and 32. In a similar way, as explained in an earlier iteration, the optimality is checked for the current allocation.

Path 1 for cell 13: 13 → 23 → 21 → 11 = 150 – 80 + 50 – 150 = –30
Path 2 for cell 22: 22 → 21 → 11 → 12 = 10 – 50 + 150 – 70 = 40
Path 3 for cell 31: 31 → 33 → 23 → 21 = 60 – 60 + 80 – 50 = 20
Path 4 for cell 32: 32 → 33 → 23 → 21 → 11 → 12 = 30 – 60 + 80 – 50 + 150 – 70 = 80

TABLE 4.17
Transportation tableau showing the entering and departing cells

Mines	Plants			Production capacity of mines
	P1	P2	P3	
M1	x_{11}=2500-θ (C_{11}= 150)	x_{12}=1500+θ (C_{12}= 70)	x_{13}=0 (C_{13} =150)	**4000 (S_1)**
M2	x_{21}=0+θ (C_{21} = 50)	x_{22}=2000-θ (C_{22} =10)	x_{23}=1000 (C_{23} = 80)	**3000 (S_2)**
M3	x_{31}=0 (C_{31}= 60)	x_{32}=0 (C_{32}= 30)	x_{33}=4000 (C_{33} = 60)	**4000 (S_3)**
Demand of plants	**2500 (D_1)**	**3500 (D_2)**	**5000 (D_3)**	

TABLE 4.18
Revised allocation after the first iteration

Mines	Plants			Production capacity of mines
	P1	P2	P3	
M1	x_{11}=500 (C_{11}= 150)	x_{12}=3500 (C_{12}= 70)	x_{13}=0 (C_{13} =150)	**4000 (S_1)**
M2	x_{21}=2000 (C_{21} = 50)	x_{22}=0 (C_{22} =10)	x_{23}=1000 (C_{23} = 80)	**3000 (S_2)**
M3	x_{31}=0 (C_{31}= 60)	x_{32}=0 (C_{32}= 30)	x_{33}=4000 (C_{33} = 60)	**4000 (S_3)**
Demand of plants	**2500 (D_1)**	**3500 (D_2)**	**5000 (D_3)**	

TABLE 4.19
Transportation tableau showing the entering and departing cells

Mines	Plants			Production capacity of mines
	P1	P2	P3	
M1	x_{11}=500-θ (C_{11}= 150)	x_{12}=3500 (C_{12}= 70)	x_{13}=0+ θ (C_{13} =150)	**4000 (S_1)**
M2	x_{21}=2000+ θ (C_{21} = 50)	x_{22}=0 (C_{22} =10)	x_{23}=1000- θ (C_{23} = 80)	**3000 (S_2)**
M3	x_{31}=0 (C_{31}= 60)	x_{32}=0 (C_{32}= 30)	x_{33}=4000 (C_{33} = 60)	**4000 (S_3)**
Demand of plants	**2500 (D_1)**	**3500 (D_2)**	**5000 (D_3)**	

In the above optimality check, a negative value still exists for cell 13 and thus the current allocation is not optimal. Thus, cell 13 is the entering cell, as shown in Table 4.19. Let the amount of allocation made to cell 13 is θ. The value of θ can be determined from the non-negativity constraints, derived from Table 4.19, as

$$500-\theta \geq 0, \text{ and } 1000-\theta \geq 0$$

From the above two constraints, the maximum value of θ = 500. The revised allocation after the second iteration is shown in transportation Table 4.20.

After the second iteration, the allocation in cell 22 is 0, and thus cell 22 is now a non-basic variable along with cells 11, 31, and 32. Again, the optimality condition is checked for the current allocation by examining the paths for all the non-basic cells.

TABLE 4.20
Revised allocation after the second iteration

Mines	Plants			Production capacity of mines
	P1	P2	P3	
M1	x_{11}=0 (C_{11}= 150)	x_{12}=3500 (C_{12}= 70)	x_{13}=500 (C_{13} =150)	**4000 (S_1)**
M2	x_{21}=2500 (C_{21} = 50)	x_{22}=0 (C_{22} =10)	x_{23}=500 (C_{23} = 80)	**3000 (S_2)**
M3	x_{31}=0 (C_{31}= 60)	x_{32}=0 (C_{32}= 30)	x_{33}=4000 (C_{33} = 60)	**4000 (S_3)**
Demand of plants	**2500 (D_1)**	**3500 (D_2)**	**5000 (D_3)**	

Path 1 for cell 11: 11 → 13 → 23 → 21 = 150 – 150 + 80 – 50 = 30
Path 2 for cell 22: 22 → 12 → 13 → 23 = 10 – 70 + 150 – 80 = 10
Path 3 for cell 31: 31 → 21 → 23 → 33 = 60 – 50 + 80 – 60 = 30
Path 4 for cell 32: 32 → 12 → 13 → 33 = 30 – 70 + 150 – 60 = 50

Since no path exhibits a negative value, this indicates that the current allocation has reached the optimality. That is, there is no scope for a further reduction in the cost.

Thus, the optimal transportation cost is = 70*3500 + 150*500 + 50*2500 + 80*500 + 60*4000
= US$7, 25,000/-

4.3.3 Solution algorithm of an unbalanced transportation model

When the supply and demand are unequal, the transportation model is called the unbalanced transportation model. The unbalanced transportation model needs to be balanced first before applying the above-mentioned algorithms. In this case, we need to add a dummy row (when demand is greater than the supply) or a dummy column (when supply is greater than the demand) and assign zero cost to each cell. After this operation, the same methods can be followed for solving the problems. The following two cases demonstrate the procedure of handling the unbalanced transportation problem.

Case 1: Supply > demand

The case is demonstrated using data shown in Example 4.2. In this example, similar to the previous example, M_1 to M_3 represents three mines, and P_1 to P_3 represents three power plants. The total supply from mines is 11,000 (= 4000 +3000 + 4000 = 11000), and the total demand of the plants is 10,000 (= 2500 +3500 + 4000 = 10000). We can see, in this case, the supply is greater than the demand; therefore, the problem is an unbalanced transportation problem. A dummy plant (P_4) needs to be added to the solution table to make the problem a balanced transportation problem. The cost of transportation from each of the mines to the dummy plant is assumed to be zero. The balanced problem after modification is represented in Table 4.21. We can also note here that the demand of the dummy plant (D_4) is added in the table with a demand of 1000 (=11000 – 10000) to balance the table. Now, this problem can be solved using the two-step approach as discussed in the previous section.

EXAMPLE 4.2

Mines	Plants			Production capacity of mines
	P1	P2	P3	
M1	150	70	150	4000
M2	50	10	80	3000
M3	60	30	60	4000
Demand of plants	2500	3500	4000	

TABLE 4.21
Balancing an unbalanced transportation problem when supply > demand

Mines	Plants				Production capacity of mines
	P1	P2	P3	Dummy Plant (P_4)	
M1	150 (x_{11})	70 (x_{12})	150 (x_{13})	0 (x_{14})	4000 (S_1)
M2	50 (x_{21})	10 (x_{22})	80 (x_{23})	0 (x_{24})	3000 (S_2)
M3	60 (x_{31})	30 (x_{32})	60 (x_{33})	0 (x_{34})	4000 (S_3)
Demand of plants	2500 (D_1)	3500 (D_2)	4000 (D_3)	1000 (D_4)	11000

Case 2: Supply < demand
The case is demonstrated using data shown in Example 4.3. In this example, similar to the previous example, M_1 to M_3 represents three mines, and P_1 to P_3 represents three power plants. The total supply from mines in 11,000 (= 4000 + 3000 + 4000 = 11000), and the total demand of the plants is 12,000 (= 2500 + 3500 + 6000 = 12000). In this case, supply < demand; therefore, a dummy row needs to be added to make it balance. The cost of transportation from each of the dummy mines to each plant is assumed to be zero. The balanced problem is represented in Table 4.22. We can also note here that the supply from the dummy mine (M_4) is added to the table with a supply of 1000

EXAMPLE 4.3

Mines	Plants			Production capacity of mines
	P1	P2	P3	
M1	150	70	150	4000
M2	50	10	80	3000
M3	60	30	60	4000
Demand of plants	2500	3500	6000	

TABLE 4.22
Balancing an unbalanced transportation problem when supply < demand

Mines	**Plants**			**Production capacity of mines**
	P1	**P2**	**P3**	
M1	150 (x_{11})	70 (x_{12})	150 (x_{13})	4000 (S_1)
M2	50 (x_{21})	10 (x_{22})	80 (x_{23})	3000 (S_2)
M3	60 (x_{31})	30 (x_{32})	60 (x_{33})	4000 (S_3)
Dummy mine (M4)	0 (x_{41})	0 (x_{42})	0 (x_{43})	1000 (S_3)
Demand of plants	2500 (D_1)	3500 (D_2)	6000 (D_3)	12000

(=12000 – 11000) to balance the table. Now, this problem can be solved using the two-step approach as discussed in the previous section.

Once the problem is balanced, a similar algorithm can be applied to solve the problems.

4.3.4 Solution algorithm of a transportation model with prohibited routes

Sometimes one or more of the routes in the transportation model are prohibited. For example, a particular mine cannot supply materials to a particular plant. When this situation occurs, we must make sure that no resources in the optimal solution are allocated to the cell representing this route. It should be noted that assigning a high relative cost (or a coefficient of M) to a cell will ensure that the variable will never be allocated on the final solution. This same principle can be used in a transportation model for the prohibited routes. An infinitely large value of M should be assigned as the transportation costs to all the cells that represent the prohibited routes. Thus, when the prohibited cells are evaluated, they will always contain a large positive cost change of M, which will keep it from being selected as an entering variable.

The algorithm is demonstrated to the same problem, as explained in **Example 4.1,** with a small change in data. Suppose transportation from Mine C to Plant 3 is prohibited. In that case, the cost of transportation from mine C to plant 3 is assigned as M. The revised transportation problem is represented in Table 4.23. M is a big positive number relative to all other transportation costs.

The same algorithms for obtaining the initial basic feasible solution (north-west corner method or matrix minimum method or VAM method) and optimal solution (MODI or Stepping stone method) can be applied. The optimal solution of the problem is shown in Table 4.24.

It can be observed from Table 4.24 that no material is allocated to prohibited routes (M3 to P3) due to higher transportation costs.

The optimal cost of transportation is

$$= 70 * 2000 + 150 * 2000 + 80 * 3000 + 60 * 2500 + 30 * 1500 = \text{US\$8, 75,000}$$

The optimal transportation cost in case of prohibited routes should be always equal or greater than the non-prohibited route transportation problems.

TABLE 4.23
Revised transportation problem

Mines	Plants			Production capacity of mines
	P1	P2	P3	
M1	150	70	150	4000
M2	50	10	80	3000
M3	60	30	M	4000
Demand of plants	2500	3500	6000	

TABLE 4.24
Optimal allocation of the revised transportation problem

Mines	Plants			
	P1	P2	P3	
M1	$x_{11}=0$ ($C_{11}=150$)	$x_{12}=2000$ ($C_{12}=70$)	$x_{13}=2000$ ($C_{13}=150$)	**4000 (S_1)**
M2	$x_{21}=0$ ($C_{21}=50$)	$x_{22}=0$ ($C_{22}=10$)	$x_{23}=3000$ ($C_{23}=80$)	**3000 (S_2)**
M3	$x_{31}=2500$ ($C_{31}=60$)	$x_{32}=1500$ ($C_{32}=30$)	$x_{33}=0$ ($C_{33}=$**M**)	**4000 (S_3)**
	2500 (D_1)	**3500 (D_2)**	**5000 (D_3)**	

4.3.5 Solution Algorithm for Degeneracy Problem

If the degeneracy exists in the initial feasible solution, a zero allocation should be made to a non-basic cell with the minimum transportation cost. Once the degeneracy problem is removed, the same algorithms (MODI or stepping stone method) can be applied to obtain the optimal solution.

4.4 ASSIGNMENT PROBLEM

The purpose of the assignment problem is to an optimal assignment of a certain number of resources (e.g., workers, machines, etc.) intended for specific jobs. The assignment problem can be represented through binary integer programming in which the goal of the objective function is to minimize the cost or time of the given assignment.

Suppose *m* jobs are to be assigned to n workers. Each worker can do each job at a specific duration with different efficiency. Let C_{ij} is the cost or time for assigning the resource *i* to job *j*. The problem is to find an optimal assignment (which job should be assigned to which worker one-to-one basis) so that the total cost or time of performing all jobs is minimal.

The mathematical form of the assignment problem becomes:

$$\text{Minimize } Z = \sum_{i=1}^{n}\sum_{j=1}^{m} C_{ij}x_{ij} \qquad \text{Objective Function}$$

Subject to,

$$\sum_{j=1}^{m} x_{ij} = 1, \qquad i = 1, 2\ldots n \qquad \text{Job constraints}$$

$$\sum_{i=1}^{n} x_{ij} = 1, \qquad j = 1, 2,\ldots, m \qquad \text{Resource/worker constraints}$$

The variable x_{ij} can be either 0 or 1, depends on the assigning and non-assigning the worker i to job j. That is,

$$x_{ij} = \begin{cases} 1, & \text{if } i^{th} \text{ worker is assigned } j^{th} \text{ job} \\ 0, & \text{if } i^{th} \text{ worker is not assigned } j^{th} \text{ job} \end{cases}$$

The model represents a special form of the linear programming model, known as 0-1 programming or binary integer linear programming. The objective function represents the total cost or time of performing all the jobs. The job constraints indicate that only one job is done by i^{th} resource (i = 1, 2,…,m). Similarly, the resource constraint indicates that only one worker should be assigned to j^{th} job (j = 1, 2,…,n).

An assignment problem can be categorized as balanced or unbalanced. For example, if the number of jobs (m) is equal to the number of resources (n), the problem is said to be a balanced assignment problem. On the other hand, if the number of jobs (m) does not equal to the number of resources (n); the problem is an unbalanced assignment problem.

In order to solve the assignment problem, the number of resources has to be equal to the number of jobs ($m = n$). The importance of the algorithm is that a large number of jobs can be assigned to different resources in order to minimize the cost or time. This special form of assignment problem can be solved efficiently using the Hungarian Assignment Method (HAM) method. The solution strategies of the general-purpose integer programming are discussed in Chapter 5.

4.4.1 The Hungarian Assignment Method (HAM)

The HAM method was first developed by Harold W. Kuhn in 1955. The HAM algorithm is largely based on the earlier works of two Hungarian mathematicians (Dénes Kőnig and Jenő Egerváry) and thus the name suggested as 'Hungarian method'. The Hungarian assignment method provides an optimal assignment for minimizing the cost or resources without examining every feasible solution. In this method, a matrix of opportunity cost is derived from the given cost/time matrix. Opportunity cost represents the relative penalties associated with assigning jobs to any resource/worker in comparison to assigning to best or least cost/resources. The detailed algorithm is explained below.

The method can be applied to a balanced assignment problem. If n resources or workers are needed for completing n jobs, the efficiency/cost matrix can be presented as:

$$
\textbf{Resources}\quad
\begin{array}{c}
\\ 1 \\ 2 \\ \\ i \\ \\ \mathbf{n}
\end{array}
\overset{\textbf{Jobs}}{\begin{bmatrix}
\mathbf{1} & \mathbf{2} & & j & & \mathbf{n} \\
C_{11} & C_{12} & \cdots & C_{1j} & \cdots & C_{1n} \\
C_{21} & C_{22} & \cdots & C_{2j} & \cdots & C_{2n} \\
\cdots & \cdots & \cdots & \cdots & \cdots & \cdots \\
C_{i1} & C_{i2} & \cdots & C_{ij} & \cdots & C_{in} \\
\cdots & \cdots & \cdots & \cdots & \cdots & \cdots \\
C_{n1} & C_{n2} & \cdots & C_{nj} & \cdots & C_{nn}
\end{bmatrix}}
$$

In the above matrix, C_{ij} represents the cost or time that i^{th} resource is assigned by j^{th} job. The optimal assignment represents the solution in which the right worker/resource is assigned by a specific job to ensure that total completion cost/time is minimal.

HAM Algorithm

Step 1: Identify the lowest cost/time element in each row of the cost/time matrix and subtract the lowest cost/time from each element of the respective row. This leads to offer at least one zero in each row of the new matrix, called the *Reduced cost matrix.*

Step 2: In the *Reduced cost matrix,* derived in Step 1, identify the smallest element in each column and subtract from every element in the respective column. The process offers at least one zero in each row and column of the new *Reduced cost matrix.*

Step 3: Cover all the 'Zero' elements in the new *Reduced cost matrix* by drawing the least number of horizontal and vertical lines. If the number of lines drawn is equal to n (the number of rows/columns), the optimal assignment is possible and proceed to step 6; else, go to step 4.

Step 4: Identify the lowest uncovered (by the lines) cost/time element, subtract this lowest element from all uncovered elements, and add this element to each value located at the intersection of any two lines.

Step 5: Steps 3 and 4 should repeat until the condition for optimal assignment mentioned in Step 3 is fulfilled.

Step 6: Assign the job to resource/worker corresponds to 'Zero' elements.

(a) Identify the rows and columns which contain only one 'Zero' element. Assign the job to the resource/worker corresponding to this element. Strike out the 'Zeros', if any, in the column or row corresponding to the element.

(b) If there is no row or column with a single 'Zero' left, then identify a row/column with multiple zero elements and choose any zero element cells arbitrarily for assigning the jobs. Strike out the remaining zeros in the selected column or row.

(c) Steps (a) and (b) should be repeated until all assignments are made.

Step 7: Calculate the total completion cost/time with reference to the original cost/time matrix table.

The steps involved in the HAM method are illustrated using Example 4.4.

Example 4.4

In an underground mine, five types of jobs have to be assigned to five operators. Each operator takes a different time period to complete different jobs, as shown in below Table 4.25. Any job can be assigned to any operator. An operator is assigned by a single job only. Find an optimum allocation of jobs in order to complete in minimum time.

TABLE 4.25
Cost or efficiency matrix showing the time spent by an individual operator to complete the different jobs

		Jobs				
		A	B	C	D	E
Operator	W1	6	5	4	7	9
	W2	10	9	8	7	6
	W3	12	10	9	6	8
	W4	9	11	6	8	11
	W5	8	9	11	12	10

TABLE 4.26
Efficiency or cost matrix showing the least time spent by each operator to complete the different jobs

		Jobs					Minimum
		A	B	C	D	E	
Operator	W1	6	5	4	7	9	$P_1 = 4$
	W2	10	9	8	7	6	$P_2 = 6$
	W3	12	10	9	6	8	$P_3 = 6$
	W4	9	11	6	8	11	$P_4 = 6$
	W5	8	9	11	12	10	$P_5 = 8$

Solution

Step 1: Locate the shortest time, or smallest cost element in each row of the **e**fficiency or cost matrix, as indicated in Table 4.26. Now subtract this smallest element from each element in that row. Let P_i be the minimum cost associated with i^{th} row. The reduced cost matrix is shown in Table 4.27.

Step 2: Locate the smallest cost element in each column in the above-reduced efficiency matrix, as shown in Table 4.27. Now subtract this smallest element from each element in that column. Let Q_j be the minimum costs of jth column.

Step 3: Draw the minimum number of lines covering all zero elements in the reduced efficiency or cost matrix, as shown in Table 4.28.

The minimum number of lines required to cover the zero elements is 5, equal to the number of rows or columns. Thus, a unique optimal assignment is possible. In the reduced cost matrix, the locations of the zero entries allow assigning unique allocations, as shown in Table 4.28. If the

TABLE 4.27
Reduced efficiency or cost matrix

		Jobs				
		A	B	C	D	E
Operator	W1	2	1	0	3	5
	W2	4	3	2	1	0
	W3	6	4	3	0	2
	W4	3	5	0	2	5
	W5	0	1	3	4	2
Minimum	$Q_1 = 0$	$Q_2 = 1$	$Q_3 = 0$	$Q_4 = 0$	$Q_5 = 0$	

TABLE 4.28
Reduced efficiency or cost matrix showing the least numbers of horizontal and vertical lines needed to cover all zero elements

		Jobs				
		A	B	C	D	E
Operator	W1	2	0	0	3	5
	W2	4	2	2	1	0
	W3	6	3	3	0	2
	W4	3	4	0	2	5
	W5	0	0	3	4	2

number of lines needed to cover all the zero element is less than the number of rows or columns, then one additional step is needed. This is explained with the same example with modified data, as shown in **Example 4.5.**

To assign the jobs to the worker, first, identify the row and column with a unique zero cell. In column A, only one zero elements exist, and hence job A should be assigned to worker W5. Once W5 is assigned by job A, the same job cannot be assigned to another, and W5 should not be assigned by any other job. Thus, all the zero elements located in row W5 and column A should be crossed. In a similar way, other jobs can be assigned, as indicated in Table 4.29.

The optimal assignment for minimizing the time is

W1->B; W2 -> E; W3 -> D; W4 -> C; W5 -> A

Therefore, the time of completion of the jobs = 5 + 6 + 6 + 6 + 8 = 31

TABLE 4.29
Reduced efficiency or cost matrix showing assigning jobs to different workers based on zero elements

		Jobs				
		A	B	C	D	E
Operator	**W1**	**2**	**0**	**0**	**3**	**5**
	W2	**4**	**2**	**2**	**1**	**0**
	W3	**6**	**3**	**3**	**0**	**2**
	W4	**3**	**4**	**0**	**2**	**5**
	W5	**0**	**0**	**3**	**4**	**2**

TABLE 4.30
Cost or efficiency matrix showing the time taken by the individual operator to complete the different jobs

		Jobs					Minimum
		A	B	C	D	E	
Operator	W1	6	5	4	7	9	P_1= 4
	W2	10	9	8	7	6	P_2= 6
	W3	12	10	9	8	6	P_3= 6
	W4	9	11	6	8	5	P_4= 5
	W5	10	9	11	12	8	P_5= 8

Example 4.5
In Example 4.4, if the time taken to complete different jobs is modified, as shown in Table 4.30, determine the optimum allocation of jobs to complete in minimum time.

Step 1: Subtract this smallest element from each element in the corresponding row. The reduced efficiency matrix is shown in Table 4.31.
Step 2: Subtract this smallest element from each element in the corresponding column of the reduced efficiency or cost matrix (Table 4.31). The resultant matrix is shown in Table 4.32.
Step 3: Draw the minimum number of lines to cover all zero elements in Table 4.32. The numbers of horizontal and vertical lines needed to cover all the zeros are shown in Table 4.33.

The minimum number of lines required to cover the zero elements is 4, which is less than the number of rows or number of columns. Therefore, the locations of the zero entries in Table 4.33 do

TABLE 4.31
Reduced efficiency or cost matrix after subtracting the smallest element from each element in the corresponding row

		Jobs				
		A	B	C	D	E
Operator	W1	2	1	0	3	5
	W2	4	3	2	1	0
	W3	6	4	3	2	0
	W4	4	6	1	3	0
	W5	2	1	3	4	0
Minimum		$Q_1 = 2$	$Q_2 = 1$	$Q_3 = 0$	$Q_4 = 1$	$Q_5 = 0$

TABLE 4.32
Reduced efficiency or cost matrix after subtracting the smallest element from each element in the corresponding column

		Jobs				
		A	B	C	D	E
Operator	W1	0	0	0	2	5
	W2	2	2	2	0	0
	W3	4	3	3	1	0
	W4	2	5	1	2	0
	W5	0	0	3	3	0

TABLE 4.33
Reduced efficiency or cost matrix showing the least numbers of horizontal and vertical lines needed to cover all zero elements

		Jobs				
		A	B	C	D	E
Operator	W1	0	0	0	2	5
	W2	2	2	2	0	0
	W3	4	3	3	1	0
	W4	2	5	1	2	0
	W5	0	0	3	3	0

not allow assigning unique jobs to the workers. That is, a unique assignment cannot be performed, and hence an additional step (Step 4 and 5 of the algorithm) need to be performed.

Step 4: Select the smallest uncovered (which are not lay on the lines) element. In this case, the smallest uncovered element is 1, as shown in Table 4.33. The reduced matrix after subtracting '1' from all uncovered elements, including itself, and add this element to each value located at the intersection of any two lines is shown in Table 4.34.

Go to Step 3: After obtaining the reduced cost matrix, Step 3 needs to be repeated. Draw the minimum number of lines to cover all zero elements in Table 4.34. The numbers of horizontal and vertical lines needed to cover all the zeros are shown in Table 4.34.

Now the minimum number of lines required to cover the zero elements is 5, equal to the number of rows or number of columns. Therefore, unique optimal assignment can be done. As discussed above, if the criterion is not satisfied after this step, similar iterative operations need to be performed. The unique assignment of jobs to a worker is indicated in Table 4.35.

TABLE 4.34
Reduced efficiency or cost matrix after subtracting the smallest uncovered element from each uncovered elements and adding with the elements of the intersection cells

		Jobs				
		A	B	C	D	E
Operator	**W1**	0	0	0	3	6
	W2	1	1	1	0	0
	W3	3	2	2	1	0
	W4	1	4	0	2	0
	W5	0	0	3	4	1

TABLE 4.35
Assigning jobs to different workers based on zero elements

		Jobs				
		A	B	C	D	E
Operator	**W1**	0	0	0	3	6
	W2	1	1	1	0	0
	W3	3	2	2	1	0
	W4	1	4	0	2	0
	W5	0	0	3	4	1

TABLE 4.36
Alternate solution of assigning jobs to different workers based on zero elements

		Jobs				
		A	B	C	D	E
Operator	**W1**	0	0	0	3	6
	W2	1	1	1	0	0
	W3	3	2	2	1	0
	W4	1	4	0	2	0
	W5	0	0	3	4	1

W1 and W5 have zero elements corresponding to both tasks A and B, and thus assigning either of the tasks A or B to W1 or W5 offers the same cost or time. This indicates an alternative solution exists for the problem. The alternative solution is shown in Table 4.36.

Optimal Assignment for minimizing the cost/time is

W1 -> A; W2 -> D; W3 -> E; W4 -> C; W5 -> B

Cost of assignment = 6 + 7 + 9 + 6 + 9 = 37

The optimal assignment for minimizing the time is

W1 -> B; W2 -> D; W3 -> E; W4 -> C; W5 -> A

Cost of assignment = 5 + 7 + 9 + 6 + 10 = 37

Corollary I: In case of unbalanced assignment problems, a dummy column/row, whichever is smaller in number, is inserted with zeros as the cost or time elements. The solution strategies for the unbalanced assignment problem are demonstrated with a new problem, as shown in **Example 4.7**.

Example 4.7

A mine manager has to assign five types of jobs to 4 contractors. Each contractor has the capability to complete each task but charges different costs in US$ (x10^3), as shown in Table 4.37. Any job can be assigned to any contractor but not more than one. Find an optimum allocation of jobs in order to complete them at a minimum cost.

Solution

The problem is an unbalanced assignment problem, as the number of rows and columns are not equal. Thus, the problem should be converted into a balanced assignment problem by adding a dummy row before applying the HAM algorithm.

In this case, the number of rows (contractor) is less than the number of columns (jobs). Therefore, a dummy row (contractor C5) with a cost equal to 'Zero' is added, as indicated in Table 4.38. For contractor C5, the cost of completing job J1 to J5 (e.g., C_{51} to C_{55}) is all set to 0.

TABLE 4.37
Cost matrix table

Contractor	Job				
	J1	J2	J3	J4	J5
C1	27	18	25	20	21
C2	31	24	21	12	17
C3	20	17	20	24	16
C4	22	28	20	16	27

TABLE 4.38
Revised cost matrix table

Contractor	Job				
	J1	J2	J3	J4	J5
C1	27	18	25	20	21
C2	31	24	21	12	17
C3	20	17	20	24	16
C4	22	28	20	16	27
Dummy (C5)	0	0	0	0	0

TABLE 4.39
Optimal assignment of jobs to contractors

Contractor	Job				
	J1	J2	J3	J4	J5
C1	5	0	3	2	3
C2	15	12	5	0	5
C3	0	1	0	8	0
C4	2	12	0	0	11
Dummy (C5)	0	4	0	4	4

TABLE 4.40
Time/cost matrix table

Workers	Tasks			
	T1	T2	T3	T4
W1	270	180	250	200
W2	310	240	210	--
W3	200	170	200	240
W4	220	280	200	160

After balancing the assignment problem, the same algorithm (HAM) can be applied for optimal assignment. The optimal assignment result is shown in Table 4.39.

The optimal assignment is

C1 -> J2; C2 -> J4; C3 -> J5; C4 -> J3; C5 -> J1

Cost of job completion = US$ (18 + 12 + 16 + 20 + 0) *1000 = US$56,000

Corollary II: If any specific job cannot be assigned to one or more workers or a worker cannot perform a job, the nature of the problem is said to be a constrained assignment problem. To solve this type of problem, the costs of performing the job by such resources or workers are assumed to be extremely large (M). The constrained assignment problem is explained with **Example 4.8**.

Example 4.8

A mine manager has to assign 4 tasks to 4 workers. The time taken in minutes by each worker to finish each task is shown in Table 4.40. Any task can be assigned to any worker but not more than one with the exception that T4 cannot be assigned to W2. Find an optimum allocation of tasks in order to complete them in a minimum time.

Solution

In this case, T4 cannot be assigned to W2, and the corresponding cell (24) should be assigned a large value, M, as shown in Table 4.41. If there are more than one restricted cell, all should be assigned with M. In this way; the restricted cell will automatically not be selected for the assignment. After this, the algorithm (HAM) can be applied to make the optimal assignment.

The optimal assignment of the different tasks to workers is shown in Table 4.42.

The optimal assignment is

W1-T2; W2-T3; W3-T1; W4-T4

Cost of completion of the jobs = (180 + 210 + 200 + 160) =. 750 minutes

That is, the minimum time required to complete all the jobs by four workers is 750 minutes.

TABLE 4.41
Revised time/cost matrix table

Contractor	Tasks			
	T1	T2	T3	T4
W1	270	180	250	200
W2	310	240	210	M
W3	200	170	200	240
W4	220	280	200	160

TABLE 4.42
Optimal assignment of tasks to workers

Contractor	Tasks			
	T1	T2	T3	T4
W1	60	0	70	20
W2	70	30	0	M
W3	0	0	30	70
W4	30	120	40	0

4.5 CASE STUDY ON THE APPLICATION OF TRANSPORTATION MODEL IN MINING SYSTEM

This case study is presented using the data from the largest iron mining company in India. The company has three iron ore mines (Mine A, Mine B, and Mines C), which are located in the same region. For confidential agreement, we are not disclosing the names and locations of the mines. Each mine has three working faces from where the daily iron ore production has been done. The production capacity of Mine A, Mine B, and Mine C is 10,000 tonnes/day, 6000 tonnes/day, and 4000 tonnes/day, respectively. The production from three faces of Mine A is 4000 tonnes/day, 3000 tonnes/day, 3000 tonnes/day respectively. Similarly, the production from three faces of Mine B is 2500 tonnes/day, 2000 tonnes/day, 1500 tonnes/day, respectively. The production from three faces of Mine C is 1500 tonnes/day, 1500 tonnes/day, 1000 tonnes/day, respectively. Mine A has two crushing/screening plants, and Mine B and Mine C both have single crushing/ screening plant. The processing capacity of two crushing/screening plants located in MINE A is 7000 tonnes/day and 5000 tonnes/day, respectively. The crushing/screening plants located at Mine B and Mine C, have 7000 tonnes/day and 5000 tonnes/day, respectively. Each mine has two stockpiles designated for supplying ores to steel plants through trucks and railways, respectively. The ores produced from three mines are solely supplied to six steel plants located in the same region. The demand of ores of six steel plants are 6000 tonne/day, 4000 tonne/day, 3000 tonne/day, 3000 tonne/day, 2000 tonne/day, 1000 tonne/

TABLE 4.43
Distance chart in km from one node to other

	Q1	Q2	R1	S1	T1	T2	U1	U2	V1	V2	P1	P2	P3	P4	P5	P6
FA1	1.0	1.5	--	--	--	--	--	--	--	--	--	--	--	--	--	--
FA2	2.0	2.5	--	--	--	--	--	--	--	--	--	--	--	--	--	--
FA3	2.5	3.0	--	--	--	--	--	--	--	--	--	--	--	--	--	--
FB1	--	--	1.8	--	--	--	--	--	--	--	--	--	--	--	--	--
FB2	--	--	2.0	--	--	--	--	--	--	--	--	--	--	--	--	--
FB3	--	--	2.5	--	--	--	--	--	--	--	--	--	--	--	--	--
FC1	--	--	--	2.0	--	--	--	--	--	--	--	--	--	--	--	--
FC2	--	--	--	2.5	--	--	--	--	--	--	--	--	--	--	--	--
FC3	--	--	--	3.0	--	--	--	--	--	--	--	--	--	--	--	--
Q1	--	--	--	--	2.0	3.0	--	--	--	--	--	--	--	--	--	--
Q2	--	--	--	--	3.0	3.5	--	--	--	--	--	--	--	--	--	--
R1	--	--	--	--	--	--	2.0	2.5	--	--	--	--	--	--	--	--
S1	--	--	--	--	--	--	--	--	3.0	4.0	--	--	--	--	--	--
T1	--	--	--	--	--	--	--	--	--	--	25	50	70	30	40	80
T2	--	--	--	--	--	--	--	--	--	--	20	46	65	32	38	82
U1	--	--	--	--	--	--	--	--	--	--	30	32	40	50	50	90
U2	--	--	--	--	--	--	--	--	--	--	25	30	35	48	48	88
V1	--	--	--	--	--	--	--	--	--	--	45	40	50	25	30	70
V2	--	--	--	--	--	--	--	--	--	--	42	35	48	24	32	68

day, and 1000 tonne/day, respectively. The distances from each face to crushing/screening plants, crushing/screening plants to stockpiles, and stockpiles to steel plants are listed in Table 4.43. Each mine supplies the uniform grade of ore, which is acceptable by each steel plant. The flows of ores are represented in Figure 4.2. That is, the movement of ore from one face to another face of the mine is not feasible. Similarly, the flows of ore from one crushing/screening plant to another crushing/screening plant and one stock pile to another stock pile are not considered.

The cost of transportation from each face of each mine to the respective crushing/screening plant is US$1 per tonne per km. The cost of transportation from each face of each crushing/screening plant to the respective stock piles is US$0.90 per tonne per km. The cost of transportation from each stock pile to the steel plant is US$0.80 per tonne per km. The goal of this case study is to determine the optimal transportations plan for minimizing the total transportation cost.

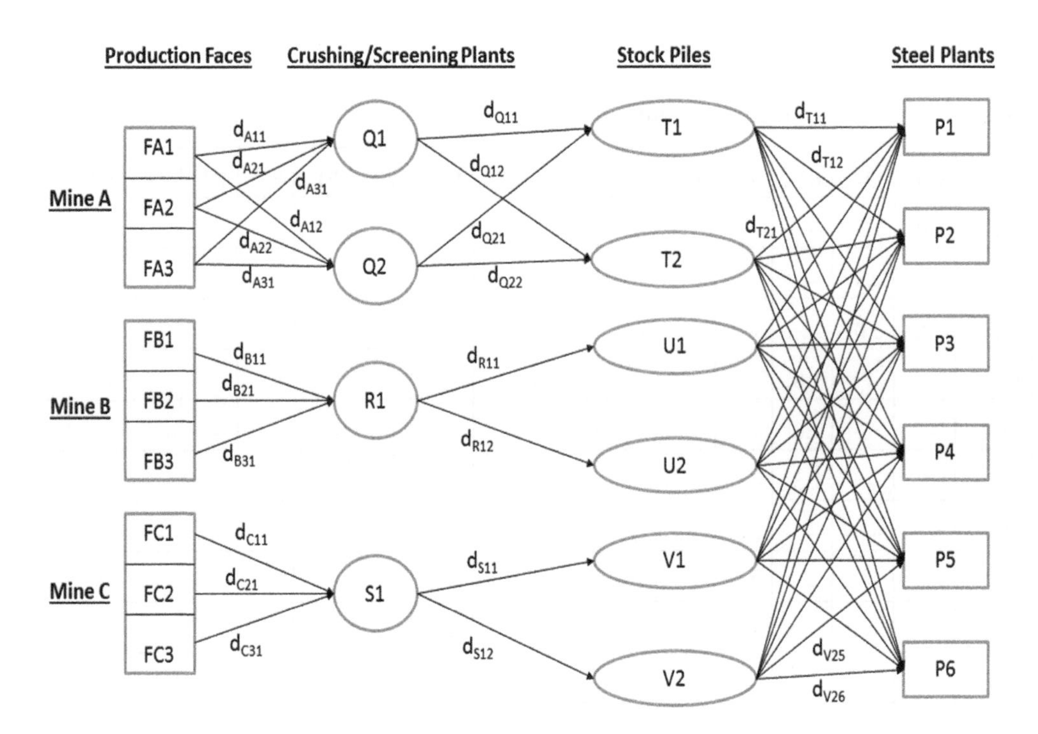

FIGURE 4.2 Feasible paths for ore flow.

Variables:

$d_{A11}, d_{A21}, d_{A31}$ are distances from 1st, 2nd, and 3rd face of Mine A to crushing/screening plant Q1.
$d_{A12}, d_{A22}, d_{A32}$ are distances from 1st, 2nd, and 3rd face of Mine A to crushing/screening plant Q2.
$d_{B11}, d_{B21}, d_{B31}$ are distances from 1st, 2nd and 3rd face of Mine B to crushing/screening plant R1.
$d_{C11}, d_{C21}, d_{C31}$ are distances from 1st, 2nd and 3rd face of Mine C to crushing/screening plant S1.
d_{Q11}, d_{Q12} are distances from crushing/screening plant Q1 to stock piles T1 and T2, respectively.
d_{Q21}, d_{Q22} are distances from crushing/screening plant Q2 to stock piles T1 and T2, respectively.
d_{R11}, d_{R12} are distances from crushing/screening plant R1 to stock piles U1 and U2, respectively.
d_{S11}, d_{S12} are distances from crushing/screening plant S1 to stock piles V1 and V2, respectively.
$d_{T1i}, d_{T2i}, d_{U1i}, d_{U2i}, d_{V1i}, d_{V2i}$ are distances from stock piles T1, T2, U1, U2, V1 and V2 to ith steel plant, respectively. $i = 1,2,3,4,5,6.$

$x_{A11}, x_{A21}, x_{A31}$ are the amount of ores transported from 1st, 2nd and 3rd face of Mine A to crushing/screening plant Q1, respectively.

$x_{A12}, x_{A22}, x_{A321}$ are the amount of ores transported from 1st, 2nd and 3rd face of Mine A to crushing/screening plant Q2, respectively.

$x_{B11}, x_{B21}, x_{B31}$ are the amount of ores transported from 1st, 2nd and 3rd face of Mine B to crushing/screening plant R1, respectively.

$x_{C11}, x_{C21}, x_{C31}$ are the amount of ores transported from 1st, 2nd and 3rd face of Mine C to crushing/screening plant S1, respectively.

x_{Q11} and x_{Q12} are the amount of ores transported from crushing/screening plant Q1 to stock piles T1 and T2, respectively.
x_{Q21} and, x_{Q22} are the amount of ores transported from crushing/screening plant Q2 to stock piles T1 and T2, respectively.
x_{R11} and x_{R12} are the amount of ores transported from crushing/screening plant R1 to stock piles U1 and U2, respectively.
x_{S11} and x_{S12} are the amount of ores transported from crushing/screening plant S1 to stock piles V1 and V2, respectively.
$x_{T1i}, x_{T2i}, x_{U1i}, x_{U2i}, x_{V1i}, x_{V2i}$ are the amount of ores transported from stock piles T1, T2, U1, U2, V1 and V2 to i^{th} steel plant, respectively. $i = 1,2,3,4,5,6$.

$C_{A11}, C_{A21}, C_{A31}$ are the unit transportation cost per unit distance from 1st, 2nd and 3rd face of Mine A to crushing/screening plant Q1.
$C_{A12}, C_{A22}, C_{A321}$ are the unit transportation cost per unit distance from 1st,2nd and 3rd face of Mine A to crushing/screening plant Q2.
$C_{B11}, C_{B21}, C_{B31}$ are the unit transportation cost per unit distance from 1st, 2nd and 3rd face of Mine B to crushing/screening plant R1.
$C_{C11}, C_{C21}, C_{C31}$ are the unit transportation cost per unit distance from 1st, 2nd and 3rd face of Mine C to crushing/screening plant S1.
C_{Q11}, C_{Q12} are the unit transportation cost per unit distance from crushing/screening plant Q1 to stock piles T1 and T2, respectively.
C_{Q21}, C_{Q22} are the unit transportation cost per unit distance from crushing/screening plant Q2 to stock piles T1 and T2, respectively.
C_{R11}, C_{R12} are the unit transportation cost per unit distance from crushing/screening plant R1 to stock piles U1 and U2, respectively.
C_{S11}, C_{S12} are the unit transportation cost per unit distance from crushing/screening plant S1 to stock piles V1 and V2, respectively.
$C_{T1i}, C_{T2i}, C_{U1i}, C_{U2i}, z_{V1i}, C_{V2i}$ are distances from stock piles T1, T2, U1, U2, V1 and V2 to ith steel plant, respectively. $i = 1,2,3,4,5,6$.

The objective of the study is to minimize the overall transportation cost of the system. The objective function represents the total cost of transportations from three different production faces of three mines to the consumer ends (steel plants). The mathematical form of the objective function is

$$\begin{aligned}
Minimize = (&C_{A11} * d_{A11} * x_{A11} + C_{A21} * d_{A21} * x_{A21} + C_{A31} * d_{A31} * x_{A31} + C_{A12} * d_{A12} * x_{A12} \\
&+ C_{A22} * d_{A22} * x_{A22} + C_{A32} * d_{A32} * x_{A32} + C_{B11} * d_{B11} * x_{B11} + C_{B21} * d_{B21} * x_{B21} \\
&+ C_{B31} * d_{B31} * x_{B31} + C_{C11} * d_{C11} * x_{C11} + C_{C21} * d_{C21} * x_{C21} + C_{C31} * d_{C31} * x_{C31}) \\
&+ (C_{Q11} * d_{Q11} * x_{Q11} + C_{Q12} * d_{Q12} * x_{Q12} + C_{Q21} * d_{Q21} * x_{Q21} + C_{Q22} * d_{Q22} * x_{Q22} \\
&+ C_{R11} * d_{R11} * x_{R11} + C_{R12} * d_{R12} * x_{R12} + C_{S11} * d_{S11} * x_{S11} + C_{S12} * d_{S12} * x_{S12}) \\
&+ \sum_{i=1}^{6} \{(C_{T1i} * d_{T1i} * x_{T1i} + C_{T2i} * d_{T2i} * x_{T2i}) + (C_{U1i} * d_{U1i} * x_{U1i} + C_{U2i} * d_{U2i} * x_{U2i}) \\
&+ (C_{V1i} * d_{V1i} * x_{V1i} + C_{V2i} * d_{V2i} * x_{V2i})\}
\end{aligned}$$

Subject to,

$x_{A11} + x_{A12} \leq 4000$ Production constraint of face 1 of Mine A
$x_{A21} + x_{A22} \leq 3000$ Production constraint of face 2 of Mine A
$x_{A31} + x_{A32} \leq 3000$ Production constraint of face 3 of Mine A

$x_{B11} \leq 2500$ Production constraint of face 1 of Mine B
$x_{B21} \leq 2000$ Production constraint of face 2 of Mine B
$x_{B31} \leq 1500$ Production constraint of face 3 of Mine B
$x_{C11} \leq 1500$ Production constraint of face 1 of Mine C
$x_{C21} \leq 1500$ Production constraint of face 2 of Mine B
$x_{C31} \leq 1000$ Production constraint of face 3 of Mine B

$x_{A11} + x_{A21} + x_{A31} \leq 7000$ Capacity constraint of Q1 in Mine A
$x_{A12} + x_{A22} + x_{A32} \leq 5000$ Capacity constraint of Q2 in Mine A
$x_{B11} + x_{B21} + x_{B31} \leq 6000$ Capacity constraint of R1 in Mine B
$x_{C11} + x_{C21} + x_{C31} \leq 4000$ Capacity constraint of S1 in Mine C

$x_{A11} + x_{A21} + x_{A31} - x_{Q11} - x_{Q12} = 0$ Ore inflow at Q1 equal to ore outflows at Q1
$x_{A12} + x_{A22} + x_{A32} - x_{Q21} - x_{Q22} = 0$ Ore inflow at Q2 equal to ore outflows at Q2
$x_{B11} + x_{B21} + x_{B31} - x_{R11} - x_{R12} = 0$ Ore inflow at R1 equal to ore outflows at R1
$x_{C11} + x_{C21} + x_{C31} - x_{S11} - x_{S12} = 0$ Ore inflow at S1 equal to ore outflows at S1

$x_{Q11} + x_{Q21} - \sum_{i=1}^{6} x_{T1i} = 0$ Ore inflow at T1 equal to ore outflows at T1

$x_{Q12} + x_{Q22} - \sum_{i=1}^{6} x_{T2i} = 0$ Ore inflow at T2 equal to ore outflows at T2

$x_{U11} - \sum_{i=1}^{6} x_{U1i} = 0$ Ore inflow at U1 equal to ore outflows at U1

$x_{U12} - \sum_{i=1}^{6} x_{U2i} = 0$ Ore inflow at U2 equal to ore outflows at U2

$x_{V11} - \sum_{i=1}^{6} x_{V1i} = 0$ Ore inflow at V1 equal to ore outflows at V1

$x_{V12} - \sum_{i=1}^{6} x_{V2i} = 0$ Ore inflow at V2 equal to ore outflows at V2

$x_{T11} + x_{T21} + x_{U11} + x_{U21} + x_{V11} + x_{V21} \geq 6000$ Demand capacity of P1
$x_{T12} + x_{T22} + x_{U12} + x_{U22} + x_{V12} + x_{V22} \geq 4000$ Demand capacity of P2
$x_{T13} + x_{T23} + x_{U13} + x_{U23} + x_{V13} + x_{V23} \geq 3000$ Demand capacity of P3
$x_{T14} + x_{T24} + x_{U14} + x_{U24} + x_{V14} + x_{V24} \geq 3000$ Demand capacity of P4
$x_{T15} + x_{T25} + x_{U15} + x_{U25} + x_{V15} + x_{V25} \geq 2500$ Demand capacity of P5
$x_{T16} + x_{T26} + x_{U16} + x_{U26} + x_{V16} + x_{V26} \geq 1500$ Demand capacity of P6

The problem is solved in Excel Solver. In the first step, the distance and cost matrices are prepared in excel for each feasible path, as shown in Figure 4.3 and Figure 4.4, respectively. The initial allocations of materials for each feasible path are also defined, as shown in Figure 4.5. After preparing the distance, cost, and initial allocation matrices, the objective function and constraints are defined in Excel solver, as shown in Figure 4.6. After defining all the constraints and objective functions, click the solve button. The solver will offer the optimal allocations, as shown in Figure 4.7.

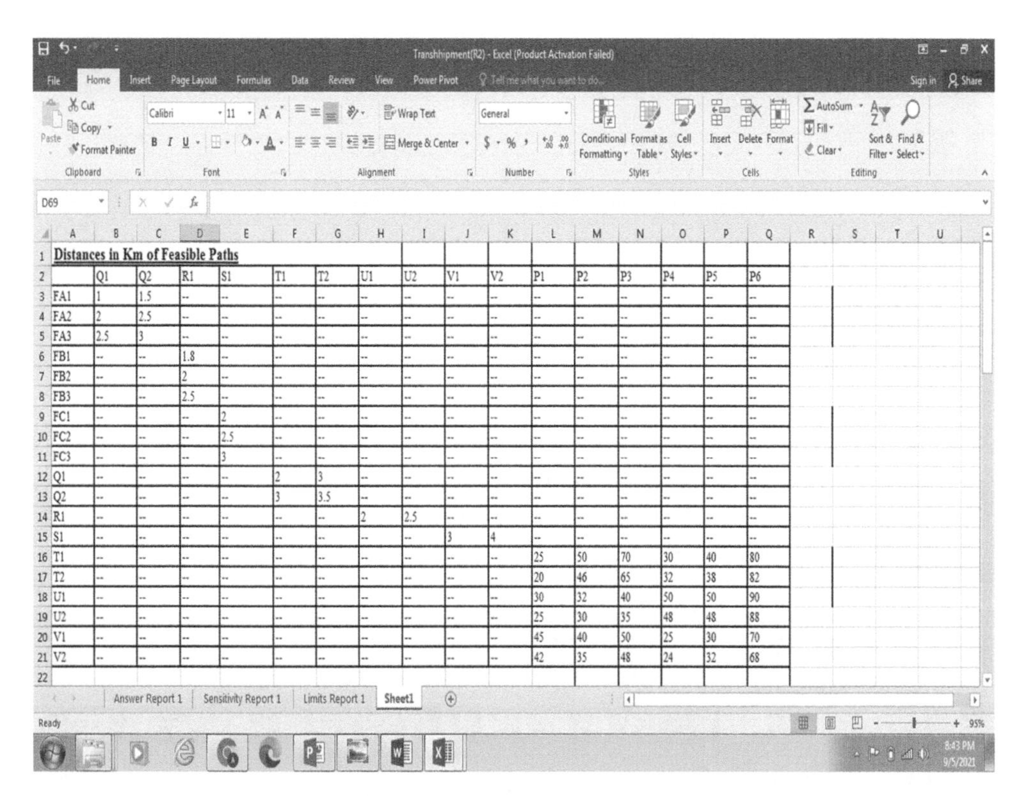

Distances in Km of Feasible Paths

	Q1	Q2	R1	S1	T1	T2	U1	U2	V1	V2	P1	P2	P3	P4	P5	P6
FA1	1	1.5	--	--	--	--	--	--	--	--	--	--	--	--	--	--
FA2	2	2.5	--	--	--	--	--	--	--	--	--	--	--	--	--	--
FA3	2.5	3	--	--	--	--	--	--	--	--	--	--	--	--	--	--
FB1	--	--	1.8	--	--	--	--	--	--	--	--	--	--	--	--	--
FB2	--	--	2	--	--	--	--	--	--	--	--	--	--	--	--	--
FB3	--	--	2.5	--	--	--	--	--	--	--	--	--	--	--	--	--
FC1	--	--	--	2	--	--	--	--	--	--	--	--	--	--	--	--
FC2	--	--	--	2.5	--	--	--	--	--	--	--	--	--	--	--	--
FC3	--	--	--	3	--	--	--	--	--	--	--	--	--	--	--	--
Q1	--	--	--	--	2	3	--	--	--	--	--	--	--	--	--	--
Q2	--	--	--	--	3	3.5	--	--	--	--	--	--	--	--	--	--
R1	--	--	--	--	--	--	2	2.5	--	--	--	--	--	--	--	--
S1	--	--	--	--	--	--	--	--	3	4	--	--	--	--	--	--
T1	--	--	--	--	--	--	--	--	--	--	25	50	70	30	40	80
T2	--	--	--	--	--	--	--	--	--	--	20	46	65	32	38	82
U1	--	--	--	--	--	--	--	--	--	--	30	32	40	50	50	90
U2	--	--	--	--	--	--	--	--	--	--	25	30	35	48	48	88
V1	--	--	--	--	--	--	--	--	--	--	45	40	50	25	30	70
V2	--	--	--	--	--	--	--	--	--	--	42	35	48	24	32	68

FIGURE 4.3 Distance matrix from one node to another for feasible paths.

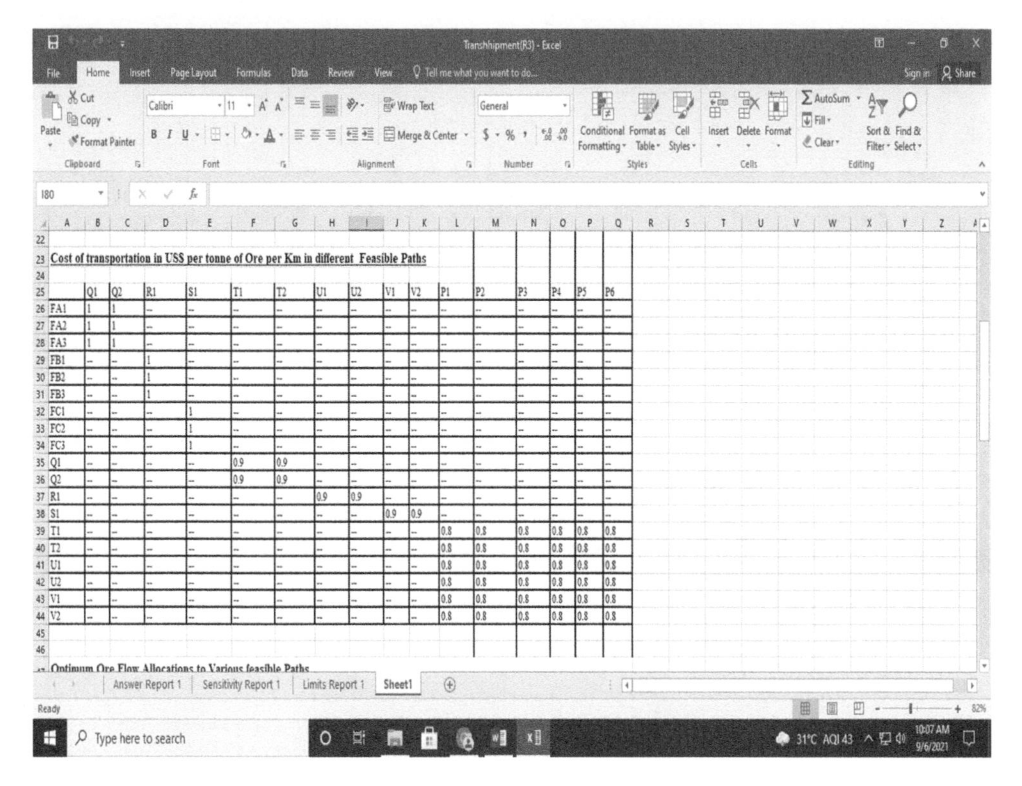

Cost of transportation in US$ per tonne of Ore per Km in different Feasible Paths

	Q1	Q2	R1	S1	T1	T2	U1	U2	V1	V2	P1	P2	P3	P4	P5	P6
FA1	1	1	--	--	--	--	--	--	--	--	--	--	--	--	--	--
FA2	1	1	--	--	--	--	--	--	--	--	--	--	--	--	--	--
FA3	1	1	--	--	--	--	--	--	--	--	--	--	--	--	--	--
FB1	--	--	1	--	--	--	--	--	--	--	--	--	--	--	--	--
FB2	--	--	1	--	--	--	--	--	--	--	--	--	--	--	--	--
FB3	--	--	1	--	--	--	--	--	--	--	--	--	--	--	--	--
FC1	--	--	--	1	--	--	--	--	--	--	--	--	--	--	--	--
FC2	--	--	--	1	--	--	--	--	--	--	--	--	--	--	--	--
FC3	--	--	--	1	--	--	--	--	--	--	--	--	--	--	--	--
Q1	--	--	--	--	0.9	0.9	--	--	--	--	--	--	--	--	--	--
Q2	--	--	--	--	0.9	0.9	--	--	--	--	--	--	--	--	--	--
R1	--	--	--	--	--	--	0.9	0.9	--	--	--	--	--	--	--	--
S1	--	--	--	--	--	--	--	--	0.9	0.9	--	--	--	--	--	--
T1	--	--	--	--	--	--	--	--	--	--	0.8	0.8	0.8	0.8	0.8	0.8
T2	--	--	--	--	--	--	--	--	--	--	0.8	0.8	0.8	0.8	0.8	0.8
U1	--	--	--	--	--	--	--	--	--	--	0.8	0.8	0.8	0.8	0.8	0.8
U2	--	--	--	--	--	--	--	--	--	--	0.8	0.8	0.8	0.8	0.8	0.8
V1	--	--	--	--	--	--	--	--	--	--	0.8	0.8	0.8	0.8	0.8	0.8
V2	--	--	--	--	--	--	--	--	--	--	0.8	0.8	0.8	0.8	0.8	0.8

Optimum Ore Flow Allocations to Various feasible Paths

FIGURE 4.4 Cost matrix from one node to another for feasible paths.

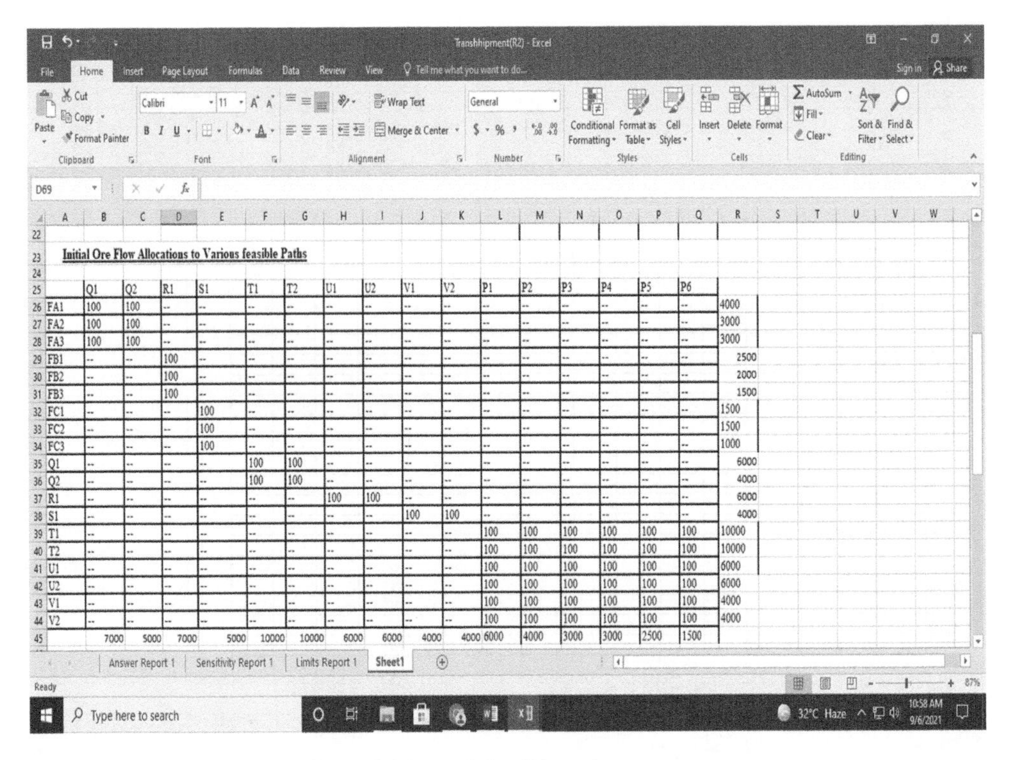

FIGURE 4.5 Initial allocation of material for each feasible path.

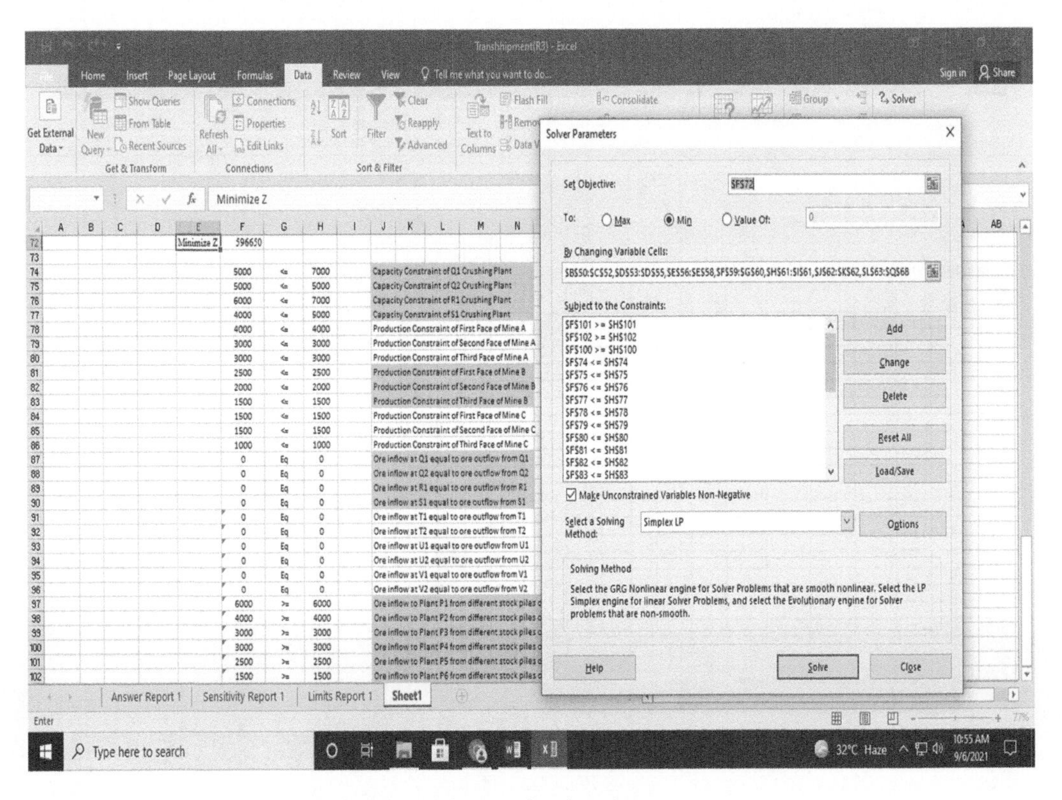

FIGURE 4.6 Defining objective function and constraints in solver.

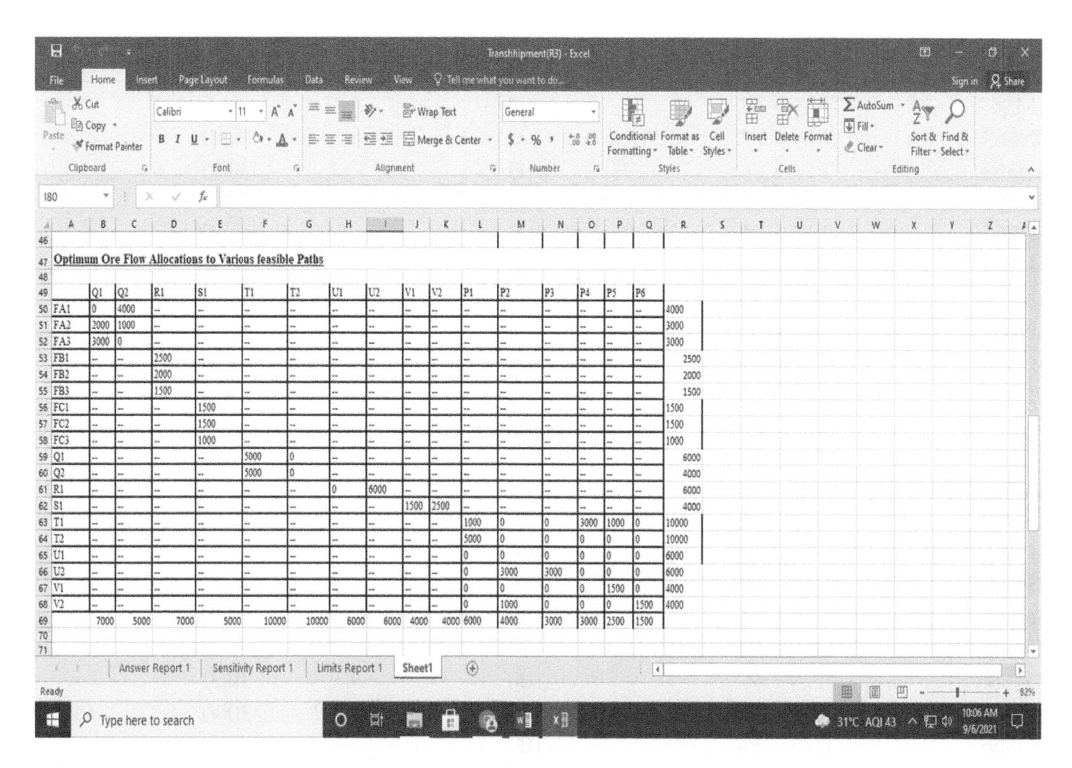

Optimum Ore Flow Allocations to Various feasible Paths

	Q1	Q2	R1	S1	T1	T2	U1	U2	V1	V2	P1	P2	P3	P4	P5	P6	
FA1	0	4000	--	--	--	--	--	--	--	--	--	--	--	--	--	--	4000
FA2	2000	1000	--	--	--	--	--	--	--	--	--	--	--	--	--	--	3000
FA3	3000	0	--	--	--	--	--	--	--	--	--	--	--	--	--	--	3000
FB1	--	--	2500	--	--	--	--	--	--	--	--	--	--	--	--	--	2500
FB2	--	--	2000	--	--	--	--	--	--	--	--	--	--	--	--	--	2000
FB3	--	--	1500	--	--	--	--	--	--	--	--	--	--	--	--	--	1500
FC1	--	--	--	1500	--	--	--	--	--	--	--	--	--	--	--	--	1500
FC2	--	--	--	1500	--	--	--	--	--	--	--	--	--	--	--	--	1500
FC3	--	--	--	1000	--	--	--	--	--	--	--	--	--	--	--	--	1000
Q1	--	--	--	--	5000	0	--	--	--	--	--	--	--	--	--	--	6000
Q2	--	--	--	--	5000	0	--	--	--	--	--	--	--	--	--	--	4000
R1	--	--	--	--	--	--	0	6000	--	--	--	--	--	--	--	--	6000
S1	--	--	--	--	--	--	--	--	1500	2500	--	--	--	--	--	--	4000
T1	--	--	--	--	--	--	--	--	--	--	1000	0	0	3000	1000	0	10000
T2	--	--	--	--	--	--	--	--	--	--	5000	0	0	0	0	0	10000
U1	--	--	--	--	--	--	--	--	--	--	0	0	0	0	0	0	6000
U2	--	--	--	--	--	--	--	--	--	--	0	3000	3000	0	0	0	6000
V1	--	--	--	--	--	--	--	--	--	--	0	0	0	0	1500	0	4000
V2	--	--	--	--	--	--	--	--	--	--	0	1000	0	0	0	1500	4000
	7000	5000	7000	5000	10000	10000	6000	6000	4000	4000	6000	4000	3000	3000	2500	1500	

FIGURE 4.7 Optimized allocations after the running solver.

Thus, the final allocations can also be obtained from the answer sheet as follows:

Cell	Name	Original Value	Final Value
\$B\$50	FA1 → Q1	100	0
\$C\$50	FA1 → Q2	100	4000
\$B\$51	FA2 → Q1	100	2000
\$C\$51	FA2 → Q2	100	1000
\$B\$52	FA3 → Q1	100	3000
\$C\$52	FA3 → Q2	100	0
\$D\$53	FB1 → R1	100	2500
\$D\$54	FB2 → R1	100	2000
\$D\$55	FB3 → R1	100	1500
\$E\$56	FC1 → S1	100	1500
\$E\$57	FC2 → S1	100	1500
\$E\$58	FC3 → S1	100	1000
\$F\$59	Q1 → T1	100	5000
\$G\$59	Q1 → T2	100	0
\$F\$60	Q2 → T1	100	5000
\$G\$60	Q2 → T2	100	0
\$H\$61	R1 → U1	100	0
\$I\$61	R1 → U2	100	6000
\$J\$62	S1 → V1	100	1500
\$K\$62	S1 → V2	100	2500
\$L\$63	T1 → P1	100	1000

Cell	Name	Original Value	Final Value
M63	T1 → P2	100	0
N63	T1 → P3	100	0
O63	T1 → P4	100	3000
P63	T1 → P5	100	1000
Q63	T1 → P6	100	0
L64	T2 → P1	100	5000
M64	T2 → P2	100	0
N64	T2 → P3	100	0
O64	T2 → P4	100	0
P64	T2 → P5	100	0
Q64	T2 → P6	100	0
L65	U1 → P1	100	0
M65	U1 → P2	100	0
N65	U1 → P3	100	0
O65	U1 → P4	100	0
P65	U1 → P5	100	0
Q65	U1 → P6	100	0
L66	U2 → P1	100	0
M66	U2 → P2	100	3000
N66	U2 → P3	100	3000
O66	U2 → P4	100	0
P66	U2 → P5	100	0
Q66	U2 → P6	100	0
L67	V1 → P1	100	0
M67	V1 → P2	100	0
N67	V1 → P3	100	0
O67	V1 → P4	100	0
P67	V1 → P5	100	1500
Q67	V1 → P6	100	0
L68	V2 → P1	100	0
M68	V2 → P2	100	1000
N68	V2 → P3	100	0
O68	V2 → P4	100	0
P68	V2 → P5	100	0
Q68	V2 → P6	100	1500

The value of the objective function is

Cell	Name	Original Value	Final Value
F72	Minimize Z --	136940	596650

Therefore, the cost of transportation is US$596650 per day with the optimal transportation plan. It is noted here that the Excel solver is used for solving this case study to demonstrate to users that there are multiple different ways to solve the same problem. It can be solved using the CPLEX that has been applied to solve other optimization problems in this book.

EXERCISE 4

Q1. Three coal mines with weekly production capacities of 60, 50, and 80 thousand tonnes, respectively, supply coal to three power plants with weekly demands of 40, 80, and 70 thousand tonnes, respectively. Coal is transported to the three power plants through roads. The transportation cost (includes handling cost) is US$1 per ton per km. The following table gives the distances between the mines and the power plants. Mine 1 is not connected to power plant 3. Determine the optimum shipping schedule to minimize the transportation cost.

		Distance Chart in km		
		Power Plants		
		P1	**P2**	**P3**
Mines	M1	12	18	-
	M2	30	10	8
	M3	20	25	12

[**Ans.** M1→P1 = 40, M1→P2 = 20, M2→P2 = 50, M3→P2 = 10, M3→P3 = 70, Cost = US$ 2430]

Q2. Three coal mines with monthly capacities of 20, 40, and 30 thousand tonnes, respectively, supply coal to power plants. The monthly demands at the three plants are estimated to be 30, 35, and 25 thousand tonnes. The cost incurred (purchase plus transportation) by each plant from different mines is listed below in the table. During peak summer, there is a 10% increase in demand at each of the three power plants. The increased demand is fulfilled by purchasing coal from another mine (M4) at a higher price of US$250 per tonne (purchasing plus transportation cost), but the mine is not connected to power plant 3. Determine the following:

a. Mathematical form of the transportation model.
b. Optimal transportation plan for minimizing the overall cost.
c. Analyse the total transportation cost for purchasing additional coals by each of the three power plants.

		Power Plants		
		P1	**P2**	**P3**
Mines	M1	US$220	US$240	US$200
	M2	US$210	US$225	US$240
	M3	US$230	US$200	US$210

[**Ans.** b. M1→P3 = 20, M2→P1 = 33, M2→P2 = 7, M3→P2 = 22.5, M3→P3 = 7.5, M4→P2 = 9; c. Revised purchasing cost = US$ 2430]

Q3. Three sand mines with daily production capacities of 600, 500, and 800 tonnes, respectively, supply three underground coal mines with daily demands of 400, 800, and 700 tonnes, respectively, for stowing. Sand is transported from the three sand mines to underground coal mines through a pipelines network. The transportation cost is US$0.5 per tonne per km. The distance between the sand mines and the coal mines is given in the following table. It is known that the pipeline between Sand mine, S2 and coal mine, M3 is not installed.

Formulate the associated transportation model and calculate the economical transportation plan for minimizing the cost.

	Mileage chart (km)			
	Mines			
	M1	M2	M3	
Sand Sources	S1	40	100	20
	S2	120	40	--
	S3	140	160	120

[**Ans**. S1→M1 = 400, S1→M3 = 200, S2→M2 = 500, S3→M2 = 300, S3→M3 = 500, Optimal shipping cost = US$ 74000]

Q4. In the above problem, suppose that the daily demand of sand at coal mine, M3 drops to 650 tonnes and surplus production at sand mines, S1 and S2 are diverted to other mines, located at 50 and 60 km from sand mine 1 and 2, respectively, by truck. The transportation cost per tonne is US$1 per tonne per km through road transport. Sand mine, S3 can divert its surplus production to other works within the mine itself. Formulate the mathematical model of the transportation problem and determine the optimum shipping cost.

[**Ans.** S1→M1 = 400, S1→M3 = 150, S1→M_O = 50, S2→M2 = 500, S3→M2 = 300, S3→M3 = 500, Optimal shipment cost = US$ 76000]

Q5. The manager of a mining workshop has to assign four jobs to 4 workers. The cost of performing a job is a function of the skill of the worker. The table summarizes the cost of assigning each individual job to an individual worker. It is known that W1 cannot do J3 and W3 cannot do J4. Determine the optimal assignment using the Hungarian method.

		Jobs			
		J1	J2	J3	J4
Workers	W1	$50	$50	--	$20
	W2	$70	$40	$20	$30
	W3	$90	$30	$50	--
	W4	$70	$20	$60	$70

[**Ans.** W1→J4, W2→J3, W3→J2, W4→J1; Optimal cost = US$ 140]

Q6. Three jobs for operating dump truck, shovel, and drill machine need to be assigned to three operators separately in a mine. The cost of completion of different jobs is different for the operators. The manager needs the work to be done at minimum cost. Cost (in US$) of completion for different workers is:

Job \ Person	A (dump truck)	B (shovel)	C (drill machine)
P	120	150	140
Q	160	80	100
R	100	120	130

Determine the optimal assignment.
[**Ans.** P→C, Q→B, R→A, Optimal Cost = US$ 320]

Q7. A panel in charge has assigned four jobs (Drilling, Blasting, Maintenance, Pump operation) to four miners. He wants the job to be done in minimum total time. As per past experience, he has the information on the time taken by the four Miners in performing these jobs, as given in the table. Time is in minutes.

Miners	**Job**			
	J1	**J2**	**J3**	**J4**
M1	135	120	153	200
M2	171	125	190	165
M3	148	156	145	192
M4	123	135	180	165

[**Ans.** M1→J2, M2→J4, M3→J3, M4→J1; Optimal cost = US$ 553]

Q8. A mining company supplies iron ores from three mines in Odisha to four steel plants located in Odisha and Jharkhand. The iron ores are shipped through rail. The supply capacity of three mines (in tonnes), the demands of ores at the four steel plants (in tonnes), and the shipping cost from each mine to each plant (in US$ per tonne). Determine the company's shipment plan for minimizing the total shipping cost.

Mines	**Plants**				**Supply (tonne)**
	P1	**P2**	**P3**	**P4**	
M1	2.40	3.00	2.70	3.20	1200
M2	2.90	1.90	2.10	3.60	850
M3	2.60	2.90	2.00	1.80	1250
Demands (tonne)	1000	750	450	800	

[**Ans.** M1→P1 = 1000, M2→P2 = 750, M3→P3 = 450, M3→P4 = 800, Optimal cost = US$ 6165]

Q9. There are three jobs A, B, and C can be assigned to P, Q, and R. A single person can finish a single job. The cost of completion of different people for the different jobs is different and is given below.

Person	**Job**		
	A	**B**	**C**
P	15	10	9
Q	9	15	10
R	10	12	8

Determine the minimum cost to finish all the jobs.
[**Ans.** P→B, Q→A, R→C, Minimum cost = US$ 27]

Q10. A mine workshop deployed four repairpersons, who needed to complete four jobs with different efficiencies. Each job can be assigned to only one repair person. The estimated processing time (in minutes) of each repair person to complete each task is given in the table below.

		Repair Person			
		W1	**W2**	**W3**	**W4**
Tasks	J1	61	92	52	72
	J2	42	49	69	85
	J3	47	59	80	71
	J4	65	70	68	72

Determine the minimum value of total processing time based on an optimal assignment.
[**Ans.** J1→W3, J2→W2, J3→W1, J4→W4, Optimal time = 220 Minutes]

Q11. A manager of a mine wants to assign five types of tasks to four different crews. The crew C4 cannot do the J4 job. The following table provides the cost (in US$) of completion by different crews to different jobs. The objective of the problem is to determine the optimum assignment of each task type for minimizing the total completion cost.

		Task type				
		J1	**J2**	**J3**	**J4**	**J5**
Crew	C1	600	400	700	800	900
	C2	500	600	800	700	700
	C3	700	500	900	600	600
	C4	800	800	600	—	700
	C5	850	750	700	820	780

[**Ans.** C1→J2, C2→J1, C3→J4, C4→J3, C5→J5, Optimal cost = US$ 2880]

Q12. Three mines are procured raw materials from three different sources. It is estimated that the penalty costs per unit of unsatisfied demand for destinations M1, M2, and M3 are $7, $6, and $4, respectively. The transportation cost for a unit item from each supply node to the destination node is given in the table below. Determine the optimum shipment plan for minimizing the total cost.

		Destination			Supply (kg)
		M1	**M2**	**M3**	
Source	S1	$ 5	$ 1	$ 7	100
	S2	$ 6	$ 4	$ 6	800
	S3	$ 3	$ 2	$ 5	150
Demand (kg)		750	200	500	

[**Ans.** S1→M1 = 100, S2→M1 = 600, S2→M2 = 100, S2→M3 = 100, S3→M1 = 150, Optimal cost = US$ 6750]

Q13. Three mines (M1, M2, and M3) supply coal to three power plants (P1, P2, and P3) on a daily basis. The storage costs of coal in different mines are $0.6, $0.5, and $0.4 per tonne for mine M1, M2, and M3, respectively. Moreover, mine M2 does have storage capacity, and thus all the produced coal should be shipped out to make space for new coal. Determine the optimal shipping schedule.

		Destination			**Supply**
		D1	**D2**	**D3**	
Source	M1	$1	$2	$1	20
	M2	$3	$4	$5	40
	M3	$2	$3	$3	30
Demand		30	20	20	

[**Ans.** M1→P3=2000, M2→P1=2000, M3→P1=1000, M3→P2=2000, Cost=US$ 17000]

Q14. The supervisor in a mining workshop needs to assign four jobs to 4 service engineers. The cost function of performing a job depends on the skill of the service engineer, as summarized in the cost data table. It is known that service engineer, E1 cannot perform Job, J3 and service engineer, E3 cannot perform job, J4. Calculate the optimal assignment.

		Jobs			
		J1	**J2**	**J3**	**J4**
Miner	E1	$60	$50	--	$30
	E2	$80	$40	$30	$40
	E3	$90	$30	$50	--
	E4	$70	$20	$60	$80

[**Ans.** E1→J4, E2→J3, E3→J2, E4→J1, Optimal cost = US$ 160]

Q15. In the above assignment problem, if an additional (fifth) service engineer, E5 becomes available for performing the four jobs at the respective costs of $70, $50, $40, and $80. Is it economical to replace one of the existing service engineer with the new one?

[**Ans.** E3 should be replaced by E5]

Q16. In the above assignment problem, if the workshop has just received the fifth job and that the respective costs of performing it by the four current workers are US$30, US$20, US$30, and US$70. Should the new job take priority over any of the four jobs already has to reduce the cost?

[**Ans.** New job should not be given priority to reduce the cost]

5 Integer linear programming for mining systems

5.1 DEFINITION

Integer linear programming (ILP) solves linear optimization models containing variables that are restricted to integer values. ILP also provides a mechanism for handling nonlinearities such as fixed costs within an LP format. In ILP, at least one of the variables is restricted to integer values. ILP problems can be classified into different types, as demonstrated in Figure 5.1.

- **Pure Integer Programming (PIP)**: In PIP, all decision variables are restricted to integer values.
- **Mixed Integer Programming (MIP)**: In MIP, some but not all of the decision variables are restricted to integer values.
- **Binary Integer Programming (BIP)**: In BIP, all decision variables are restricted to binary values, i.e., 0 or 1.
- **Integer Knapsack (IK)**: The knapsack problem is a problem in combinatorial optimization in which finding an optimal solution is done from a finite set of alternatives.

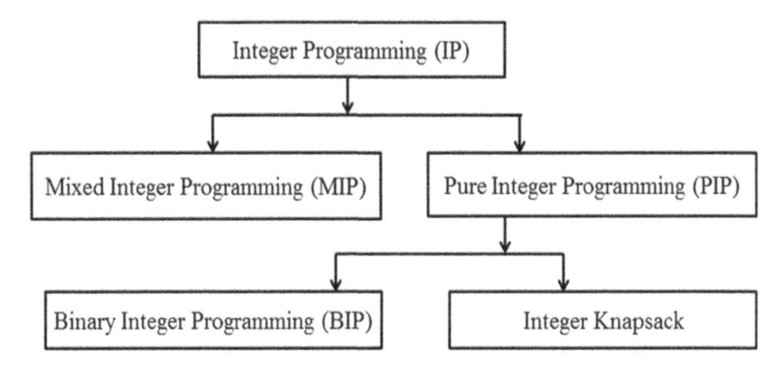

FIGURE 5.1 Classification of Integer Programming.

5.2 FORMULATION OF ILP

The linear-programming models that have been discussed in Chapter 3 are continuous in the sense that decision variables are allowed to be non-integer or fractions. But on many occasions, one or more of the decision variables are restricted to an integer value. The formulation of integer programming is explained with the following example.

Example: Capital Budgeting Problem for Selecting a Mining Project

There are four possible mining projects, which each run for five years and have the following characteristics.

DOI: 10.1201/9781003200703-5

Project	Return ($m)	Capital Requirements ($m)				
		Year 1	Year 2	Year 3	Year 4	Year 5
1	2	4	3	2	1	1
2	3	5	4	2	2	1
3	4	6	4	2	2	1
4	5	6	5	3	1	1
Capital Availability		20	13	7	5	2

Formulate the problem as ILP to select the projects that offer a maximum total return.
An ILP can be formulated using the same approach as that used in LP. The problem has three components like decision variables, constraints, and objective.

Variables
The objective of the problem is to decide whether to choose a project or not. In contrast to LP, the formulation of ILP requires introducing variables that take the integer values of the decision variable. In this case, a binary integer variable (0 or 1) is introduced to represent *binary* decisions like select a project or not selecting a project. That is,

- the positive decision (the project is selected) being represented by the value 1; and
- the negative decision (the project is not selected) being represented by the value 0.

The variables are defined as

$$x_{j=1\,to\,4} = \begin{cases} 1, & \text{if project } j \text{ is selected} \\ 0, & \text{if project } j \text{ is not selected} \end{cases}$$

Constraints
The constraints relating to the availability of capital funds each year are

$$4x_1 + 5x_2 + 6x_3 + 6x_4 \le 20 \quad \text{Capital availability constraint for year 1}$$

$$3x_1 + 4x_2 + 4x_3 + 5x_4 \le 13 \quad \text{Capital availability constraint for year 2}$$

$$2x_1 + 2x_2 + 2x_3 + 3x_4 \le 7 \quad \text{Capital availability constraint for year 3}$$

$$1x_1 + 2x_2 + 2x_3 + 1x_4 \le 5 \quad \text{Capital availability constraint for year 4}$$

$$1x_1 + 1x_2 + 1x_3 + 1x_4 \le 2 \quad \text{Capital availability constraint for year 5}$$

Objective function
The objective function for maximizing the total return is given by

$$Maximize\ Z = 2x_1 + 3x_2 + 4x_3 + 5x_4$$

Thus, the complete ILP can be represented as

$$Maximise\ Z = 2x_1 + 3x_2 + 4x_3 + 5x_4$$

subject to,

$$4x_1 + 5x_2 + 6x_3 + 6x_4 \le 20$$

$$3x_1 + 4x_2 + 4x_3 + 5x_4 \le 13$$

$$2x_1 + 2x_2 + 2x_3 + 3x_4 \leq 7$$

$$1x_1 + 2x_2 + 2x_3 + 1x_4 \leq 5$$

$$1x_1 + 1x_2 + 1x_3 + 1x_4 \leq 2$$

$$x_j = 0 \text{ or } 1, \qquad j = 1 \text{ } to \text{ } 4$$

5.3 SOLUTION ALGORITHMS OF AN ILP

Unlike LP, *no* unique computationally effective algorithms exist for solving IP problems. That is, IP is difficult to solve. The solution methods for IP can also be categorized as *optimal and heuristic.* An optimal algorithm (mathematically) *guarantees* to find the optimal solution but is difficult to solve large problems. On the other hand, a heuristic algorithm offers a feasible solution close to the optimal solution. Therefore, a heuristic algorithm, in general, can be used to solve large problems. In this section, only optimal solution algorithms are discussed.

ILP is very much required when non-integer decision variables have little meaning in decision-making problems. However, since LP solution procedures such as the simplex method treat variables as continuous, this method cannot be helpful for the optimization solution for ILP. Predominantly, two algorithms are widely used to solve an ILP: **cutting plane method or Gomory's cut method** and **branch and bound (B&B) algorithm**.

5.3.1 CUTTING PLANE METHOD OR GOMORY'S CUT METHOD

The cutting plane method is one of the optimization methods that iteratively adds valid inequalities to the original problem to narrow down the search area enclosed by the constraints or cuts while retaining the feasible points. The step-wise solution algorithm of the cutting plane method is explained below:

Algorithm of cutting plane method

Step 1: Solve the linear programming (LP) relaxation.
Step 2: If the resulting optimal solution, X, is an integer, the optimal solution is found, and the algorithm stops. On the other hand, if the solution is a non-integer, go to Step 3.
Step 3: Generate a cut using the Gomory cutting plane algorithm, i.e., a constraint satisfied by all feasible integer solutions but not by X.
Step 4: Add this new constraint, re-solve the problem, and go back to Step 2. This terminates after a finite number of iterations, and the resulting X is an integer and optimal.

The detailed explanations of the solution algorithm are given while solving the following integer programming.

Example 5.1

Maximize $z = 4x_1 + 5x_2$

Subject to, $3x_1 + 2x_2 \leq 12$ (Constraint 1)

$x_1 + 2x_2 \leq 6$ (Constraint 2)

$x_1, x_2 \geq 0$, and Integer (Non-negativity and integer constraint)

Solution

Step 1: In the first step, the problem is solved by considering only the non-negativity constraint of the decision variables, i.e., $x_1, x_2 \geq 0$, and ignoring the integer restrictions. In that case, the linear programming problem can be solved using the simplex method, as the variables are continuous. However, introducing $x_1, x_2 \geq 0$, and Integer criteria in decision variables, makes the problem more challenging to solve. The method of ignoring the integer constraints is called LP relaxation.

The process of obtaining the solution to the given ILP problem by ignoring the integer constraints is called LP relaxation. However, the rounded solution can't be an optimum. Thus, the reduced problem is represented by LP1

LP1: *Maximize* $z = 4x_1 + 5x_2$

Subject to, $3x_1 + 2x_2 \leq 12$

$x_1 + 2x_2 \leq 6$

$x_1, x_2 \geq 0$

The standard form of the LP1 is

Maximize $z = 4x_1 + 5x_2$

Subject to, $3x_1 + 2x_2 + x_3 = 12$ (5.1)

$x_1 + 2x_2 + x_4 = 6$ (5.2)

$x_1, x_2, x_3, x_4 \geq 0$

In the standard form of the problem, x_3 *and* x_4 are slack variables, which are introduced in two constraints to make them equality constraints. The slack variables should be non-negative. The reduced LP problem (LP1) is solved using the simplex method, as explained in Section 3.3.2 of Chapter 3. The following tables (Table 5.1, Table 5.2, and Table 5.3) show the iterative process to generate solutions using the simplex methods. To follow the steps of the simplex method, readers can consult with examples in Chapter 3.

TABLE 5.1
Simplex table showing starting solution or initial basic feasible solution

	Basis	Coefficient of				Solution	Ratio	
		x_1	x_2	x_3	x_4	RHS		
Row 1	x_3	3	2	1	0	12	12/2 = 6	
Row 2	x_4	1	2	0	1	6	6/2 = 3	← Departing Variable
Row z	z	–4	–5	0	0	0		
			↑ Entering Variable					

TABLE 5.2
Simplex table showing solution after the first iteration

	Basis	Coefficient of				Solution	Ratio	
		x_1	x_2	x_3	x_4	RHS		
Row 1	x_3	2	0	1	–1	6	6/2 = 3	← Departing Variable
Row 2	x_2	1/2	1	0	1/2	3	3/0.5 = 6	
Row z	z	–3/2	0	0	5/2	15		
		↑ Entering Variable						

TABLE 5.3
Simplex table showing solution after the second iteration

	Basis	Coefficient of				Solution
		x_1	x_2	x_3	x_4	RHS
Row 1	x_1	1	0	1/2	–1/2	3
Row 2	x_2	0	1	–1/4	3/4	3/2
Row z	z	0	0	3/4	7/4	39/2

In the second iteration, the solution reached the optimal stage as all the coefficients in the z-row under the basic variables are non-negatives.

Thus, the optimal solution to the problem using the simplex method is

$$x_1 = 3, \quad x_2 = 3/2$$

$$z = 4x_1 + 5x_2 = 4*3 + 5*3/2 = 19.5$$

The graphical representation of the solution is shown in Figure 5.2.

Step 2: The optimal solution is non-integer type as one of the decision variables is non-integer. It violates the constraint of integer, and thus an integer solution needs to be searched by introducing additional constraint, as explained in Step 3.

Step 3: This can be done by adding a new constraint using Gomory's cut method in the final solution, as obtained in Step I. A Gomory's cut is a linear constraint added to an LP problem, which does not exclude any feasible integer solution of the problem under consideration. It is used in conjunction with the simplex method to generate optimal integer solutions.

The optimal simplex table (Table 5.3) gives

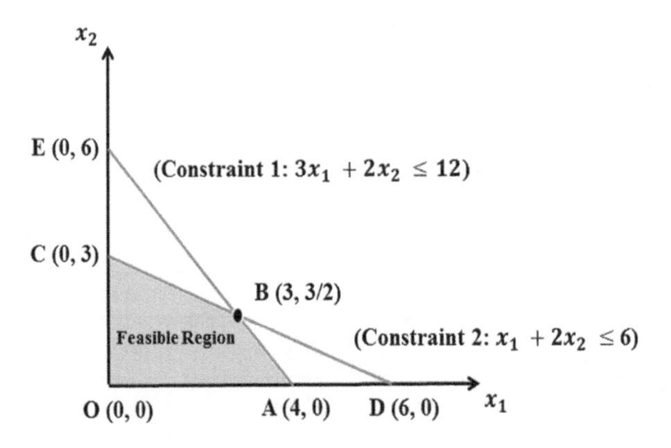

FIGURE 5.2 Graphical representation of the feasible solution space for the problem.

$$z + \frac{3}{4}x_3 + \frac{7}{4}x_4 = \frac{39}{2} \quad [\text{z row}] \tag{5.3}$$

$$x_1 + \frac{1}{2}x_3 - \frac{1}{2}x_4 = 3 \quad [\text{Row 1}] \tag{5.4}$$

$$x_2 - \frac{1}{4}x_3 + \frac{3}{4}x_4 = \frac{3}{2} \quad [\text{Row 2}] \tag{5.5}$$

The constraint equations above derive Gomory's cut, provided its right-hand side is fractional. Out of three constraint equations (Eq. 5.5–5.7), the right-hand side of Eq. 5.4 is non-fractional and thus cannot be used for generating a Gomory's cut. The generation of the Gomory's cut from the other constraints (Eq. 5.3 and Eq. 5.5) is explained below.

Eq. (5.3) can be written as:

$$z + \frac{3}{4}x_3 + \left(1 + \frac{3}{4}\right)x_4 = 19 + \frac{1}{2}$$

Moving all the integer components to the left-hand side and the fractional component to the right-hand side, we have,

$$z + x_4 - 19 = \frac{1}{2} - \frac{3}{4}x_3 - \frac{3}{4}x_4 \tag{5.6}$$

Since x_3 and x_4 are non-negative, both the fractions in the Eq. (5.6) are positive. Therefore, the right-hand side of the above equation should follow:

$$\frac{1}{2} - \frac{3}{4}x_3 - \frac{3}{4}x_4 \leq \frac{1}{2}$$

Also, the left-hand side of Eq. (5.6) is an integer, and hence the right-hand side should be an integer value. To satisfy the condition, we have the associated cut as

$$\frac{1}{2} - \frac{3}{4}x_3 - \frac{3}{4}x_4 \leq 0 \quad \textbf{Cut 1} \tag{5.7}$$

The above equation holds good with the justification that an integer value less than 1/2 should be less and equal to zero. The above constraint represents a cut, which may provide an integer solution.

Similarly, Eq. (5.5) can be written as:

$$x_2 + \left(-1 + \frac{3}{4}\right)x_3 + \left(0 + \frac{3}{4}\right)x_4 = 1 + \frac{1}{2}$$

$$\Rightarrow x_2 - x_3 - 1 = \frac{1}{2} - \frac{3}{4}x_3 - \frac{3}{4}x_4$$

The associate cut is

$$\frac{1}{2} - \frac{3}{4}x_3 - \frac{3}{4}x_4 \le 0 \qquad \textbf{Cut}\,2 \tag{5.8}$$

In the cutting plane algorithm, the cut, which has the highest fractional value on the right-hand side, should be used first for obtaining the integer solution. In this case, both the cuts have the same fractional value (=1/2), and thus either of the cut can be used by breaking the tie. The arbitrary selection of the cut 2, we have,

$$-\frac{3}{4}x_3 - \frac{3}{4}x_4 + s_1 = -\frac{1}{2}[s_1 \ge 0 \,\text{and integer}] \tag{5.9}$$

In Eq. (5.9), S_1 is a slack variable.

The revised simplex table after adding the new constraint (Cut 2) can be derived from the optimal simplex table (Table 5.3) by adding a new row (s_1-row). The revised table is shown in Table 5.4.

The above simplex table is optimal as all the coefficients in the z-row are non-negative but infeasible as one of the elements in the RHS column is negative. Thus, a dual simplex algorithm is applied to obtain the solution. For a detailed explanation, readers should refer to the dual simplex algorithm explained in Section 3.5.2 of Chapter 3. The optimal feasible simplex table is shown in Table 5.5.

TABLE 5.4
Revised simplex table after adding the new constraint

	Basis	**Coefficient of**						
		x_1	x_2	x_3	x_4	s_1	**RHS**	
Row x_1	x_1	1	0	1/2	–1/2	0	3	
Row x_2	x_2	0	1	–1/4	3/4	0	3/2	
Row s_1	s_1	0	0	–3/4	–3/4	1	–1/2	**LV**
Row z	z	0	0	3/4	7/4	0	39/2	
Ratio = z/Row s_1 (for -ve coefficients of s_1 row only)				1	7/3			
				EV				

TABLE 5.5
Revised optimal feasible simplex table

	Basis	Coefficient of					Solution
		x_1	x_2	x_3	x_4	s_1	RHS
Row x_1	x_1	1	0	0	-1	2/3	8/3
Row x_2	x_2	0	1	0	1	−1/3	5/3
Row x_3	x_3	0	0	1	1	−4/3	2/3
Row z	z	0	0	0	1	1	19

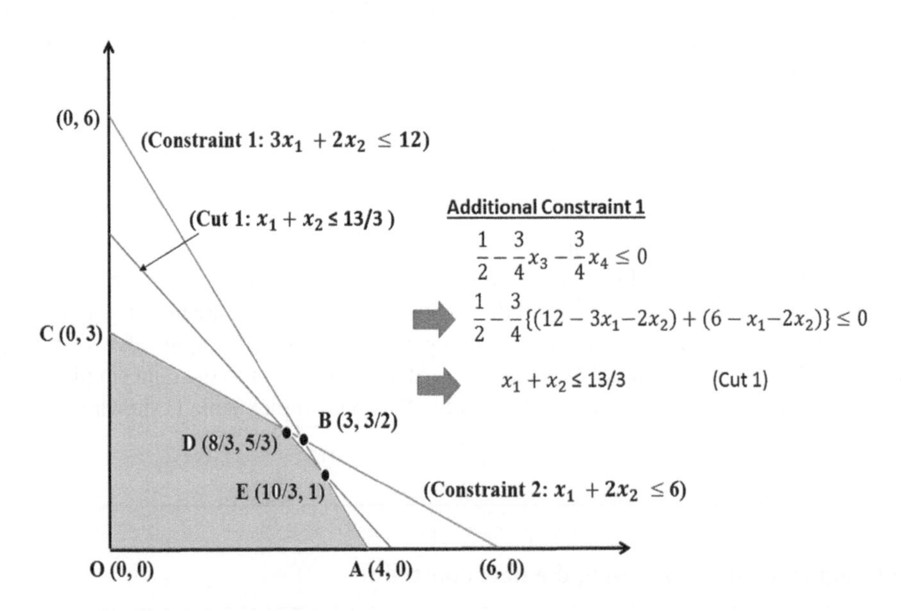

FIGURE 5.3 Resultant optimal feasible solution after introducing a new constraint.

Thus, the optimal solution after introducing the new constraint (cut 2) is $x_1 = 8/3$, $x_2 = 5/3$ and z = 19. That is, both the decision variables are still non-integer.

The optimal solution after introducing the new constraint (cut 1) can also be derived using the graphical method, as there are only two decision variables. The new constraint can be represented in decision variables x_1 and x_2 by putting the values of x_3 and x_4 from Eq. (5.1) and Eq. (5.2), as represented in Figure 5.3. After adding a new constraint in the feasible region, which is shown in Figure 5.2, the resultant feasible region is narrowed, and a new optimal solution (D: $x_1 = 8/3$, $x_2 = 5/3$ and z = 19) was obtained. The modified feasible region is shown in Figure 5.3. The same optimal solution is obtained as that of the simplex method.

In the optimal simplex table (Table 5.5), the right-hand side of all the three rows (Row x_1, Row x_2, and Row x_3) are fractions, and thus either of the three can be used as a source row for generating the cut. The fractional components of x_1-row, Row x_2-row, and x_3-row are 2/3 (= 2 + 2/3), 2/3 (1 + 2/3), and 2/3, respectively. The fractional component of each row is equal, and thus either one can be chosen to derive the new constraint. In this time, Row x_3 is used as the source row for generating the cutting plane. The following equation can be derived from row x_3 of Table 5.5.

$$x_3 + x_4 - \frac{4}{3}s_1 = \frac{2}{3} \tag{5.10}$$

$$\Rightarrow x_3 + x_4 + \left(-2 + \frac{2}{3}\right)s_1 = \frac{2}{3}$$

In a similar way, as explained above, moving all the integer components to the left-hand side and the fractional component to the right-hand side, we have,

$$x_3 + x_4 - 2s_1 = \frac{2}{3} - \frac{2}{3}s_1$$

The associate cut is

$$\frac{2}{3} - \frac{2}{3}s_1 \le 0 \tag{5.11}$$

$$\Rightarrow -\frac{2}{3}s_1 + s_2 = -\frac{2}{3} \quad [s_2 \ge 0 \text{ and integer}] \tag{5.12}$$

In Eq. (5.12), S_2 is a slack variable.

The revised simplex table after adding the new constraint [Eq. (5.12)] is derived from the optimal simplex table (Table 5.5) by adding a new row (s_2-row). The revised simplex table is shown in Table 5.6.

The simplex table is optimal as all the coefficients in the z-row are non-negative but infeasible as one of the elements in the RHS column is negative. Again, a dual simplex algorithm is applied to

TABLE 5.6
Revised simplex table after adding the new constraint

		Coefficient of						**Solution**	
	Basis	x_1	x_2	x_3	x_4	s_1	s_2	**RHS**	
Row x_1	x_1	1	0	0	−1	2/3	0	8/3	
Row x_2	x_2	0	1	0	1	−1/3	0	5/3	
Row x_3	x_3	0	0	1	1	−4/3	0	2/3	
Row s_2	s_2	0	0	0	0	−2/3	1	−2/3	**LV**
Row z	z	0	0	0	1	1	0	19	
Ratio = z/Row s_1 (for −ve coefficients of s_1 row only)						3/2			
						EV			

obtain the solution. The resultant optimal simplex table after applying the dual simplex algorithm is shown in Table 5.7.

Thus, the optimal solution, after introducing the two cuts is $x_1 = 2$, $x_2 = 2$, is z = 18. That is, both the decision variables are now integers.

The optimal solution after introducing the new cut can also be derived using the graphical method. The new constraint [Eq. (5.12)] can be represented in terms of decision variables x_1 and x_2 by putting the values of s_1 from Eq. (5.1), Eq. (5.2), and Eq. (5.10), as represented in Figure 5.4. After adding a new constraint in the feasible region, shown in Figure 5.3, the resultant feasible region is narrowed, and a new optimal solution (E: $x_1 = 2$, $x_2 = 2$ and z = 18) is obtained, as shown in Figure 5.4. The same optimal solution is obtained as that of the simplex method.

TABLE 5.7
Revised optimal feasible simplex table

	Basis	Coefficient of						Solution
		x_1	x_2	x_3	x_4	s_1	s_2	RHS
Row x_1	x_1	1	0	0	−1	0	1	2
Row x_2	x_2	0	1	0	1	0	−1/2	2
Row x_3	x_3	0	0	1	1	0	2	2
Row s_2	s_1	0	0	0	0	1	−3/2	1
Row z	z	0	0	0	1	0	3/2	18

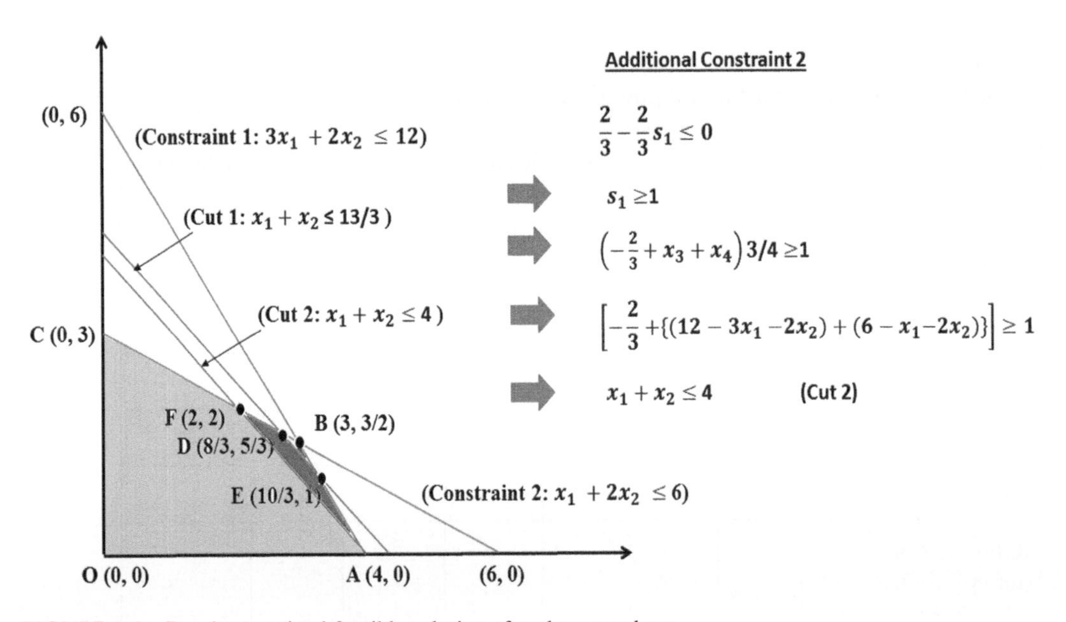

FIGURE 5.4 Resultant optimal feasible solution after the second cut.

Thus, the optimal solution to the problem is

$$x_1 = 2, \quad x_2 = 2$$

$$z = 4x_1 + 5x_2 = 4*2 + 5*2 = 18$$

5.3.2 Branch and bound (B&B) algorithm

Branch and bound (B&B), also called tree search, solves combinatorial optimization problems. The B&B method employs a diagram consisting of nodes and branches as a framework for deriving the solution. In this algorithm, all the solutions may need to explore to obtain an optimal solution. Land and Doig (1960) had first used this algorithm for discrete programming and it has become the most commonly used tool for solving optimization problems. In this method, the optimal integer solution can be identified without searching all the feasible solutions. Moreover, the method can still guarantee to offer the optimal integer solution. The steps involved in the B&B method for obtaining an optimal integer solution for a model is as follows:

Step 1: The optimal solution of the ILP problem is obtained by relaxing the integer constraint.
Step 2: The optimal solution should be assigned as the upper bound (UB), and the rounded-down integer solution should be assigned as the lower bound (LB).
Step 3: The decision variable that has the highest fractional value should be selected for branching. Two new constraints are derived by selecting the nearby integers with ≤ and ≥ sign, respectively.
Step 4: Two new relaxed LPP are formed by adding two constraints.
Step 5: Determine the optimal solutions of the relaxed LPPs.
Step 6: The optimal solution at each node represents the UB, and the existing maximum integer solution (at any node) represents the LB.
Step 7: If the branching produces a feasible integer solution with the highest UB value of any ending node, the optimal integer solution has been reached, and further branching is stopped. If a feasible integer solution does not found, branch from the node with the greatest UB.
Step 8: Go to Step 3 and repeat the process.

The B&B algorithm to solve an IP model is explained using the same example (Example 5.1).

Maximize $z = 4x_1 + 5x_2$

Subject to,	$3x_1 + 2x_2 \le 12$	Constraint (1)
	$x_1 + 2x_2 \le 6$	Constraint (2)
	$x_1, x_2 \ge 0$, and Integer	Non-negativity and integer constraint

Initially, the problem is solved by relaxing the integer constraint. That is, all the variables are assumed to be continuous rather than discrete. The revised problem is represented as LP1. That is, any LP method [Graphical method (only for two-variable problems) or simplex method (for any number of decision variables)] can be used to obtain the optimal solution without ensuring that the solution is an integer type. For a detailed explanation of the graphical method or simplex method, readers should refer to Section 3.2 in Chapter 3.

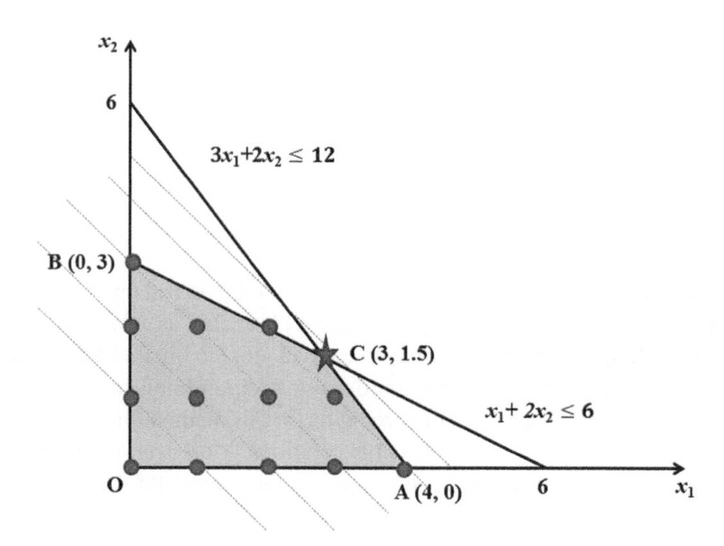

FIGURE 5.5 Solution Space for original (LP1) with the ignorance of integer constraint.

LP1 : *Maximize* $z = 4x_1 + 5x_2$

$$\text{Subject to, } 3x_1 + 2x_2 \leq 12$$
$$x_1 + 2x_2 \leq 6$$
$$x_1, x_2 \geq 0$$

In this case, the LP1 has only two decision variables and thus uses a graphical method to derive the optimal solution, as shown in Figure 5.5. The optimal solution of LP1 is obtained at C ($x_1 = 3, x_2 = 1.50$), and the optimal value of the objective function is 19.5. Out of two decision variables, one decision variable (x_2) is a non-integer, and thus a non-integer solution is obtained. That is, the optimal solution does not satisfy the integer constraint. The integer solutions located in the feasible region are shown using small circles in Figure 5.5. Out of all, the optimal integer solution needs to be searched. The searching for an optimal integer solution is a challenging task, and thus the B&B algorithm can be applied. Thus, the problem represented in LP1 is further branched into two problems (LP2 and LP3) with new constraints in the LP1 problem. As x_2 (=3/2) is a non-integer in the LP1 solution, the new constraints are derived by branching x_2 to a nearby integer. That is, $\boldsymbol{x_2 \leq 1}$ and $\boldsymbol{x_2 \geq 2}$ are the two constraints, which are added in LP2 and LP3, respectively. If both the variables are non-integer types, either of them can be branched to derive the branched problems. The branched problems (LP2 and LP3) derived from LP1 by adding new constraints are as follows:

LP2:	**LP3:**
Maximize $z=4x_1+5x_2$ subject to, $3x_1+2x_2 \leq 12$ $x_1+2x_2 \leq 6$ $\boldsymbol{x_2 \leq 1}$ $x_1, x_2 \geq 0$, and Integer	*Maximize* $z=4x_1+5x_2$ subject to, $3x_1+2x_2 \leq 12$ $x_1+2x_2 \leq 6$ $\boldsymbol{x_2 \geq 2}$ $x_1, x_2 \geq 0$, and Integer

It can be represented using a tree search diagram, as shown in Figure 5.6.

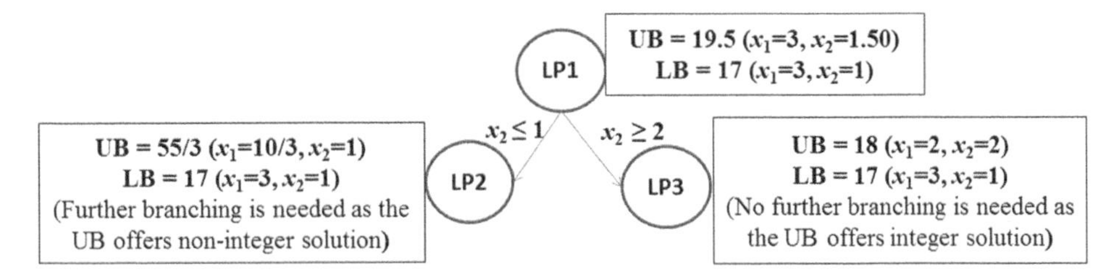

FIGURE 5.6 Solution subsets of the sub-problems derived after first branching.

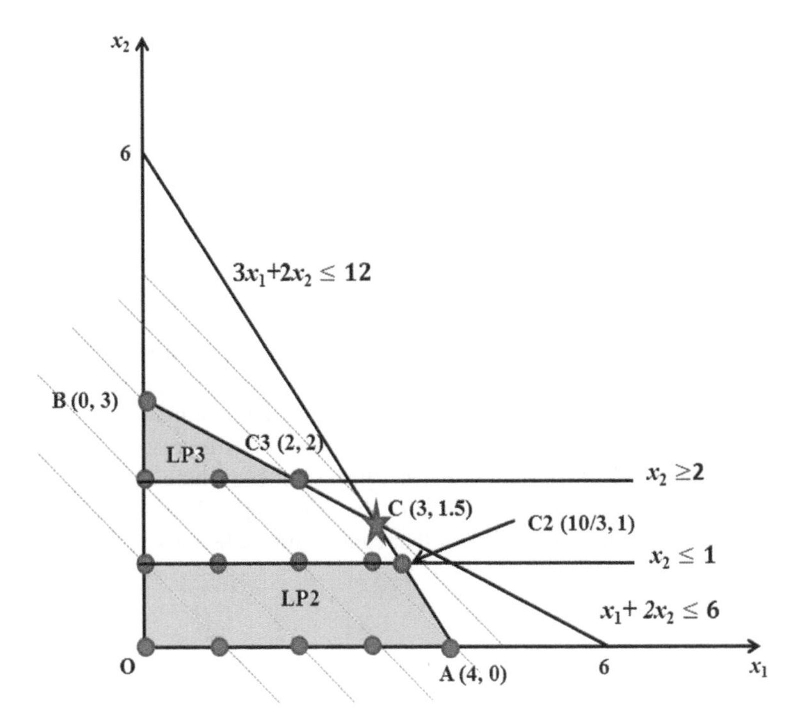

FIGURE 5.7 Optimal solutions of LP2 and LP3.

The optimal solution points for LP2 and LP3 are C2 (x_1=10/3, x_2=1) and C3 (x_1=2, x_2=2), respectively, as shown in Figure 5.7. It is observed that LP3 is pruned, and the solution can be the candidate for the optimal solution of the original ILP until a better solution is obtained. Therefore, no further branching of LP3 is required. Unlike LP3, the solutions of LP2 indicate that one decision variable is an integer, but the other one is still a non-integer value. Thus, the problem LP2 is further branches into two new problems (LP4 and LP5) by adding additional constraints. In B&B, before making a decision about the optimal solution of the original problem, the lower and upper bound of all the branched problems need to be examined.

As x_1(=10/3) is a non-integer in the LP2 solution, the new constraints are derived by branching x_1 to a nearby integer. That is, $x_1 \leq 3$ and $\boldsymbol{x_1 \geq 4}$ are the two constraints, which are added in LP4 and LP5, respectively. The branched problems (LP4 and LP5) derived from LP2 by adding new constraints are as follows:

LP4:	**LP5:**
Maximize $z=4x_1+5x_2$	*Maximize* $z=4x_1+5x_2$
subject to,	subject to,
$3x_1+2x_2 \le 12$	$3x_1+2x_2 \le 12$
$x_1+2x_2 \le 6$	$x_1+2x_2 \le 6$
$\boldsymbol{x_2 \le 1}$	$\boldsymbol{x_2 \ge 2}$
$\boldsymbol{x_1 \le 3}$	$\boldsymbol{x_1 \ge 4}$
$x_1, x_2 \ge 0$, and Integer	$x_1, x_2 \ge 0$, and Integer

The optimal solution points for LP4 and LP5 are C4 (x_1=3, x_2=1) and C5 (x_1=4, x_2=0), respectively, as shown in Figure 5.8.

It is observed that both the problems, LP4 and LP5, are pruned, and no further branching is needed as both the solutions are integer types. If LP4 or/and LP5 are offer non-integer solutions, the problem requires branching further by adding a new constraint. Out of all the solution points (C2, C3, C4, and C5), the best solution needs to be identified based on the value of the objective function of the original ILP.

The objective function value at C3, C4, and C5 are

$$z = \begin{cases} 4*2+5*2=18, & \text{at } C3(2,2) \\ 4*3+5*1=17, & \text{at } C4(3,1) \\ 4*4+5*0=16, & \text{at } C5(4,0) \end{cases}$$

The above results indicate that all the solutions are integer. The optimal solution is found at C3 (2, 2), and the maximum value of the objective function is 18, as represented in Figure 5.9. LP3 is fathomed as this cannot yield a better solution. No further branching is required for LP3. But, we cannot say LP3 offers the optimum solution for the original problem because LP may offer the better solution. Any unexamined sub-problem that cannot offer a better solution than the LB must

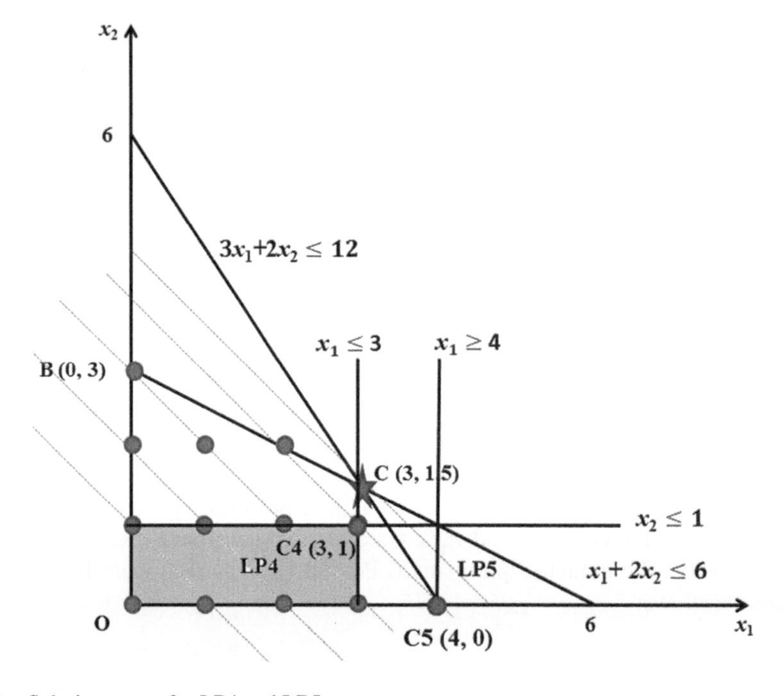

FIGURE 5.8 Solution space for LP4 and LP5.

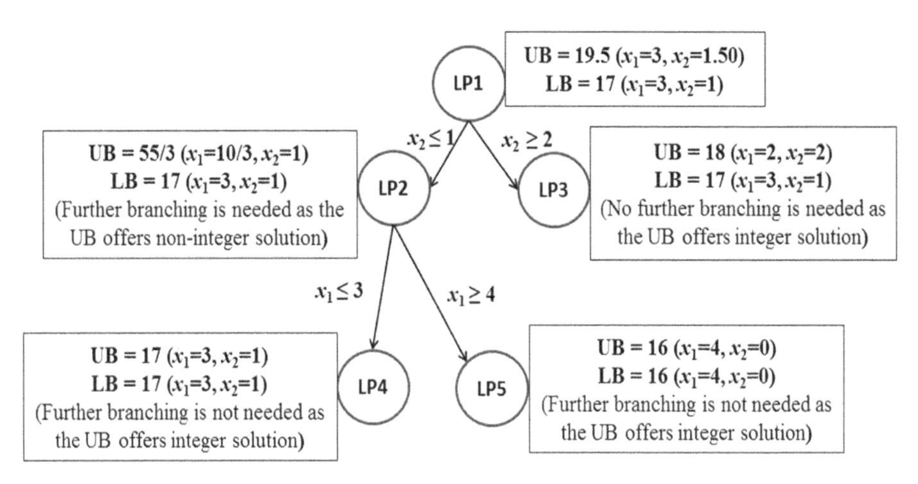

FIGURE 5.9 Solution subsets of the subproblems derived after the second level of branching.

be discarded as non-promising. On the other hand, if any unexamined sub-problem produces a better integer solution, then the LB must be updated accordingly.

The B&B algorithm can also be used to solve the binary integer and mixed-integer problem in a similar approach as used in solving the pure integer problem. The application of the B&B algorithm in solving Binary Integer Programming is explained using the following Assignment Problem, as given in Example 5.2.

Example 5.2

A mine manager has to assign four types of jobs to four contract persons. Each contractor has the capability to complete each task but charges different costs in US$ (x10^6), as shown below in the cost matrix table (Table 5.8). Any job can be assigned to any contractor but not more than one. Suggest an optimum allocation of jobs to complete at a minimum cost using a binary integer algorithm.

Formulation of the Problem

Let x_{ij} is the binary variable, which indicates that j^{th} job is assigned to i^{th} contractor.

The binary integer linear programming model for this problem is formulated as follows:

$$\begin{aligned} Min\,Z = {} & 27(x_{11})+18(x_{12})+25(x_{13})+20(x_{14})+31(x_{21})+24(x_{22})+21(x_{23}) \\ & +40(x_{24})+20(x_{31})+17(x_{32})+20(x_{33})+24(x_{34})+22(x_{41})+28(x_{42}) \\ & +20(x_{43})+16(x_{44}) \qquad [\textbf{Objective function}] \end{aligned}$$

Subject to,

$$x_{11}+x_{12}+x_{13}+x_{14} \leq 1 \quad \text{Constraint (1)}$$
$$x_{21}+x_{22}+x_{23}+x_{24} \leq 1 \quad \text{Constraint (2)}$$
$$x_{31}+x_{32}+x_{33}+x_{34} \leq 1 \quad \text{Constraint (3)}$$
$$x_{41}+x_{42}+x_{43}+x_{44} \leq 1 \quad \text{Constraint (4)}$$
$$x_{11}+x_{21}+x_{31}+x_{41} \leq 1 \quad \text{Constraint (5)}$$
$$x_{12}+x_{22}+x_{32}+x_{42} \leq 1 \quad \text{Constraint (6)}$$
$$x_{13}+x_{23}+x_{33}+x_{43} \leq 1 \quad \text{Constraint (7)}$$

TABLE 5.8
Cost matrix for completion of tasks by different contractors

Contractor	Job			
	J1	J2	J3	J4
C1	**27**	**18**	**25**	**20**
C2	31	24	21	40
C3	20	17	20	24
C4	22	28	20	16

$$x_{14} + x_{24} + x_{34} + x_{44} \leq 1 \qquad \text{Constraint (8)}$$

$$x_{ij} = 0, 1 \qquad \text{Binary constraint}$$

Constraints (1)–(4) represents only one job is assigned to one contractor, and Constraints (5)–(8) represent one contractor can get only one job.

There are two methods to determine the cost function:

- For each contractor, the unassigned jobs with minimum cost must be assigned.
- For each job, the unassigned contractor charges the lowest cost must be assigned.

Though for a small problem, it can be easily assigned the jobs with the lowest total minimum cost. But, for a larger problem, it is difficult to assign the tasks, and thus B&B algorithms can be applied to obtain the solution.

Before applying the B&B algorithm, the lower bound (LB) and upper bound (UB) of the solution needs to be determined. To determine the LB, the lowest cost in each row is selected, and the sum of all these lowest costs represents the LB of the solution. Similarly, the sum of the highest cost in a row represents the UB of the solution. Since the objective function of the problem is to determine the minimum cost and thus the algorithm is focused on updating the LB in each level rather than focussing on the UB. In this problem, the lower bound of the starting solution is 72 ($C_1 \rightarrow J_2$, $C_2 \rightarrow J_3$, $C_3 \rightarrow J_2$, $C_4 \rightarrow J_4$), and the UB is 119 ($C_1 \rightarrow J_1$, $C_2 \rightarrow J_4$, $C_3 \rightarrow J_4$, $C_4 \rightarrow J_2$). That is, the total cost cannot be lower than US$ 72 million and higher than US$ 119 million. Here, the LB is updated every time if a change occurs. This problem is represented by LP1.

Let's start the solution by assigning J_1 to C_1 (LP2). Once J_1 is assigned to C_1, the updated LBs are determined by considering the lowest cost in the rest of the rows. The lowest cost in C_2-row is 21, and thus J_3 is assigned to C_2. After assigning J_1 to C_1 and J_3 to C_2, the rest of the jobs are assigned to C_3 and C_4. In C3-row, the lowest cost excluding J_1 and J_3 columns (already assigned) is 17, which is corresponds to J_2 column. Therefore, job J2 is assigned to C_3. Finally, job J_4 is unassigned, and hence the same is assigned to C4, as shown in Figure 5.10. The total cost incurs in completing the task is 81 (27+21+17+16). This is higher than the earlier LB, and thus the value of LB is updated. That is, if C_1 is assigned to J_1, the minimum cost incurs to complete all the tasks is 81 million US$. The job assignment made in this are represented by blue cells, as shown in Figure 5.10(a).

Similarly, if the job J_1 is assigned to C_2, C_3, and C_4, the respective allocations of the rest of the jobs are represented in Figure 5.10(b), Figure 5.10(c), and Figure 5.10(d). The cost incurs in three

(a) $C_1 \rightarrow J_1$

Contractor	Job			
	J_1	J_2	J_3	J_4
C_1	27	18	25	20
C_2	31	24	21	40
C_3	20	17	20	24
C_4	22	28	20	16

Contractor	Job			
	J_1	J_2	J_3	J_4
C_1	27	18	25	20
C_2	31	24	21	40
C_3	20	17	20	24
C_4	22	28	20	16

Contractor	Job			
	J_1	J_2	J_3	J_4
C_1	27	18	25	20
C_2	31	24	21	40
C_3	20	17	20	24
C_4	22	28	20	16

Contractor	Job			
	J_1	J_2	J_3	J_4
C_1	27	18	25	20
C_2	31	24	21	40
C_3	20	17	20	24
C_4	22	28	20	16

(b) $C_1 \rightarrow J_2$

Contractor	Job			
	J_1	J_2	J_3	J_4
C_1	27	18	25	20
C_2	31	24	21	40
C_3	20	17	20	24
C_4	22	28	20	16

Contractor	Job			
	J_1	J_2	J_3	J_4
C_1	27	18	25	20
C_2	31	24	21	40
C_3	20	17	20	24
C_4	22	28	20	16

Contractor	Job			
	J_1	J_2	J_3	J_4
C_1	27	18	25	20
C_2	31	24	21	40
C_3	20	17	20	24
C_4	22	28	20	16

Contractor	Job			
	J_1	J_2	J_3	J_4
C_1	27	18	25	20
C_2	31	24	21	40
C_3	20	17	20	24
C_4	22	28	20	16

(c) $C_1 \rightarrow J_3$

Contractor	Job			
	J_1	J_2	J_3	J_4
C_1	27	18	25	20
C_2	31	24	21	40
C_3	20	17	20	24
C_4	22	28	20	16

Contractor	Job			
	J_1	J_2	J_3	J_4
C_1	27	18	25	20
C_2	31	24	21	40
C_3	20	17	20	24
C_4	22	28	20	16

Contractor	Job			
	J_1	J_2	J_3	J_4
C_1	27	18	25	20
C_2	31	24	21	40
C_3	20	17	20	24
C_4	22	28	20	16

Contractor	Job			
	J_1	J_2	J_3	J_4
C_1	27	18	25	20
C_2	31	24	21	40
C_3	20	17	20	24
C_4	22	28	20	16

(d) $C_1 \rightarrow J_4$

Contractor	Job			
	J_1	J_2	J_3	J_4
C_1	27	18	25	20
C_2	31	24	21	40
C_3	20	17	20	24
C_4	22	28	20	16

Contractor	Job			
	J_1	J_2	J_3	J_4
C_1	27	18	25	20
C_2	31	24	21	40
C_3	20	17	20	24
C_4	22	28	20	16

Contractor	Job			
	J_1	J_2	J_3	J_4
C_1	27	18	25	20
C_2	31	24	21	40
C_3	20	17	20	24
C_4	22	28	20	16

Contractor	Job			
	J_1	J_2	J_3	J_4
C_1	27	18	25	20
C_2	31	24	21	40
C_3	20	17	20	24
C_4	22	28	20	16

FIGURE 5.10 Assignment of jobs with new constraints (a) $C_1 \rightarrow J_1$(b) $C_1 \rightarrow J_2$ (c) $C_1 \rightarrow J_3$ (d) $C_1 \rightarrow J_4$.

cases ($C_1 \rightarrow J_2$ (LP3), $C_1 \rightarrow J_3$ (LP4), $C_1 \rightarrow J_4$ (LP5)) are respectively 75 million US$, 85 million US$, and 80 million US$. In the next iteration, only the lowest cost node is branched to obtain the solution. The node that showed a higher value could not offer a better solution, and thus, branching is not done for those nodes. Thus, in this case, only LP2 is further branched into three constraints to obtain the optimal solution. That is, if C_1 is assigned by J_2, J_3, J_4, the assignments are represented by a blue cell in Figure 5.10(b), Figure 5.10(c), and Figure 5.10(d), respectively.

Once all the options for assigning the jobs to contractor C_1, the lowest cost option is further branched. This is because the options with a higher cost at the first level cannot offer a better solution. In this case, the assignment of job J_2 to C_1 offers the lowest cost (= 75 million US$). Thus, the node with $C_1 \rightarrow J_2$ is further branched for examining the cost by assigning $C_2 \rightarrow J_1$ (LP6), $C_2 \rightarrow J_3$ (LP7), and $C_2 \rightarrow J_4$ (LP8). Since J_2 is already assigned to C_1, it cannot be further assigned. The assignment with all these cases is shown in Figure 5.11(a)–(c). The costs incurred with additional restriction are 85 million US$, 75 million US$, and 98 million US$ respectively for LP6, LP7, and LP8. That is the lowest cost offers by LP7 ($C_1 \rightarrow J_2$ and $C_2 \rightarrow J_3$). LP7 is further branched with additional restrictions as $C_3 \rightarrow J_1$ (LP9) and $C_3 \rightarrow J_4$ (LP10). The jobs J_2 and J_4 cannot be further assigned. These assignments are shown in Figure 5.12(a)–(b). The cost of completion in the

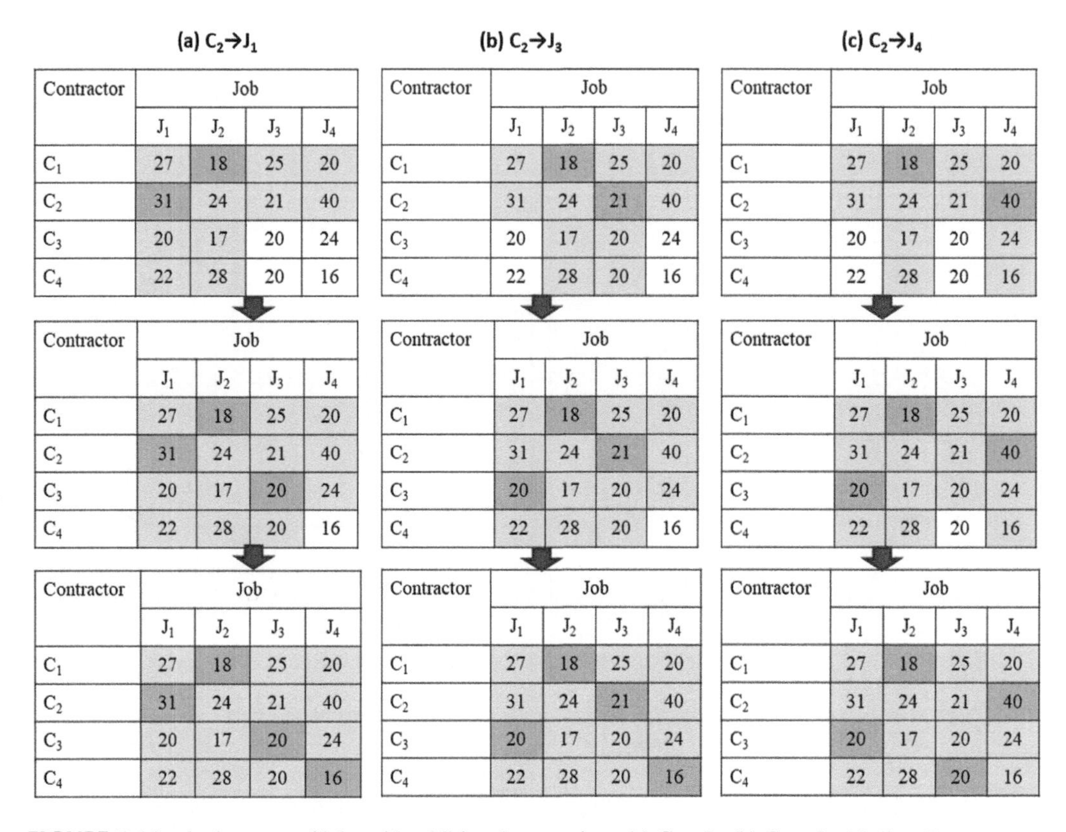

(a) $C_2 \rightarrow J_1$

Contractor	Job			
	J_1	J_2	J_3	J_4
C_1	27	18	25	20
C_2	31	24	21	40
C_3	20	17	20	24
C_4	22	28	20	16

Contractor	Job			
	J_1	J_2	J_3	J_4
C_1	27	18	25	20
C_2	31	24	21	40
C_3	20	17	20	24
C_4	22	28	20	16

Contractor	Job			
	J_1	J_2	J_3	J_4
C_1	27	18	25	20
C_2	31	24	21	40
C_3	20	17	20	24
C_4	22	28	20	16

(b) $C_2 \rightarrow J_3$

Contractor	Job			
	J_1	J_2	J_3	J_4
C_1	27	18	25	20
C_2	31	24	21	40
C_3	20	17	20	24
C_4	22	28	20	16

Contractor	Job			
	J_1	J_2	J_3	J_4
C_1	27	18	25	20
C_2	31	24	21	40
C_3	20	17	20	24
C_4	22	28	20	16

Contractor	Job			
	J_1	J_2	J_3	J_4
C_1	27	18	25	20
C_2	31	24	21	40
C_3	20	17	20	24
C_4	22	28	20	16

(c) $C_2 \rightarrow J_4$

Contractor	Job			
	J_1	J_2	J_3	J_4
C_1	27	18	25	20
C_2	31	24	21	40
C_3	20	17	20	24
C_4	22	28	20	16

Contractor	Job			
	J_1	J_2	J_3	J_4
C_1	27	18	25	20
C_2	31	24	21	40
C_3	20	17	20	24
C_4	22	28	20	16

Contractor	Job			
	J_1	J_2	J_3	J_4
C_1	27	18	25	20
C_2	31	24	21	40
C_3	20	17	20	24
C_4	22	28	20	16

FIGURE 5.11 Assignment of jobs with additional constraints: (a) $C_2 \rightarrow J_1$, (b) $C_2 \rightarrow J_3$, (c) $C_2 \rightarrow J_4$.

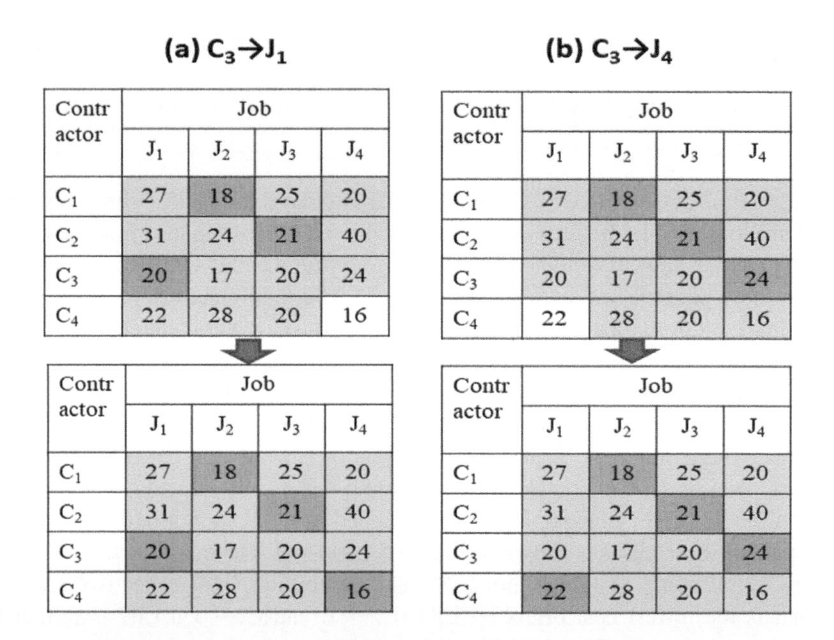

(a) $C_3 \rightarrow J_1$

Contr actor	Job			
	J_1	J_2	J_3	J_4
C_1	27	18	25	20
C_2	31	24	21	40
C_3	20	17	20	24
C_4	22	28	20	16

Contr actor	Job			
	J_1	J_2	J_3	J_4
C_1	27	18	25	20
C_2	31	24	21	40
C_3	20	17	20	24
C_4	22	28	20	16

(b) $C_3 \rightarrow J_4$

Contr actor	Job			
	J_1	J_2	J_3	J_4
C_1	27	18	25	20
C_2	31	24	21	40
C_3	20	17	20	24
C_4	22	28	20	16

Contr actor	Job			
	J_1	J_2	J_3	J_4
C_1	27	18	25	20
C_2	31	24	21	40
C_3	20	17	20	24
C_4	22	28	20	16

FIGURE 5.12 Assignment of jobs with additional constraints: (a) $C_3 \rightarrow J_1$ and (b) $C_3 \rightarrow J_4$.

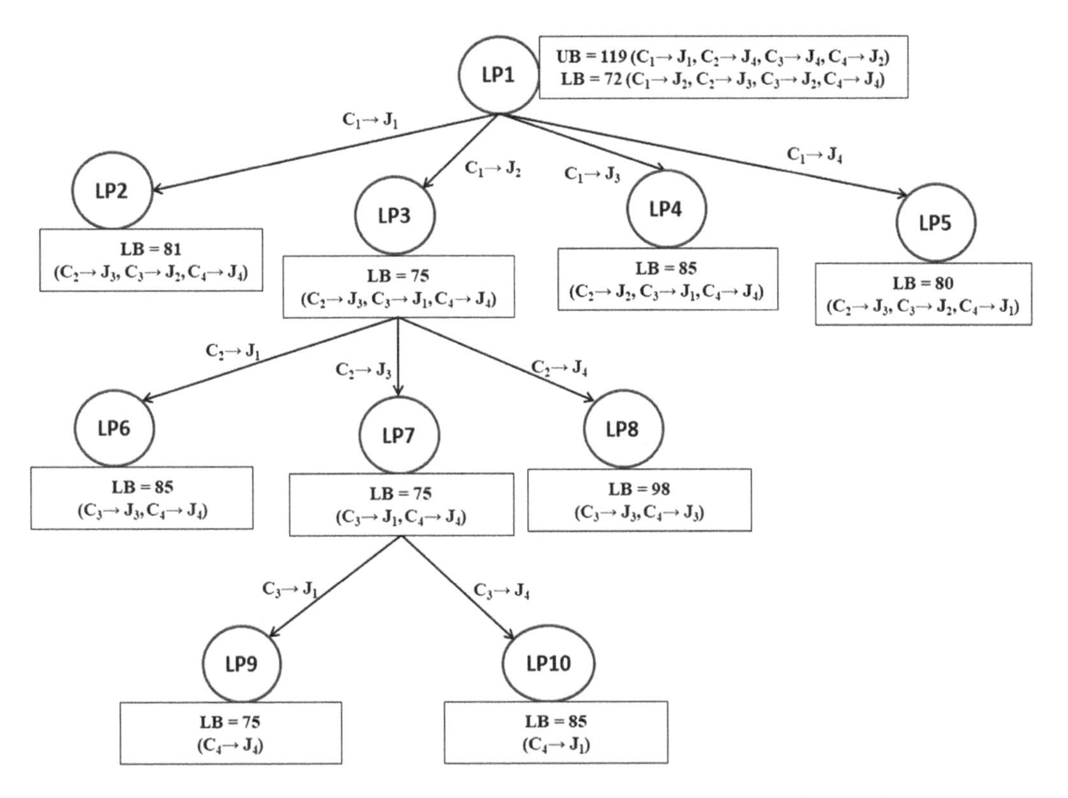

FIGURE 5.13 Searching of solution subsets of the sub-problems derived using B&B algorithm.

two alternative assignments, shown in Figure 5.12(a) and Figure 5.12(b) are US$75 million and US$85 million, respectively. Finally, the low-cost alternative should be chosen. The final solution with all the branches is represented in Figure 5.13.

Thus, the minimum z is obtained for the following assignments:

$$C_1 \rightarrow J_2$$
$$C_2 \rightarrow J_3$$
$$C_3 \rightarrow J_1$$
$$C_4 \rightarrow J_4$$

The value of the objective function is $z = 75$.

5.4 CASE STUDY OF THE APPLICATION OF MIXED INTEGER PROGRAMMING (MIP) IN PRODUCTION SCHEDULING OF A MINE

The present case study demonstrates the application of MIP in production scheduling for an open-pit iron ore mine. The iron ore mine is situated in the south-eastern part of India. This is a hilly deposit with a highly undulating ground level. The highest point of the deposit is 1269 m above the mean sea level (MSL) and the lowest point of mineralization is at 950 m above MSL. Most of the area of the deposit is covered by green vegetation. The geological study of the deposit revealed that this iron ore deposit was formed during the Precambrian age. This series of ore consists of iron ore, unenriched banded iron formation rocks (Banded Hematite Quartzite), shale, tuff, and quartzite. The major iron ore bodies occur along the top of the range and generally at the bottom of the underlying shale. The deposit is situated in the southern ridge of the range. There is a number of folds, faults present in the mine, which indicates that the deposit is highly disturbed in nature. There were 77 borehole data available from the mine for conducting this study. The boreholes are located in a grid pattern; however, the spacing of the boreholes varies from 200–250 metres. The average length of the boreholes is about 100 metres. Samples from the boreholes were collected in cores of less than 1-metre. The mine has seven different lithologies, namely Steel Grey Hematite (SGH), Blue hematite (BH), Laminated Hematite (LH), Laterite (L), Blue Dust (BD), Shale (SHL), and Banded Hematite Quartzite (BHQ). The high-grade iron ore is associated with steel grey, blue and laminated hematite. The average depth of the mine is 160 m. The iron ore mine is equipped with a good quality control structure to control the quality of the ore from the extraction point to the dispatch point. The mine has an inbuilt crushing and screening system. There is no requirement of beneficiation because of high ore quality; only ore processing is required to get the product sizes. The entire ore flow diagram of the iron ore mine is shown in Figure 5.14.

In this case study mine, only one sample attribute (Fe) is available from borehole data. This is because the iron produced by this mine is of very high grade. There is very little chance that impurities like silica, phosphorus, and sulfur exceed their maximum allowable limit defined by the customer. That is the reason why the mines authority is not maintaining the data of the other attributes.

For resource modelling, the borehole information collected from the case study mine was composited. The compositing was done based on lithology. A total number of 1456 composited data was available for resource evaluation purposes. The geotechnical study of the mine shows that the optimum bench height for the mine is 10 m. Therefore, the resource modelling was performed considering a block size of 10 m along Z-direction. The 20 m × 20 m × 10 m block size was selected

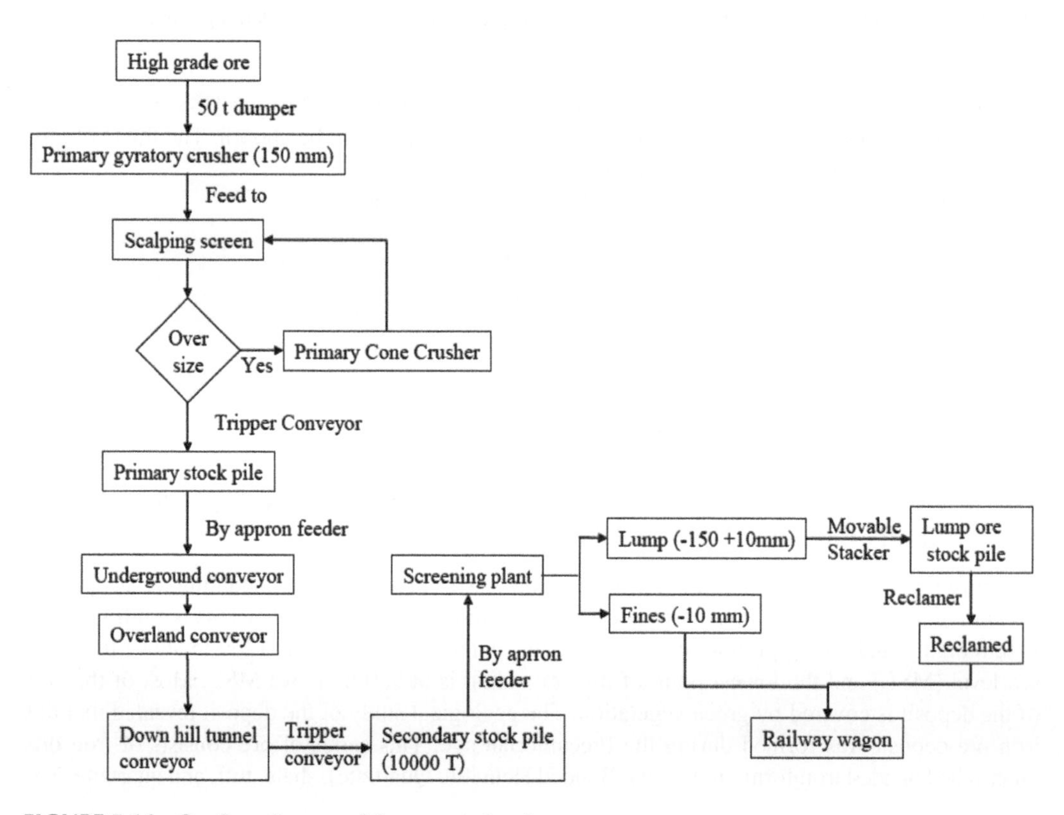

FIGURE 5.14 Ore flow-diagram of the case study mine.

TABLE 5.9
Descriptive statistics of Fe

Statistics	Sample Number	Mean (%)	Variance (%²)	Coefficient of variation	Skewness	Kurtosis
Iron (Fe)	1456	63.85	64.52	12.58	–2.74	8.39

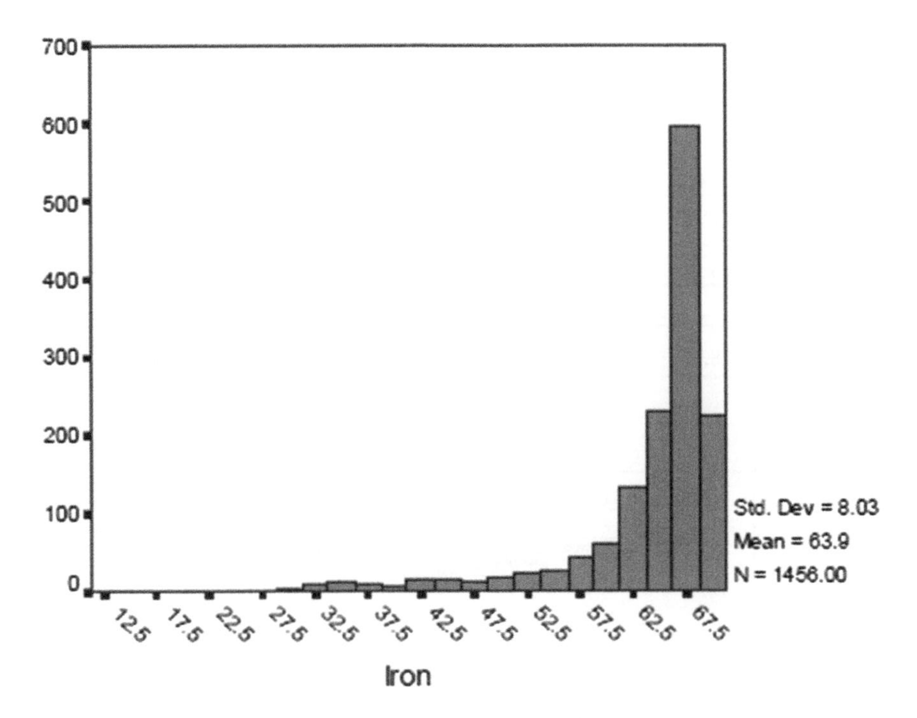

FIGURE 5.15 Histogram of the composited Fe data from the study mine.

for resource estimation using the block kriging algorithm. However, before resource modelling, a detailed statistical analysis of composited Fe samples was carried out. Table 5.9 shows the descriptive statistics of the composited data. It is seen from Table 5.9 that the grade of iron is highly skewed towards the right side, but it has comparatively less variance. The frequency distribution of the data set was also prepared using the histogram plot (Figure 5.15), which clearly shows the skewness property of the data set. It is seen from the descriptive statistics that the average grade of Fe is 63.8 (%), and the variance is 64.52 (%²). The Skewness (–2.74) of the data indicates that the distribution is highly skewed. The K-S goodness of fit test was carried out for the normality checking of the data at 5% level of significance. The result of the K-S test shows that the data are not normally distributed.

This skewed distribution of the data leads to exploring the reason for the skewness of the histogram. It may be due to the number of lithologies present in the study area. Therefore, statistical analysis of the lithological data was carried out to show the statistical variability within the lithology. The descriptive statistics of the seven lithotypes are given in Table 5.10. The results show that the lithology-based classification reduces the data variance. The maximum variance was found in BHQ lithology with 235.19, and minimum variance has found in steel grey hematite that is 1.5. But the distribution pattern has not been changed; all are showing the negatively skewed distribution. Box plot was prepared to visualize the variation lithology-wise for the Fe attribute. Figure 5.16 presents

TABLE 5.10
Lithology wise descriptive statistics of Fe

Statistics	Mean (%)	Variance (%^{2})	Coefficient of variation	Skewness	Kurtosis
Steel grey hematite (SGH)	68.74	1.5	1.78	−5.78	68.85
Blue hematite (BH)	67.41	6.9	3.89	−12.36	277.15
Laminated Hematite (LH)	65.54	12.45	5.38	−9.7	160
Blue dust (BD)	63.57	16.82	6.45	−1.66	6.52
Laterite (L)	66.59	6.29	3.76	−2.1	11.87
Banded Hematite Quartz (BHQ)	52.8	235.19	29.04	−0.935	−0.171
Shale (SHL)	48.89	137.15	23.95	−0.896	0.081

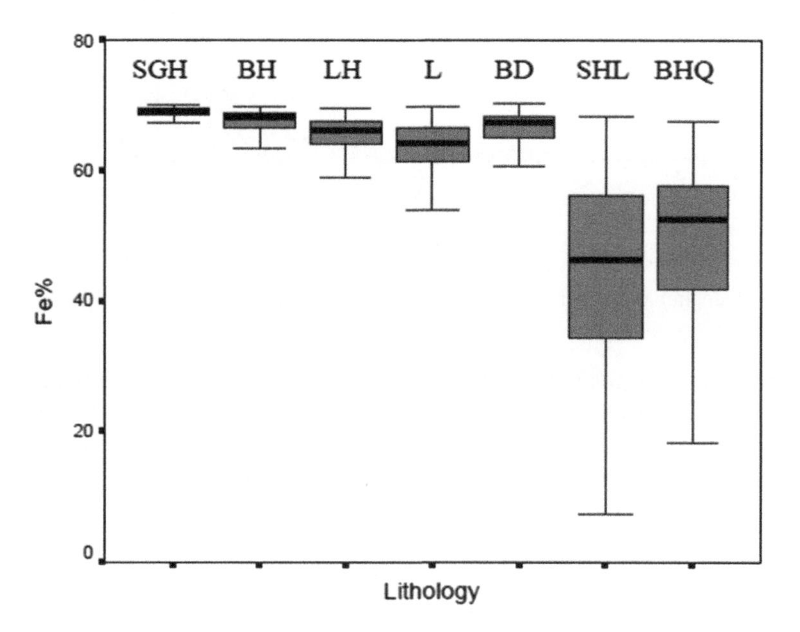

FIGURE 5.16 Lithology wise box plot of Fe for the case study mine.

the lithology-wise box plots of variable Fe. The box plot results revealed that the four ore-forming lithologies, i.e., SGS, BH, LH, and L, have less variance compare to the other three lithologies. This lithology-wise statistical analysis revealed the importance of lithology for grade estimation. Therefore, resource estimation was performed independently for each lithology after preparing lithology-wise solid modelling. All analysis was carried out using Surpac software (www.3ds.com/products-services/geovia/products/surpac/).

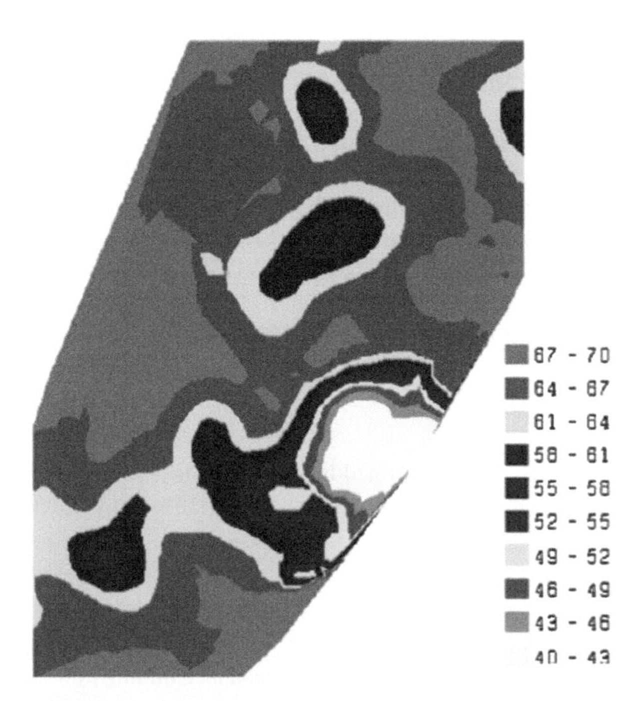

FIGURE 5.17 Ore grade map of 1172 m bench by ordinary kriging.

The ordinary kriging was used for the grade estimation. The spatial correlation of the data was measured using the variogram analysis. Both the omnidirectional and direction variograms were studied for checking the anisotropy. A spherical variogram model was fitted for modelling the spatial continuity of all lithology. A cross-validation exercise optimally chose variogram's parameters. After fitting the variograms, the search for anisotropy was carried out. The variograms were calculated at eight different directions with 22.50 tolerance. The directional difference was not obtained for the anisotropic modelling. Therefore, the isotropic model for ordinary kriging was used for modelling purposes. The blocks of the deposit for all seven lithologies were estimated using the ordinary kriging estimation technique. The ore grade map of the deposit was estimated bench-wise. A total number of 14 ore grade maps were generated for attribute Fe. Figure 5.17 shows the grade maps of Fe for the 1172 m bench. The number of blocks within the deposit that could fall within the ultimate pit of the deposit is 49603, and the total volume, tonnage, and average grade are 198,412,000 m^3, 567.46 Million Tonne (MT), and 64.86%. The generated orebody model was used for production scheduling.

The open-pit production schedule assigns the extraction sequence of the blocks from the estimated orebody model to maximize the net present value over the mine life from the deposit. The production schedule of open-pit mine can either be an integer programming problem or mixed-integer programming problem, which can be solved using the methods discussed in this chapter. However, due to the geometry of the deposit and the slope constraints requirement, the integer programming formulation may generate an infeasible solution (Chatterjee and Dimitrakopoulos, 2020). Therefore, mixed integer programming (MIP) formulation is used in this study. In the MIP formulation, three resource constraints (mining, processing, and metal) are considered; however other constraints like grade blending constraints can also be used. In the MIP, the resource constraints are considered soft constraints and can be violated using the penalty terms.

The mathematical representation of objective function for the open pit production schedule can be written as:

$$Z = Max\sum_{t=1}^{T}\sum_{i=1}^{N}c_{it}x_{i,t} - \sum_{t=1}^{T}\left(v_t^{o-}d_t^{o-} + v_t^{o+}d_t^{o+} + v_t^{m-}d_t^{m-} + v_t^{m+}d_t^{m+}\right) \quad (5.13)$$

The first part of the objective function maximizes the cumulative net profit from the mine throughout the life of the mine. The second part of the objective function penalizes for over- and under-production of ore and metal quantity from target limits. The deviation of total material production (ore plus waste) is not allowed because the case study mine doesn't have any option to hire extra equipment and labour for the over-production of mining, and they also don't allow idle time of equipment for under-production at any circumstances. However, in general, any constraints other than slope and reserve constraints can be penalized if those constraints can be allowed to deviate from target limits.

The production schedule MIP formulation is subject to the following constraints:

Reserve constraints: A block can be mined only once during its period, which is defined as

$$\sum_{t=1}^{T} x_{it} \leq 1 \qquad i = 1,\ldots,N \tag{5.14}$$

Slope constraints: A block cannot be mined before its predecessors. To access a given block, a set of overlying blocks needs to be exacted. Slope constraints are written as

$$x_{it} - \sum_{\tau=1}^{t} x_{p\tau} \leq 0 \qquad p \in P_i, \qquad t = 1,\ldots,T \tag{5.15}$$

Mining constraints: The amount of materials (ore plus waste) mined during each period should be at least equal to a minimum value to avoid an unbalanced mining flow throughout the periods. On the other hand, it should not exceed the mining equipment capacity available during that period. In this case study, the violations of these constraints are also not acceptable; therefore, the same mining constraints equations as the previous model are used here.

$$\begin{aligned} \underline{MC} &\leq \sum_{i=1}^{N} mc_i * x_{it} \qquad t = 1,\ldots,T \\ \overline{MC} &\geq \sum_{i=1}^{N} mc_i * x_{it} \qquad t = 1,\ldots,T \end{aligned} \tag{5.16}$$

Processing constraints: The modified processing constraints with allowable violation occurs production capacity of the plant and the minimum production requirement, these upper and lower bounds are necessary to ensure a smooth feed of ore to the mill, but when the limit of exceeded for either of the lower or upper limit bounds then penalties are added in the objective function. In any case, these violations should not be more than d_t^{o-}, d_t^{o+}, for a specific period. The modified processing constraints for both upper and lower bounds can be presented as:

$$\begin{aligned} \underline{PC} &\leq \sum_{i=1}^{N} pc_i * mc_i * x_{it} + d_t^{o-} \qquad t = 1,\ldots,T \\ \overline{PC} &\geq \sum_{i=1}^{N} pc_i * mc_i * x_{it} - d_t^{o+} \qquad t = 1,\ldots,T \end{aligned} \tag{5.17}$$

Metal production constraints: Same as processing constraints, the metal production constraints are also allowed some amount of violation from their target bounds, and their associated penalties are added in the objective function. The maximum allowable violations are d_t^{m-}, d_t^{m+}, for lower and upper limit bounds. The modified constraints are presented as:

$$\begin{aligned} \underline{MP} &\leq \sum_{i=1}^{N} mp_i * pc_i * x_{it} + d_t^{m-} \qquad t = 1,\ldots,T \\ \overline{MP} &\geq \sum_{i=1}^{N} mp_i * pc_i * x_{it} - d_t^{m+} \qquad t = 1,\ldots,T \end{aligned} \tag{5.18}$$

Decision variables are

$$x_{it} = 0 \; or 1 \qquad i = 1,\ldots,N, \;\; t = 1,\ldots,T$$

$$d_t^{o-}, d_t^{o+}, d_t^{m-}, \; d_t^{m+} \geq 0 \qquad t = 1,\ldots,T$$

where

$v_t^{o-} = \dfrac{c^{o-}}{(1+d_2)^t}$ = Unit shortage of ore that can be associated with failure meet $\underline{PC}$ during period t.

(c^{o-} is the undiscoundted unit shortage cost, and d_2 represent the risk discount rate)

$v_t^{o+} = \dfrac{c^{o+}}{(1+d_2)^t}$ = Unit surplus cost incurred if the total weight of the ore blocks mined during period t exceeds $\overline{PC}$.

$v_t^{m-} = \dfrac{v^{m-}}{(1+d_2)^t}$ = Unit shortage cost associated with failure to meet $\underline{\text{MP}}$ during period t.

$v_t^{m+} = \dfrac{v^{m+}}{(1+d_2)^t}$ = Unit surplus cost incurred if the metal production during period t exceeds $\overline{MP}$.

d_t^{o-} = Shortage of amount of ore at a discounted cost in time period t.

d_t^{o+} = Surplus of the amount of ore at a discounted cost in time period t.

d_t^{m-} = Shortage of metal for selling at discounted time period t.

d_t^{m+} = Surplus of metal for selling at discounted time period t.

In this model, the x_{it} is the first stage decision variables which can only take the binary value; however, $d_t^{o-}, d_t^{o+}, d_t^{m-}, d_t^{m+}$ are second-stage decision variables that can take any real positive values. In this case study, these variables are allowed to make any real positive value up to infinity for which the objective function is penalized. However, one can restrict the amount of maximum deviation if required. Practically, the optimizer selects positive values for these four decision variables during the optimization process. The optimizer selects these decision variables values as zero if allowing the deviation is actually reduced the objective function values.

Although the MIP formulation is easy to implement, it is the most intractable, and finding a feasible solution is NP-hard (Lerchs and Grosmann, 1965). Since the production scheduling problem is NP-hard, we have discretized it into a set of small sub-problems, and each sub-problem was solved sequentially, as discussed in Chatterjee and Dimitrakopulos (2020) and Chowdhury and Chatterjee (2014). Finally, the results of all sub-problems were combined to get the final solution of the main problem. Each sub-problem was solved sequentially by branch-and-cut algorithm, which is a combined method of branch-and-bound and cutting plane, discussed in this Chapter. The large optimization problem was written using ZIMPL software (https://zimpl.zib.de/), which creates the *lp* file, which can be used in any standard optimization solver. The IBM CPLEX solver (www.ibm.com/analytics/cplex-optimizer) is used in this case study; however, any other commercial or free solvers can be used for solving the problem. The list of available solvers can be found here (https://neos-server.org/neos/solvers/). Appendix 5 shows the steps involved in solving the MIP problem using ZIMPL and CPLEX.

For decision-making of the blocks to be mined, an economic block model was created first from the resource block model. This was performed by considering the current production from mine and process costs and commodity price's current economic conditions. The geotechnical and economic

TABLE 5.11
Details of different parameters values and constraints limits used for MIP

Description	Values
Total number of blocks	49603
Slope angle (degree)	45°
Block dimensions (m × m × m)	20 × 20 × 10
Specific gravity of rock (tonne/m^3)	2.86
Recovery (%)	0.74
Discount (%)	10
Selling price of metal (US $/tonne)	40
Selling cost of ore (US $/tonne)	3.6
Processing cost of ore (US $/tonne)	12
Mining cost of rock (US $/ tonne)	3
Block mass or tonnage (tonne)	11440
Mining constraints upper bound (Million tonne)	25
Mining constraints lower bound (Million tonne)	10
Processing constraints upper bound (Million tonne)	10
Processing constraints lower bound (Million tonne)	6
Metal production constraints upper bound (Million tonne)	4
Metal production constraints lower bound (Million tonne)	2

parameters and the different constraints limits that are used in this case study are presented in Table 5.11.

To generate production schedule, the sequential branch-and-cut algorithm was implemented. Results demonstrated that a total of 1150 blocks are scheduled in period 1. The remaining blocks 48453 (i.e., 49603–1150) were then used for the production schedule of period 2 using the same method. Finally, the results show that the mine can produce profitably for ten years by extracting a total number of 16532 blocks. The accuracy of the process is evaluated by calculating the tolerance gap of optimal solution and solution time. Table 5.12 shows the calculated result by the branch-and-cut method. Figure 5.18 presents an East-West section of the production schedule of the study mine. As seen from Figure 5.18, slope constraints are properly respected in this proposed study.

The scheduling of the case study presented here is not possible to solve optimally using any commercial solver. Therefore, it is difficult to calculate the optimality gap of our proposed method for the real mining problem. However, in literature, it has been shown with several problems and comparison with the upper bound solution that the sequential branch-and-cut is very close to the upper bound solution (Chatterjee and Dimitrakopoulos, 2020; Lamghari and Dimitrakopoulos, 2012). Therefore, it can be considered that the generated production schedule for the case study mine is near the optimum solution. Also, the total time required for solving the 23-year production schedule

TABLE 5.12
Production schedule of the case study mine

Period (T)	No of blocks (N)	No of block extracted in period	Solution time in second (t)	Gap in percentage (%)	Iterations for solving the problem
1	49603	2185	107.69	0.01	35
2	47418	2185	116.20	0.00	731
3	45233	2185	124.22	0.00	147
4	43048	2185	123.79	0.00	57
5	40863	2185	106.94	0.00	44
6	38678	2185	60.19	0.00	45
7	36493	2185	64.47	0.00	59
8	34308	2185	86.75	0.00	70
9	32123	2185	110.64	0.00	17767
10	29938	2185	74.40	0.01	1018
11	27753	2185	72.03	0.01	320
12	25568	2185	62.84	0.00	53
13	23383	2185	310.12	0.00	83701
14	21198	2185	925.09	0.01	126678
15	19013	2185	574.47	0.00	101180
16	16828	2185	119.70	0.00	34083
17	14643	2185	38.95	0.01	1787
18	12458	2185	13.68	0.00	55
19	10273	2185	22.79	0.01	865
20	8088	2185	15.91	0.01	706
21	5093	2185	5.63	0.01	869
22	3718	2121	2.96	0.00	8314
23	1597	1597	0.23	0.00	567

is 3139.69 seconds; thus, the branch-and-cut is computationally efficient to solve large mine scheduling problems.

Figure 5.19 presents material production (i.e., mining capacity) from the deposit during production periods. The results show that the optimizer tries to maximize the production to generate more cashflow. The results also show that no deviation from the production target was observed.

Similar to material production, ore and metal productions were also calculated. From the ore production data, it was observed that the ore productions are not within the ore production limits of the mine over the production periods. During initial periods of time, the ore production is higher than

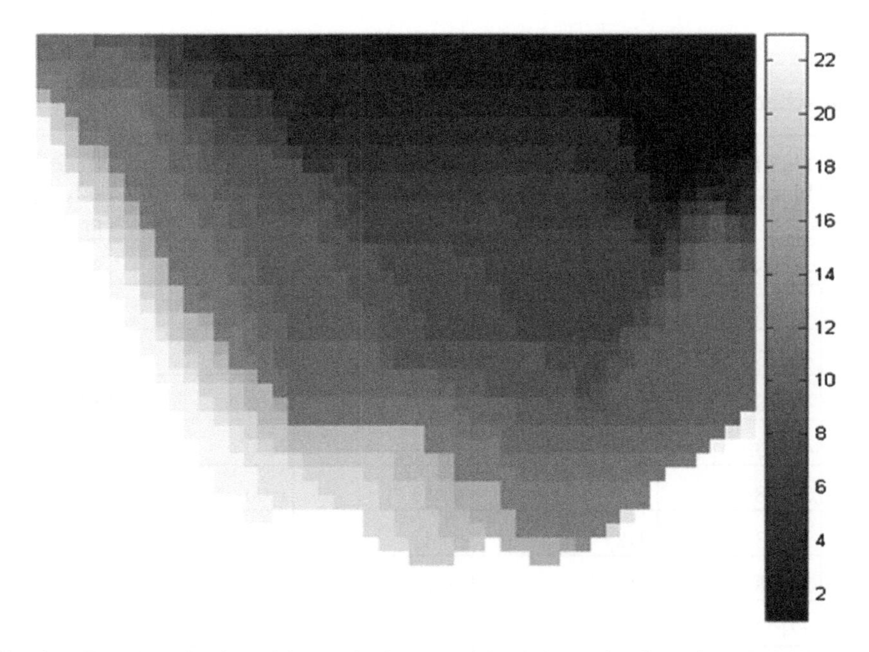

FIGURE 5.18 East-West Section of the production schedule of the study mine using Model 2.

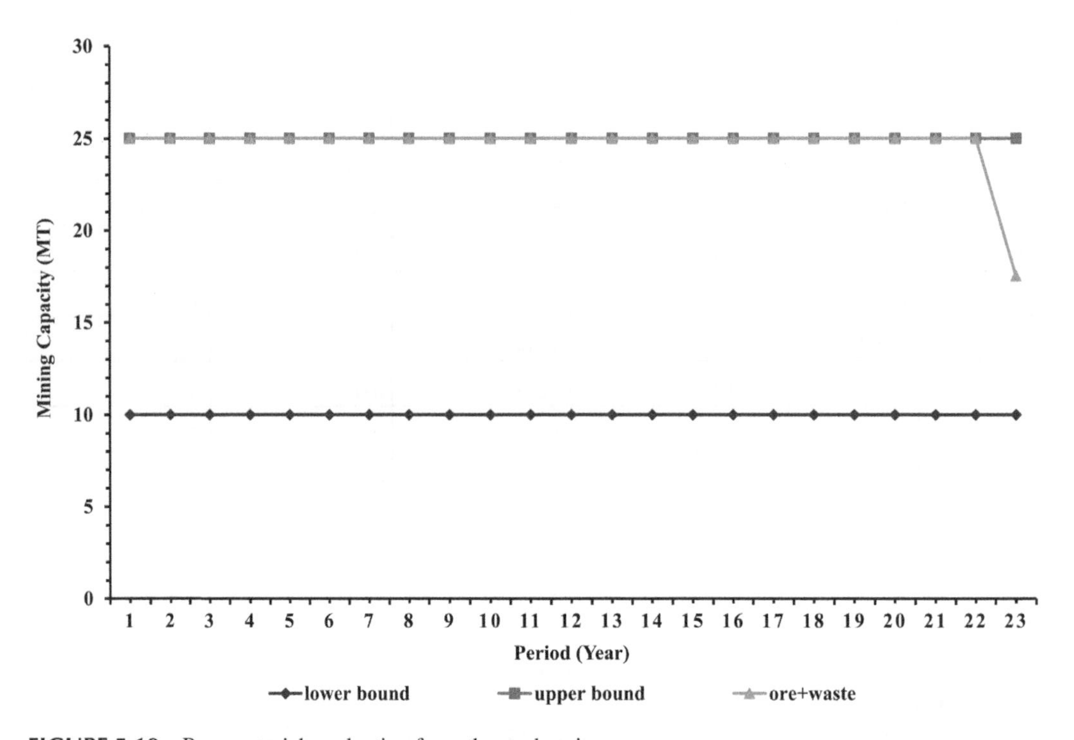

FIGURE 5.19 Raw material production from the study mine.

the target bounds, and at the later periods, the ore production is lower than the target bounds. This is because the ore production constraint was relaxed with a penalty in the objective function, and therefore this constraint was not the hard constraint for production scheduling formulation. The penalty of the over-production at the initial periods was compensated by producing more metals from the initial period, which in turn produces more cash inflow at the initial periods. Since the penalty due to the underproduction cannot be compensated, the optimizer tries to delay the underproduction violation and underproduction violation was only observed in the last three periods of the mine life.

Similar to ore production, it was also observed that the metal productions are also not within the target limits in the generated schedule. As observed in ore production, during initial production periods, the metal production is higher than the target bounds, which generates significantly high cash inflow during early mine life. The optimizer pushes the under-production for the extremely later stage of the mine life. The goal of the production schedule is to maximize the sum of discounted cashflow; therefore, the optimizer tries to generate as much cashflow as possible during the first few years, deferring the risk of not meeting the target for the later periods. The risk profile of metal production also shows that the bounds of the risk is also very tight and tighter in the initial periods than in later periods.

After assigning blocks to different production periods, the discounted cashflow and NPV of the deposit were calculated. The expected NPV of the case study mine is 898 Million US dollars (M$). The total average amount of ore within the ultimate pit is 265 Mt, and the average amount of metal is 129 Mt. The total amount of materials within the ultimate pit is 568 Mt.

APPENDIX 5

This appendix shows how to use ZIMPL and CPLEX together to solve open pit production scheduling by MIP. The two input data that are required for solving this problem are the economic block value of the ore body model and the slope file. If the metal production constraints or grade blending constraints are considered, then a grade data file is also required. Users can create the economic block value data from the resource model (grade) by simple arithmetic and using the economic parameters. Later, in Chapter 6, a Matlab code is provided to calculate the economic block model from the resource model. If users need ready-to-use economic block value data and slope data, they can check the Minelib webpage (http://mansci-web.uai.cl/minelib/). The block economic value data, grade data, and slope data should be saved in text format (.txt).

The block economic value data should have two columns, where the first column represents the block index, and the second column represents the block economic value. The positive block economic value represents the ore block, and the negative block economic value represents the waste block. The data file should not have any header row. The structure of the block economic value data 'cp_bev.txt' will look like Figure 5A1. The grade data will have a similar structure as the block economic value date. The only difference is that the second column will have grade information in place of economic block value.

Like block economic value data, slope data also needs to be saved in text format (.txt). The slope file will also have two columns. The first column represents the index of the block under consideration, and the second column represents the index of the block that needs to be mined before mining the index of the respective row of the first column block. These index numbers are associated with the block economic value data. For example, the first block in the block economic value file (first row in cp_bev.txt file) will have index 1, the 100th block in the block economic value file (100th row in cp_bev.txt file) will have index 100, etc. The structure of the file is very similar to Figure 5A2.

Once we have these two files ready, we can create the optimization formulation. The standard file format for CPLEX solver is .lp format. The ZIMPL open source software can be used for creating the .lp file. The ZIMPL executable file and installation guide can be found here (https://zimpl.zib.de/). The users' guide of ZIMPL software can be found here (https://zimpl.zib.de/download/zimpl.pdf). ZIMPL can be run from command prompt by calling the executable, or double-clicking the executable file in the Windows operating system. Creating the .lp file in ZIMPL requires parameter file (.par format) and input file (.zpl format), along with the block economic value file and slope file, as discussed before. Both the .par and .zpl files can be created using any text editor and saved the

cp_bev.txt - Notepad

File Edit Format View Help

```
1         4.43E+05
2         4.28E+05
3         4.29E+05
4         4.55E+05
5         4.44E+05
6         3.94E+05
7         3.84E+05
8         3.86E+05
9         4.26E+05
10        4.27E+05
11        3.69E+05
12        3.53E+05
13        3.12E+05
14        3.32E+05
15        3.55E+05
16        3.94E+05
17        3.55E+05
18        2.91E+05
19        2.92E+05
20        3.09E+05
21        2.86E+05
22        3.16E+05
23        4.06E+05
24        4.33E+05
25        4.56E+05
26        4.43E+05
27        4.44E+05
28        4.56E+05
```

FIGURE 5A1 Structure of block economic value data file.

slope - Notepad

File Edit Format View Help

```
1       15
2       16
3       17
4       18
5       19
6       20
7       21
8       24
9       25
10      26
11      27
12      28
13      29
14      30
15      51
16      52
17      53
18      54
19      55
20      56
21      57
22      58
23      59
24      62
25      63
26      64
27      65
28      66
```

FIGURE 5A2 Structure of slope file for the MIP model for open-pit production schedule.

```
a_par - Notepad
File  Edit  Format  View  Help
# parameters for sequential production scheduling

P          1  # yr
I          49603 # block
S          265098  # slope

BLOCK_MASS        11440 # in tons
DISCOUNT          0.10 # discount factor

MAX_MINING        25000000 # in tons
MIN_MINING        10000000 # in tons

MAX_PROC1         10000000 # maximum ore limit in ton
MIN_PROC1         6000000 # minimum ore limit in ton

MAX_MET1          4000000 # upper limit of metal production in ton

MIN_MET1          2000000 # lower limit of metal production in ton
```

FIGURE 5A3 File format of parameter file for ZIMPL software.

files in the respective file format. The ZPL file contains the set of equations to formulate a mixed-integer linear programming problem. Inputs for any study are the block economic value (bev.txt) and slope file (slope.txt).

The parameter file (.txt file format) contains the input parameters needed for the study. It contains information regarding the number of years, number of blocks in the block model (number of rows in bev.txt file), number of slope constraints (number of rows in slope.txt file), block mass in case mass is constant for all blocks, mining, processing, metal production constraints, etc. Since the case study uses a sequential branch-and-bound approach to solve the problem, the parameter file (Ex: a_par.txt) for the first period will look like Figure 5A3. In Figure 5A3, the parameters P, I, and S represent number of periods, number of blocks, and number of slope constraints, respectively. The remaining other values in the parameter file are obtained from Table 5.11.

Assuming all files, i.e., cp_bev.txt, cp_grd.txt (grade data file), slope.txt, and a_par.txt, and the ZIMPL executable file zimpl.exe are located in the same folder E:\Example; the ZPL file can be written as follows and be saved as 'prod_schedule.zpl'.

```
# Constants
set PAR :=
{"I","P","S","BLOCK_MASS","DISCOUNT","MAX_MINING","MIN_MINING","MAX_
PROC1","MIN_PROC1","MAX_MET1","MIN_MET1"};

# Reading parameter file
param params[PAR] := read " E:\Example\a_par.txt" as "<1s> 2n" comment "#";

# Define variables
param I := params["I"];
param P := params["P"];
param S := params["S"];
param BLOCK_MASS := params["BLOCK_MASS"];
param DISCOUNT := params["DISCOUNT"];
param MAX_MINING := params["MAX_MINING"];
```

```
param MIN_MINING := params["MIN_MINING"];
param MAX_PROC1 := params["MAX_PROC1"];
param MIN_PROC1 := params["MIN_PROC1"];
param MAX_MET1 := params["MAX_MET1"];
param MIN_MET1 := params["MIN_MET1"];

#Define set
set Blocks := {1 to I};
set Years := {1 to P};
set Slopes := {1 to S};

# Read block economic value
param INCOME1[Blocks] := read "E:\Example\cp_bev.txt" as "<1n> 2n";

# Read grade data
param GR1[Blocks] := read "E:\Example\cp_grd.txt" as "<1n> 2n";

# Read slope data
param slope1[Slopes] := read "E:\Example\slope.txt" as "1n";
param slope2[Slopes] := read "E:\Example\slope.txt" as "2n";

# Define integer variables
var x[Years * Blocks] binary; # Extract

# Define linear variables
var mx_met1[Years];
var mn_met1[Years];
var mx_d1[Years];
var mn_d1[Years];

# Objective function
maximize npv :
       sum <t,i> in Years * Blocks: (1 / (1 + DISCOUNT)^(t-1)) * (INCOME1[i]
* x[t,i]) - sum <t> in Years: (15*mn_d1[t]+ 20*mn_met1[t]+15*mx_d1[t]+
20*mx_met1[t]);

# Reserve constraints
subto reserve_x :
    forall <i> in Blocks do
      sum <t> in Years :  x[t,i] <= 1;

# Maximum mining capacity
subto mining_max :
    forall <t> in Years do
      sum <i> in Blocks : x[t,i] * BLOCK_MASS <= MAX_MINING;

# Minimum processing capacity
subto proc_min1:
    forall <t> in Years do
      sum <i> in Blocks with INCOME1[i]>=0: (x[t,i] * BLOCK_MASS) + (mn_
d1[t]) >= MIN_PROC1;
```

```
# Maximum processing capacity
subto proc_max1:
   forall <t> in Years do
     sum <i> in Blocks with INCOME1[i]>=0: (x[t,i] * BLOCK_MASS) - (mx_
d1[t])<= MAX_PROC1;

# Maximum metal production capacity
subto metal_max1 :
   forall <t> in Years do
     sum <i> in Blocks with INCOME1[i]>=0:( x[t,i] * ((GR1[i]*0.9)/
100)*BLOCK_MASS) - (mx_met1[t]) <= MAX_MET1;

# Minimum metal production capacity
subto metal_min1 :
   forall <t> in Years do
     sum <i> in Blocks with INCOME1[i]>=0: (x[t,i] * ((GR1[i]*0.9)/
100)*BLOCK_MASS) + (mn_met1[t]) >= MIN_MET1;

# Slope constraints
subto slope1 :
forall <t> in Years do
   forall <s> in Slopes do
             x[t,slope1[s]]- sum <t> in Years: x[t,slope2[s]]<=0;
```

To run the ZPL file in ZIMPL using the command prompt, we need to call the ZIMPL executable file and the 'prod_schedule.zpl' file, as shown in Figure 5A4. It should be noted here that there should be space in between 'C:\Example\zimple.exe' and 'C:\Example\prod_schedule.zpl' in the command prompt. It will create a 'prod_schedule.lp' file that can be used in CPLEX for the optimization of the problem.

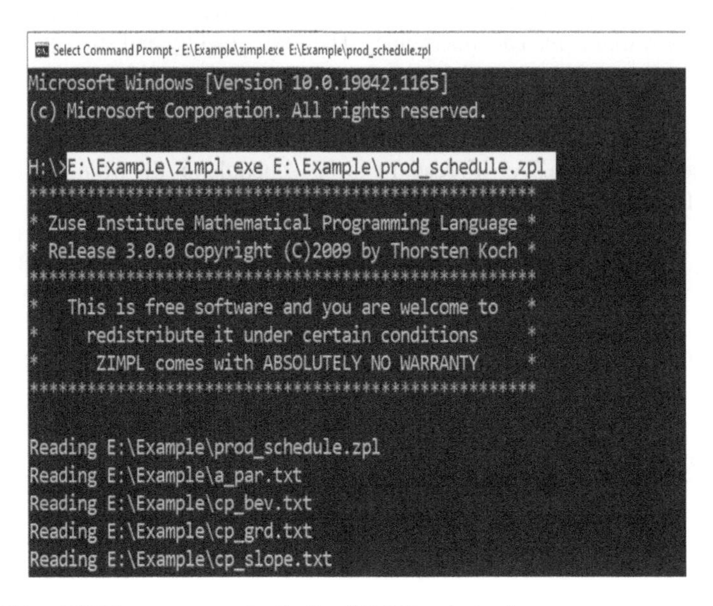

FIGURE 5A4 Using ZIMPL to create an lp file for CPLEX optimization.

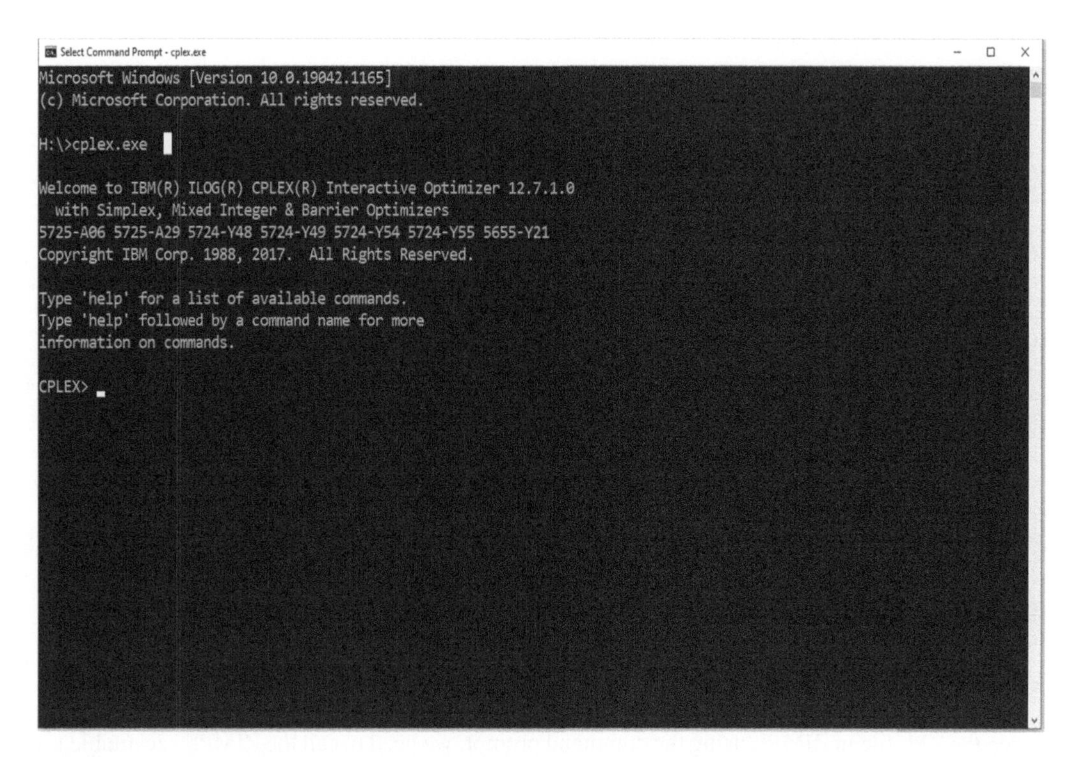

FIGURE 5A5 Using CPLEX from the command prompt.

To run CPLEX, we need to have CPLEX installed on the computer. The installation guide and user guide of CPLEX can be found here: www.ibm.com/analytics/cplex-optimizer. It is noted here that although CPLEX is commercial software, it can be used free of cost using IBM academic initiative program (www.ibm.com/academic/home). Therefore, faculty and students can use it for free if they have university credentials and follow some terms and conditions mentioned in the IBM academic initiative program.

Once CPLEX is installed, users can call the executable file from the command prompt. By types 'cplex.exe', as shown in Figure 5A5. To read the 'prod_schedule.lp' file from CPLEX, users need to write 'READ E:\Example\prod_schedule.lp' in the command prompt.

After CPLEX reads the problem, the command "optimize" needs to use for the optimization. Based on the lp formulation and decision variable types, CPLEX will automatically select the optimization algorithm. By writing 'optimize' on the CPLEX prompt and hitting enter KEY, the user can run the optimization in CPLEX (Figure 5A6).

After getting the optimal solution, we can write the solution using the command "write" followed by the address and name where we want to save the file (Figure 5A7). The extension .sol is used to save the solution file. The CPLEX solution file can be open in notepad and other text editing software. The solution file will look like Figure 5A8.

```
Select Command Prompt - cplex.exe
Microsoft Windows [Version 10.0.19042.1165]
(c) Microsoft Corporation. All rights reserved.

H:\>cplex.exe

Welcome to IBM(R) ILOG(R) CPLEX(R) Interactive Optimizer 12.7.1.0
  with Simplex, Mixed Integer & Barrier Optimizers
5725-A06 5725-A29 5724-Y48 5724-Y49 5724-Y54 5724-Y55 5655-Y21
Copyright IBM Corp. 1988, 2017.  All Rights Reserved.

Type 'help' for a list of available commands.
Type 'help' followed by a command name for more
information on commands.

CPLEX> READ E:\Example\prod_schedule.lp
Problem 'E:\Example\prod_schedule.lp' read.
Read time = 0.14 sec. (14.43 ticks)
CPLEX> optimize
CPXPARAM_Read_APIEncoding                         "UTF-8"
Found incumbent of value -1.0590000e+008 after 0.00 sec. (4.46 ticks)
Tried aggregator 5 times.
MIP Presolve eliminated 32055 rows and 0 columns.
MIP Presolve modified 2 coefficients.
Aggregator did 49 substitutions.
Reduced MIP has 126152 rows, 16487 columns, and 296349 nonzeros.
Reduced MIP has 16483 binaries, 0 generals, 0 SOSs, and 0 indicators.
Presolve time = 4.22 sec. (5257.98 ticks)
Probing fixed 1828 vars, tightened 0 bounds.
Probing changed sense of 1397 constraints.
Probing time = 2.28 sec. (812.15 ticks)
```

FIGURE 5A6 Optimize MIP using CPLEX software.

```
Select Command Prompt - cplex.exe
Root node processing (before b&c):
  Real time             =   19.53 sec. (16793.16 ticks)
Parallel b&c, 8 threads:
  Real time             =    0.00 sec. (0.00 ticks)
  Sync time (average)   =    0.00 sec.
  Wait time (average)   =    0.00 sec.
                          ------------
Total (root+branch&cut) =   19.53 sec. (16793.16 ticks)

Solution pool: 11 solutions saved.

MIP - Integer optimal, tolerance (0.0001/1e-006):  Objective = 1.1953024114e+008
Current MIP best bound = 1.1953322672e+008 (gap = 2985.59, 0.00%)
Solution time =   19.55 sec.  Iterations = 11321  Nodes = 0 (1)
Deterministic time = 16793.21 ticks  (859.12 ticks/sec)

CPLEX> write E:\Example\prod_schedule.sol
Incumbent solution written to file 'E:\Example\prod_schedule.sol'.
CPLEX> _
```

FIGURE 5A7 Writing CPLEX solution on a computer.

prod_schedule.sol - Notepad

File Edit Format View Help

```
<?xml version = "1.0" encoding="UTF-8" standalone="yes"?>
<CPLEXSolution version="1.2">
 <header
   problemName="E:\Example\prod_schedule.lp"
   solutionName="incumbent"
   solutionIndex="-1"
   objectiveValue="119530241.1392"
   solutionTypeValue="3"
   solutionTypeString="primal"
   solutionStatusValue="102"
   solutionStatusString="integer optimal, tolerance"
   solutionMethodString="mip"
   primalFeasible="1"
   dualFeasible="0"
   MIPNodes="0"
   MIPIterations="11321"
   writeLevel="1"/>
 <quality
   epInt="1.0000000000000001e-05"
   epRHS="9.9999999999999995e-07"
   maxIntInfeas="0"
   maxPrimalInfeas="0"
   maxX="4707.6930400000183"
   maxSlack="12610480"/>
 <linearConstraints>
  <constraint name="reserve_x_1" index="0" slack="1"/>
  <constraint name="reserve_x_2" index="1" slack="1"/>
  <constraint name="reserve_x_3" index="2" slack="1"/>
```

Ln 1, Col 1 100% Windows (CRLF) UTF-8

FIGURE 5A8 CPLEX solution file structure.

EXERCISE 5

Q1: Three mines supply coal to three power plants. The capacity of the plants, the coal supply by three mines, and the unit transportation cost (US$) from the mines to plants are given in the following table.

Mine	**Power Plant**				**Supply (tonne)**
	P1	**P2**	**P3**	**P4**	
M1	15	13	16	20	1500
M2	17	14	16	18	1900
M3	14	15	18	15	2200
Demand (tonne)	1600	1800	1000	1200	

The fixed costs for the set-up of the three power plants are US$12 million, US$15 million, and US$25 million, respectively. Formulate an ILP to determine the optimum solution.

[**Ans.** M1→P1=600, M1→P2= 900, M2→P2=900, M2→P3= 1000, M3→P1= 1000, M3→P4 = 1200, Cost = US$ 148300]

Q2: In an underground mine, miners work five consecutive days and have the next two days off. The number of workers required each day are as follows:

Day	Number of Miners
Monday	50
Tuesday	55
Wednesday	60
Thursday	52
Friday	55
Saturday	40
Sunday	30

Determine the minimum number of workers needed for the successful operation of the mine.

[**Ans.** Monday = 29, Tuesday = 9, Wednesday = 10, Thursday = 4, Friday = 5, Saturday = 12, Sunday =0, Total Worker = 69]

Q3. A mining company wants to invest US$24 million. The company has six investment options. The following table lists the investment required in each project and the NPV of the respective projects. Determine the optimal investment decision for maximizing the net present value.

Project	Investment required (million US$)	NPV of the Return (million US$)
1	6	9
2	8	12
3	5	7
4	4	5
5	7	10
6	6	8

[**Ans.** Project 1, 2, 5, and 7 should be selected. Maximum NPV= 35]

Q4. The following table gives the expected returns and associated yearly investment of five projects over a 5-year planning horizon.

Projects	Year					Returns (million $)
	1	2	3	4	5	
1	8	6	5	5	3	20
2	12	10	9	8	5	40
3	6	6	4	4	2	20
4	5	6	2	2	1	15
5	10	8	8	5	4	30
Available fund (million $)	30	30	30	30	30	

Determine the type of projects for which investments are made over the 5-year horizon to maximize the return.

[**Ans.** Project 2, 3, and 5 should be selected. Maximum return = 90]

Q5. In Q1, a new mine, M4 with a production capacity of 3200 tonnes and fixed cost for production is US$ 30000 can start the production. M4 supply coal to four plants with a transportation cost of US$ 12, US$ 14, US$ 16, and US$ 15, respectively. Find the optimal solution.

[**Ans.** M1→P2=1500, M2→P2= 300, M2→P3=800, M4→P1= 1600, M4→P3= 200, M4→P4 = 1200, Cost = US$ 145900]

6 Dynamic programming for mining systems

6.1 INTRODUCTION

Dynamic programming (DP) often has been used to optimize mine systems engineering to maximize the mine operations' economic returns. DP is a mathematical optimization method in which a dynamic system of equations is solved numerically. In DP, problems are divided into *stages,* and decisions on the optimal solution are taken in each stage to obtain an integrated solution.

The range of problems that can be solved is extremely wide and encompasses natural resources management and many other industrial applications (Kennedy, 1986). The method has also been applied in different types of decision-making problems of mine systems. However, decision-making of mine systems through the optimal solution is not a trivial process. This is because of the complexity of the different operational and economic variables that need to be considered and because these decisions need to be made over time, i.e., considering the future, which is uncertain. The application of dynamic programming can be used for making optimal decisions over time.

The viability of a mining project is generally evaluated through optimization techniques. Sometimes, the decision-making in the mines is done based on the optimal solution of a specific mining problem, but other times a near-optimal solution is good enough for decision making.

Richard Bellman introduced DP in the 1940s to make a stage-wise optimal decision of a problem. The method was refined in the year 1953 for breaking a large or complex decision-making problem into nested smaller decision problems (Dreyfus, 2002). In DP, the optimization problem is solved in recursive form. In every context, it simplifies a complicated problem by breaking it down into simpler sub-problems in a recursive manner. If sub-problems are nested recursively inside larger problems, then there is a relation between the value of the larger problem and the values of the sub-problems (Cormen et al., 2001).

Dynamic programming is adaptable to both linear and non-linear models. In addition, integer variables and functions cause no particular difficulties with DP. The required model structure for DP is challenging to describe in general terms because DP models can have a variety of mathematical forms. However, problems suitable for DP can usually be interpreted as a sequential allocation process.

6.2 SOLUTION ALGORITHM OF DYNAMIC PROGRAMMING

The decision-making problem of mine systems can be solved by using either of forward-recursion or backward-recursion algorithm of DP. The key idea of DP is that the complex problem can be divided into multiple sub-problems, and the optimization is carried out in stages, where each sub-problem is one stage. In the *forward-recursion* algorithm, the optimization process starts from the first stage and moves towards the last stage. On the other hand, in the b*ackward-recursion* method, the solution starts from the last stage and moves toward the first stage. According to Kennedy (1986), the forward-recursion is more suitable for problems involving uncertain time horizons. A backward recursion is appropriate for solving problems that contain options with the same time horizon.

There is no generalized structure of the dynamic programming, as this can apply to both the linear and non-linear problems. Therefore, the algorithms are explained using case examples.

DOI: 10.1201/9781003200703-6

Here is an overview of the method:

Step 1: *The defined problem is divided into smaller sub-problems. Each sub-problem is defined as stage. A stage consists of a number of states or options, and the optimal solution of each stage needs to be identified.*

Step 2: *The solution of one sub-problem is extended to the next stage to obtain the optimal solution for the next stage.* During computation of the current stage, the optimal decision for each of state does not require any additional computations of the previous stages. This is known as the **principle of optimality** for dynamic programming.

Step 3: *Continue with Step 2 until the optimal solution of the original problem is obtained.* The principle of optimality allows the solution of the problem stage by stage **recursively**.

The algorithm of dynamic programming is demonstrated by three examples.

6.3 EXAMPLE 1: MAXIMISING PROJECT NPV

In a large open-pit mine, the production scheduling is a challenging task as this influences the cost of operations. The long-term planning of open-pit mine is a dynamic process and should be reviewed periodically for enhancing the net present value (NPV) and life of mine. The objective of the optimum production scheduling is to maximize the net present value (NPV) of the mining project. Assuming the life of the mine of the project is four years (or the production period of the mine is four), as shown in Figure 6.1. The cash flow (CF) generated in a specific production period depends on the production schedule selected, as represented in Table 6.1. In each production period, only one activity can be selected to generate a valid production schedule. The number of available activities in each period depends on mining geotechnical constraints. The annual discounting factor of cash flow is 10%. Determine the optimum production schedule for maximizing NPV using dynamic programming.

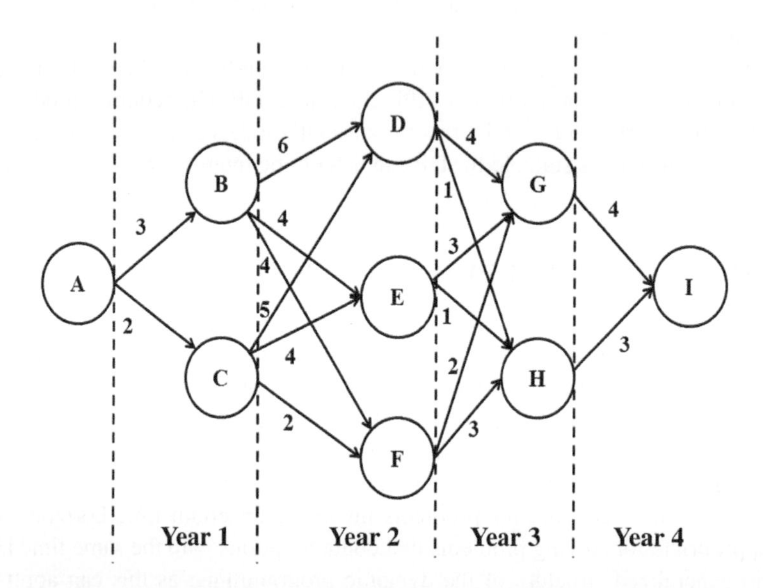

FIGURE 6.1 Production periods in a mine.

TABLE 6.1
The available activities and their associated cash flows from the mining production schedule

Production Plans

Year 1	CF (10^6US$)	Year 2	CF (10^6US$)	Year 3	CF (10^6US$)	Year 4	CF (10^6US$)
AB	3	BD	6	DG	4	GI	4
AC	2	BE	4	DH	1	HI	3
		BF	4	EG	3		
		CD	5	EH	1		
		CE	4	FG	2		
		CF	2	FH	3		

Solution

In this case, the mine planner has multiple options for production scheduling to extract the entire ores in four years. This is because the production rate in each year is different and has different NPV. The estimated CFs for different production plan in each year are shown in Table 6.1. In addition, the production plan options for each year are graphically represented in Figure 6.2. Therefore, the mine planner needs to decide the best production schedule plan for the mine to maximize the CF.

To solve the problem using the DP technique, the problem is divided into stages. In this case, the production plan of four years needs to be optimized and thus divided into four stages. In DP, the optimization solution can be derived in two ways, viz. Backward recursion method and forward recursion method. In the backward recursion method, the process starts at the end of the life of mine (node I) and goes backward, optimizing each production period until arriving in the beginning node (node A). On the other hand, in the forward recursion method, the process starts at the first node (node A) and goes forward till it reaches to end node (node I). A detailed explanation of both the algorithms is given below in subsequent sub-sections.

6.3.1 Backward Recursion Algorithm

The problem is solved **recursively by working backward** in the network

Let P_{ij} be the return in the selection of the production schedule represented by path connecting nodes i and j.

And, $f_t(S_i)$ be the optimum value in stage t for ith node.

Each stage represents the CF in a specific year in different production plan.

In the backward recursion method, the computation will start from the end node (node I) and move towards the start node (node A) through Stage 4, Stage 3, Stage 2, and Stage 1.

Stage 4 computations

The fourth stage has only two states or two options to reach the start node from the end node of Stage 4. The start node of Stage 4 is the end node of Stage 3.

$$f_4(S_G) = P_{GI} = 4$$

$$f_4(S_H) = P_{HI} = 3$$

That is, the returns for selecting the production plan GI and HI are US$ 4 million and US$ 3 million, respectively.

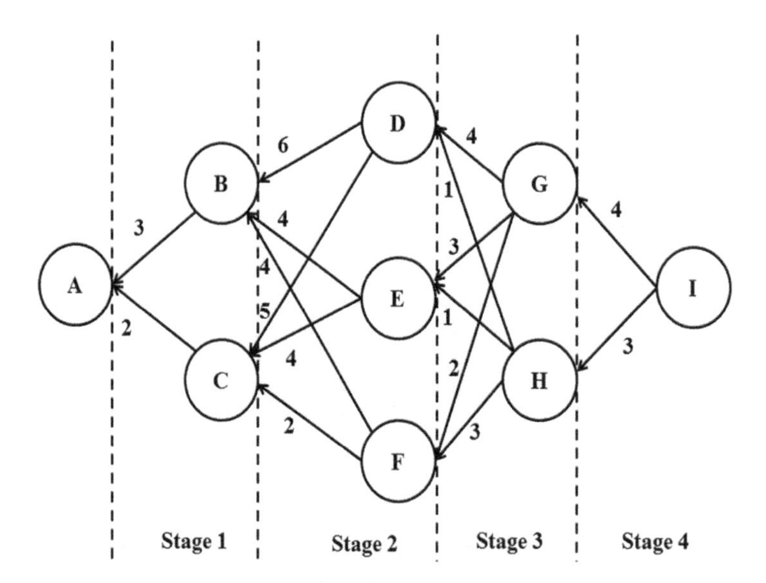

FIGURE 6.2 Graphical representations of production stages.

Stage 3 computations

To move backward from the end node to the start node of Stage 3, six options (DG, EG, FG, DH, EH, and FH) are available (three options from each end node). The return function needs to be calculated for each production plan to determine the optimal value for Stage 3. The start node of Stage 3 can be reached from the end node of Stage 4 via node G or H.

The optimal return values to reach different nodes (D, E, and F) of Stage 3 from the end node, I, are as follows:

$$f_3(S_D) = max\begin{cases} P_{DG} + f_4(S_G) \\ P_{DH} + f_4(S_H) \end{cases} = max\begin{cases} 4+4=8^* \\ 3+3=6 \end{cases} = 8$$

$$f_3(S_E) = max\begin{cases} P_{EG} + f_4(S_G) \\ P_{EH} + f_4(S_H) \end{cases} = max\begin{cases} 3+4=7^* \\ 1+3=4 \end{cases} = 7$$

$$f_3(S_F) = max\begin{cases} P_{FG} + f_4(S_G) \\ P_{FH} + f_4(S_H) \end{cases} = max\begin{cases} 2+4=6^* \\ 3+3=6^* \end{cases} = 6$$

In each case, the value of $f_t(S_i)$ should be considered the optimal value (maximum for maximization problem and minimum for minimization problem). In this problem, $f_4(S_G)$ and $f_4(S_H)$ have only one value.

Stage 3 computations indicate that the maximum value to reach nodes D, E, and F from the end node, I, are US$ 8 million, US$7 million, and US$6 million, respectively. The maximum value can be achieved by selecting the production schedule paths from end node to nodes D, E, and F: DGI, EGI, and FGI/FHI.

Stage 2 computations

Stage 2 has two states, i.e., states B and C. To move backward from end node to start node of Stage 2, six options (BD, CD, BE, CE, BF, and CF) are available (two options from each end node of

Stage 2). Therefore, the return function needs to be calculated for each production plan in this stage to determine the optimal value for Stage 2.

The optimal return values to reach different nodes (B and C) of Stage 2 from the end node, I, are as follows:

$$f_2(S_B) = max\begin{cases} P_{BD} + f_3(S_D) \\ P_{BE} + f_3(S_E) \\ P_{BF} + f_3(S_F) \end{cases} = max\begin{cases} 6+8=14^* \\ 4+7=11 \\ 4+6=10 \end{cases} = 14$$

$$f_2(S_C) = max\begin{cases} P_{CD} + f_3(S_D) \\ P_{CE} + f_3(S_E) \\ P_{CF} + f_3(S_F) \end{cases} = max\begin{cases} 5+8=13^* \\ 4+7=11 \\ 2+6=8 \end{cases} = 13$$

Stage 2 computations indicate that the maximum value to reach nodes B and C from the end node, I, are US$14 million and US$13 million, respectively. The maximum value can be achieved by selecting the production schedule paths from end node to nodes B and C are BDGI and CDGI.

Stage 1 computations

To move backward from the end node to the start node of Stage 1, two options (AB and AC) are available (one option from each end node of Stage 1). The return function needs to be calculated for each production plan in this stage to determine the optimal value for Stage 1.

The optimal return values to reach the start node (node A) of Stage 1 from end node, I, are as follows:

$$f_1(S_A) = max\begin{cases} P_{AB} + f_2(S_B) \\ P_{AC} + f_2(S_C) \end{cases} = max\begin{cases} 3+14=17^* \\ 2+13=15 \end{cases} = 17$$

Stage 1 computations indicate that the maximum value to reach starting node, A from the end node, I, is US$17 million.

It is clear from the above results that the maximum value obtained in Stage 1 is US$ 17 million, and the corresponding production schedule path is AB. The node connecting by B will only be considered in the next stage as this will have the maximum return value. The maximum return path connecting node B to the next stage is BD (=US$12 million). Thus, in the next stage, the paths connecting node D is the candidate for the maximum return path. The results show that the production schedule path DG has the maximum return value (=US$ 8 million). There is only one path that exist, which connects the end node I to node G. Therefore, the production schedule path, which offers maximum returns, is I→G→D→B→A (shown with red arrows in Figure 6.3).

If the discounting factor is 10%, then the net present value (NPV) is given by

$$NPV = \frac{CF_{AB}}{(1+i)} + \frac{CF_{BD}}{(1+i)^2} + \frac{CF_{DG}}{(1+i)^3} + \frac{CF_{GI}}{(1+i)^4}$$

$$\Rightarrow NPV = \frac{3}{(1+0.1)} + \frac{6}{(1+0.1)^2} + \frac{4}{(1+0.1)^3} + \frac{4}{(1+0.1)^4} = \$13.42\,\text{million}$$

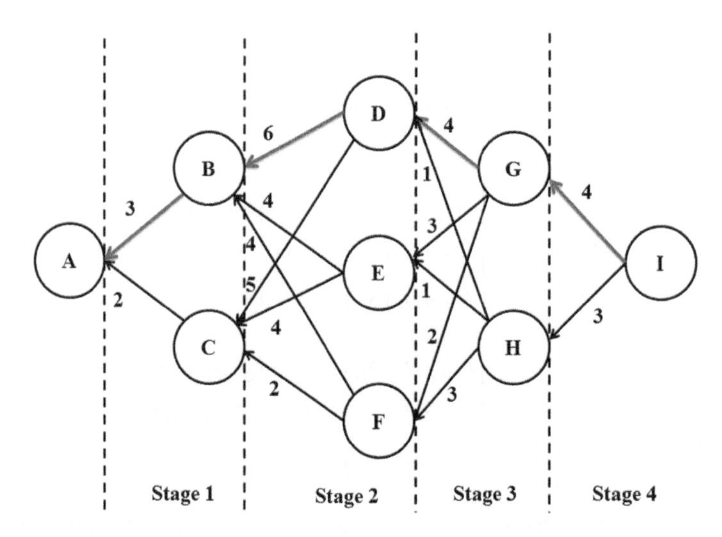

FIGURE 6.3 Production plan showing optimal cash flow path in backward recursion.

6.3.2 Forward Recursion Algorithm

The same problem can be solved using the forward-recursion method. In the forward-recursion method, the recursion moves in a forward direction from starting node, A, to end node, I, as shown in Figure 6.4.

Stage 1 computations

The first stage has only two states, following two paths (i.e., path AB and path AC), to reach the 2nd stage.

$$f_1(S_B) = P_{AB} = 3$$

$$f_1(S_C) = P_{AC} = 2$$

That is, the returns for selecting the production plan AB and AC are US$ 3 million and US$ 2 million, respectively.

Stage 2 computations

To move forward from the start node to the end node of Stage 2, six options (BD, BE, BF, CD, CE, and CF) are available (three options from each end node). The return function needs to be calculated for each production plan to determine the optimal value for Stage 2. The end node of Stage 2 can be reached from the start node of Stage 1 via node B or C.

The optimal return values to reach different end nodes (D, E, and F) of Stage 2 from starting node, A, are as follows:

$$f_2(S_D) = max\begin{cases} P_{BD} + f_1(S_B) \\ P_{CD} + f_1(S_C) \end{cases} = max\begin{cases} 6+3=9^* \\ 5+2=8 \end{cases} = 9$$

$$f_2(S_E) = max\begin{cases} P_{BE} + f_1(S_B) \\ P_{CE} + f_1(S_C) \end{cases} = max\begin{cases} 4+3=7^* \\ 4+2=6 \end{cases} = 7$$

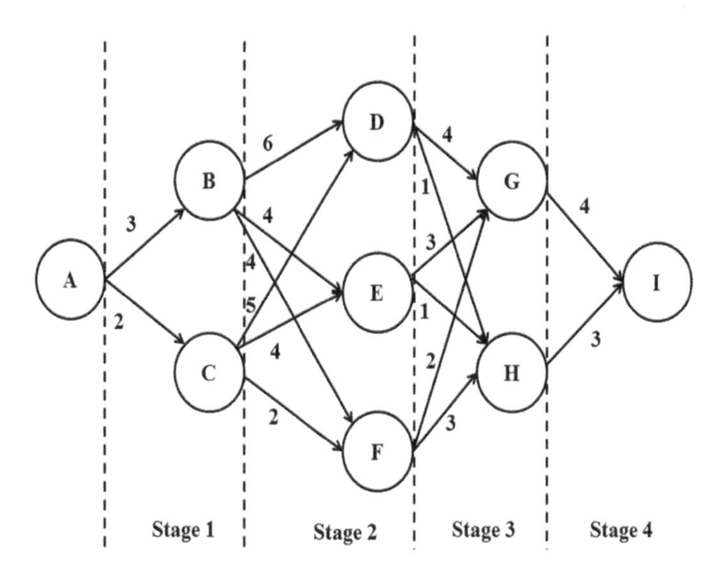

FIGURE 6.4 Movement in a forward recursion method.

$$f_2(S_F) = max\begin{cases} P_{BF} + f_1(S_B) \\ P_{CF} + f_1(S_C) \end{cases} = max\begin{cases} 4+3=7^* \\ 2+2=4 \end{cases} = 7$$

In each case, the value of $f_t(S_i)$ should be considered the optimal value (maximum for maximization problem and minimum for minimization problem). In this problem, $f_1(S_B)$ and $f_1(S_C)$ have only one value.

Stage 2 computations indicate that the maximum value to reach nodes D, E, and F from the start node, A, are US$ 9 million, US$ 7 million, and US$ 7 million, respectively. The maximum value can be achieved by selecting the production schedule paths from start node, A to nodes D, E, and F are ABD, ABE, and ABF.

Stage 3 computations

Stage 3 has two states, i.e., states G and H. To move forward from the start node to the end node of Stage 3, six options (DG, DH, EG, EH, FG, and FH) are available (two options from each end node of Stage 2). The return function needs to be calculated for each production plan to determine the optimal value for Stage 2.

The optimal return values to reach different end nodes (G and H) of Stage 3 from start node, A, are as follows:

$$f_3(S_G) = max\begin{cases} P_{DG} + f_2(S_D) \\ P_{EG} + f_2(S_E) \\ P_{FG} + f_2(S_F) \end{cases} = max\begin{cases} 4+9=13^* \\ 3+7=10 \\ 2+7=9 \end{cases} = 13$$

$$f_3(S_H) = max\begin{cases} P_{DH} + f_2(S_D) \\ P_{EH} + f_2(S_E) \\ P_{FH} + f_2(S_F) \end{cases} = max\begin{cases} 1+9=10^* \\ 1+7=8 \\ 1+7=8 \end{cases} = 10$$

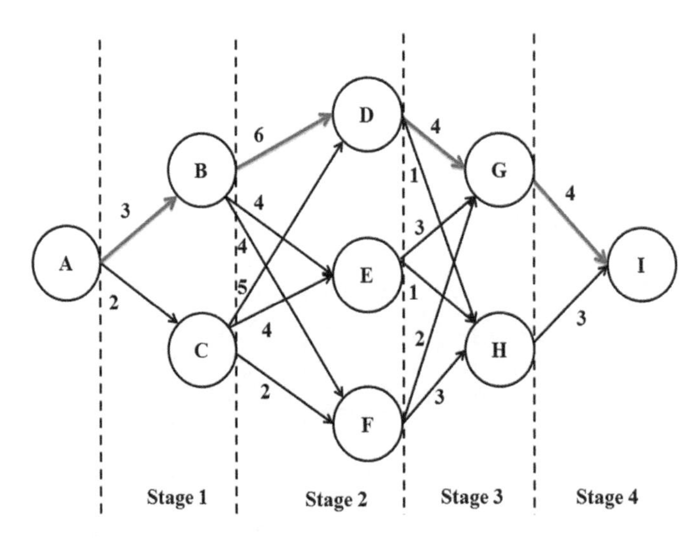

FIGURE 6.5 Production plan showing optimal cash flow path in the forward recursion.

Stage 3 computations indicate that the maximum value to reach nodes G and H from the start node, A, are US$13 million and US$10 million, respectively. The maximum value can be achieved by selecting the production schedule paths from the start node to nodes G and H are ABDG and ABDH.

Stage 4 computations

To move forward from the start node to the end node of Stage 4, two options (GI and HI) are available (one option from each start node of Stage 4).

The optimal return values to reach the end node (node I) of Stage 4 from start node, A, is as follows:

$$f_4(S_I) = max\begin{cases} P_{GI} + f_3(S_G) \\ P_{HI} + f_3(S_H) \end{cases} = max\begin{cases} 4+13=17^* \\ 3+10=13 \end{cases} = 17$$

Stage 4 computations indicate that the maximum value to reach end node, I from the start node, A, is US$ 17 million. Therefore, the production schedule path, which offers maximum returns, is A→B→D→G→I (shown with red arrows in Figure 6.5).

It can be observed that the optimal solution remains the same irrespective of the use of the forward recursion and backward recursion method.

6.4 EXAMPLE 2: DECISION ON ULTIMATE PIT LIMIT (UPL) OF TWO-DIMENSIONAL (2-D) BLOCKS

A vertical section of ore/waste blocks with their economic values ($B_{r,c}$) is shown in Figure 6.6. The blocks can be extracted with a maximum slope of 1:1. The assigned block value is the selling price of the recoverable mineral contained in the block less the costs for mining, processing, transportation, and required marketing to give the saleable product per block.

The objective of this example is to determine the ultimate pit limit for maximising return using dynamic programming.

Solution

In the given blocks, a row of "dummy blocks" is added on the top of the section with '0' economic block value. It is given that the block value $B_{r,c}$ represents the value of the selling price of

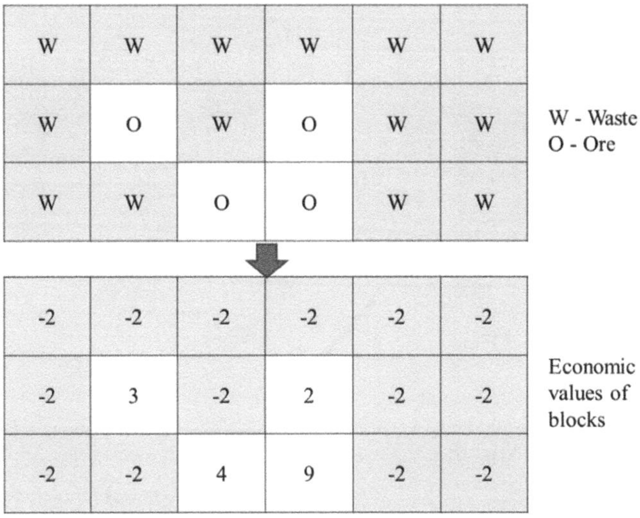

W	W	W	W	W	W
W	O	W	O	W	W
W	W	O	O	W	W

W - Waste
O - Ore

-2	-2	-2	-2	-2	-2
-2	3	-2	2	-2	-2
-2	-2	4	9	-2	-2

Economic values of blocks

FIGURE 6.6 Vertical section of ore/waste blocks with their economic values.

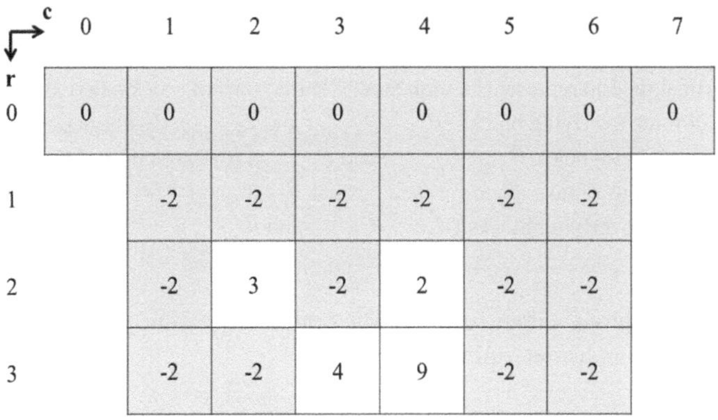

r \ c	0	1	2	3	4	5	6	7
0	0	0	0	0	0	0	0	0
1		-2	-2	-2	-2	-2	-2	
2		-2	3	-2	2	-2	-2	
3		-2	-2	4	9	-2	-2	

FIGURE 6.7 A vertical section represents the economic values ($B_{r,c}$) of individual blocks.

the mineral contained in the block less the mining cost, processing cost, transportation cost, and required marketing cost. All the blocks are uniquely defined with row (r) and column number (c), as shown in Figure 6.7. To extract a block in the level r, requires the prior mining of the blocks directly above this block in the same column and all the blocks within the minimum removal cone consistent with the slope stability requirements of the concerned block.

Now, considering any block $B_{r,c}$ in column c in relation to the neighboring column c-1, a maximum slope of 1:1 stipulates that the block $B_{r,c}$ can only be mined jointly with the block $B_{r-1,c-1}$, block $B_{r-1,c}$, and block $B_{r-1,c+1}$ as showed in Figure 6.8(a). Subsequently, the block $B_{r-1,c-1}$ can only be mined jointly with the block $B_{r-2,c-2}$, block $B_{r-2,c-1}$, and block $B_{r-1,c}$ as showed in Figure 6.8(b). Similarly, the block $B_{r-1,c+1}$ can only be mined jointly with the block $B_{r-2,c}$, block $B_{r-2,c+1}$ and block $B_{r-2,c+2,}$ as shown in Figure 6.8(c).

Therefore, to extract the block, $B_{r,c}$, the overlying blocks ($B_{r-1,c}$, $B_{r-1,c-1}$, $B_{r-1,c+1}$, $B_{r-2,c}$, $B_{r-2,c-1}$, $B_{r-2,c-2}$, $B_{r-2,c+1}$, and $B_{r-2,c+2}$) need to be extracted. Thus, the economic value of extracting any block is the sum of the block values that need to be extracted. The objective of extracting block $B_{r,c}$ is to

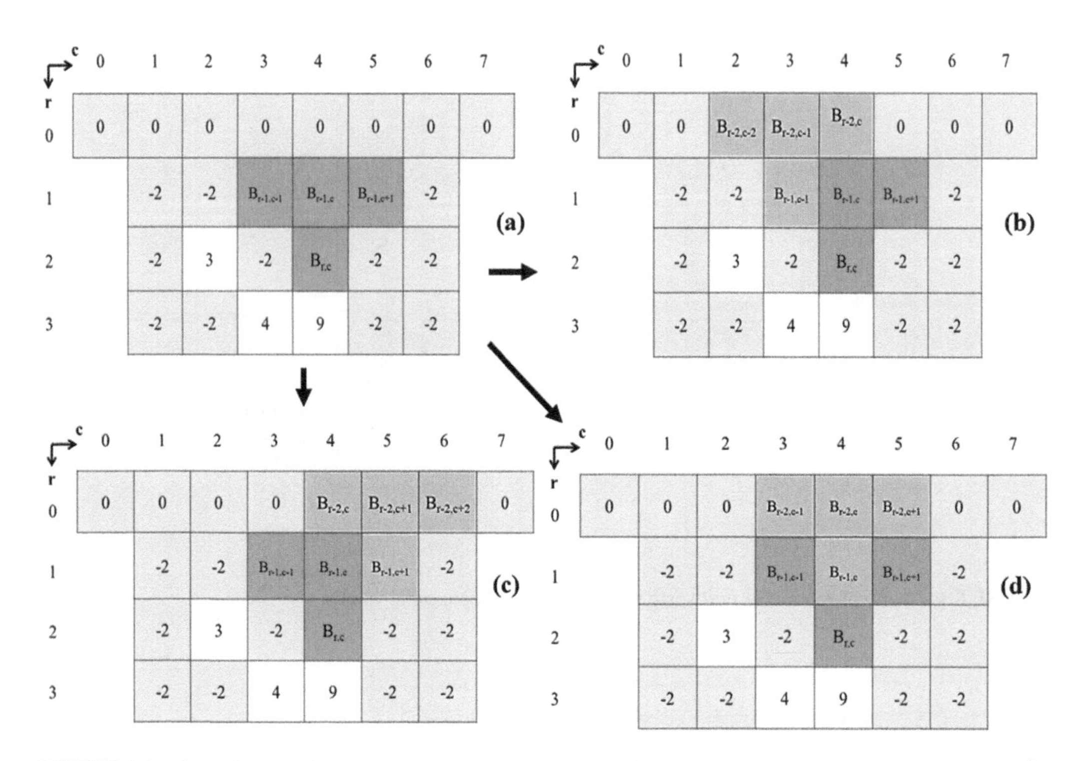

FIGURE 6.8 A vertical section represents a sequence of extractions of blocks: (a) extraction of the block, $B_{r,c}$ requires extractions of three overlying blocks ($B_{r-1,c}$, $B_{r-1,c-1}$, and $B_{r-1,c+1}$), (b) extraction of block, $B_{r-1,c}$ require extractions of three overlying blocks ($B_{r-2,c}$, $B_{r-2,c-1}$, and $B_{r-2,c+1}$), (c) extraction of the block, $B_{r-1,c-1}$ requires extractions of three overlying blocks ($B_{r-2,c-1}$, $B_{r-2,c-2}$, and $B_{r-2,c}$), and (d) extraction of the block, $B_{r-1,c+1}$ require extractions of three overlying blocks ($B_{r-2,c}$, $B_{r-2,c+1}$, and $B_{r-1,c+2}$).

maximize the sum of the block values for a pit. The optimal $P_{r,c}$ value of mining the pit down to and including the block $B_{r,c}$, can be determined as:

$$P_{r,c} = \sum_{r} B_{r,c} + max \begin{cases} P_{r-1,\, c-1} \\ P_{r,\, c-1} \\ P_{r+1,\, c-1} \end{cases} \tag{6.1}$$

It is apparent that Eq. (6.1) is a recursion formula giving a relation applicable to any block in the section. There are three possible options to move from block $B_{r,c}$ as shown in Figure 6.8. Therefore, it is possible to use Eq. (6.1) to derive all the permissible pit boundaries on the section to identify the most optimal pit configuration. The practice is to start from one end of the cross-section (e.g., with the block B_{07}) and systematically work through the block model applying Eq. (6.1) to each block in turn. The determination of the block pit value $P_{r,c}$ for each block is performed during the forward pass in which the blocks are examined in the order: level-by-level within each column, followed by column-by-column in the sections. At the end of the forward pass, each block will be characterized by its pit value $P_{r,c,}$ and an arrow showing its optimum neighbour.

For r = 2, and c = 4, Eq. (6.1) becomes

$$P_{2,4} = \left(B_{0,4} + B_{1,4} + B_{2,4}\right) + max \begin{cases} P_{1,3} \\ P_{2,3} \\ P_{3,3} \end{cases}$$

$$= (0 - 2 + 2) + max \begin{cases} P_{1,3} \\ P_{2,3} \\ P_{3,\,3} \end{cases}$$

That is, for block $B_{2,4}$, three possible movements are possible. The propagation of block extraction can move towards blocks $B_{1,3}$, $B_{2,3}$, or $B_{3,3,}$ depending on the pit value of the blocks.

To determine the pit value for each block, dynamic programming is used. The problem is divided into stages based on the number of columns of blocks. In this case, eight columns are considered, and thus the number of stages is also eight.

Computation of Stage 1

The first stage has only one state to reach the 2nd stage. The pit value for each block of the first stage is determined as

$$P_{00} = \sum_{r=0} B_{r0} + max \begin{cases} P_{r-1,\,c-1} \\ P_{r,\,c-1} \\ P_{r+1,\,c-1} \end{cases}$$

For the first row of blocks, only horizontal and diagonally downward movement is allowed, and the above equation becomes

$$P_{00} = \sum_{r=0} B_{r0} + max \begin{cases} P_{r,\,c-1} \\ P_{r+1,\,c-1} \end{cases} = B_{00} = 0$$

For starting block (B_{00}), no previous pit value exists. The pit values (P_{rc}) of blocks of Stage 1 are shown in Figure 6.9.

r \ c	0	1	2	3	4	5	6	7
0	0 (0)	0	0	0	0	0	0	0
1		-2	-2	-2	-2	-2	-2	
2		-2	2	-2	2	-2	-2	
3		-2	-2	4	9	-2	-2	

Stage 1

FIGURE 6.9 Pit values (P_{rc}) of blocks (represented in the orange color in parenthesis) determined in Stage 1.

Computation of Stage 2

All the blocks in the first column are categorized as waste, and thus economic block values are negative. Blocks $B_{2,1}$ and $B_{3,1}$ cannot be extracted to maintain the 1:1 pit slope. The pit values (P_{r1}) for Stage 2 blocks are as follows:

$$P_{01} = \sum_{r=0} B_{r1} + max \begin{cases} P_{00} \\ P_{10} \end{cases} = B_{01} + P_{00} = 0 + 0 = 0$$

$$P_{11} = \sum_{r=1} B_{r1} + max \begin{cases} P_{0,0} \\ P_{1,0} \\ P_{2,0} \end{cases} = \left(B_{01} + B_{11}\right) + P_{00} = \left(0 - 2\right) + 0 = -2$$

For block B_{11}, only one feasible path (towards block B_{00}) exists as blocks B_{10} and B_{20} are not defined due to negative economic block values.

$$P_{21} = \sum_{r=2} B_{r1} + max \begin{cases} P_{10} \\ P_{20} \\ P_{30} \end{cases} = \left(B_{01} + B_{11} + B_{21}\right) = \left(0 - 2 - 2\right) = -4$$

$$P_{31} = \sum_{r=31} B_{r1} + max \begin{cases} P_{20} \\ P_{30} \\ P_{40} \end{cases} = \left(B_{01} + B_{11} + B_{21} + B_{21}\right) = \left(0 - 2 - 2 - 2\right) = -6$$

The pit value of block B_{40} is also assumed to be zero. The pit values (P_{rc}) of blocks of Stage 2 and arrows to optimum neighbours are shown in Figure 6.10.

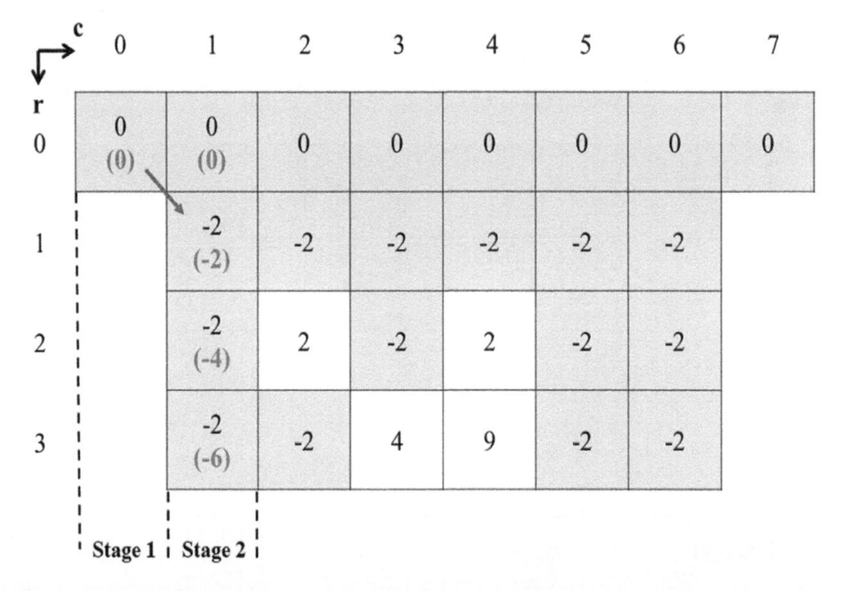

FIGURE 6.10 Pit values (P_{rc}) of blocks (represented in the orange color in parenthesis) determined in Stage 2.

Computation of Stage 3

Unlike Stage 2, the economic block value of all blocks in the third stage (second column) is not negative. Block $B_{3,2}$ cannot be extracted to maintain the 1:1 pit slope. The pit values (P_{r2}) for Stage 3 blocks are as follows:

$$P_{02} = \sum_{r=0} B_{r2} + max \begin{cases} P_{01} \\ P_{11} \end{cases} = B_{02} + P_{01} = 0 + 0 = 0$$

$$P_{12} = \sum_{r=1} B_{r2} + max \begin{cases} P_{01} \\ P_{11} \\ P_{21} \end{cases} = (B_{02} + B_{12}) + P_{01} = (0 - 2) + 0 = -2$$

$$P_{22} = \sum_{r=2} B_{r2} + max \begin{cases} P_{11} \\ P_{21} \\ P_{31} \end{cases} = (B_{02} + B_{12} + B_{12}) + P_{11}$$
$$= (0 - 2 + 3) + (-2) = -1$$

$$P_{32} = \sum_{r=3} B_{r3} + max \begin{cases} P_{21} \\ P_{31} \\ P_{41} \end{cases} = (B_{02} + B_{12} + B_{22} + B_{32}) + P_{21}$$
$$= (0 - 2 + 2 - 2) + (-4) = -6$$

Block $B_{4,1}$ is not defined, and thus no feasibility to move beyond block $B_{3,2}$. The pit values (P_{rc}) of blocks of Stage 3 and arrows to optimum neighbors is shown in Figure 6.11.

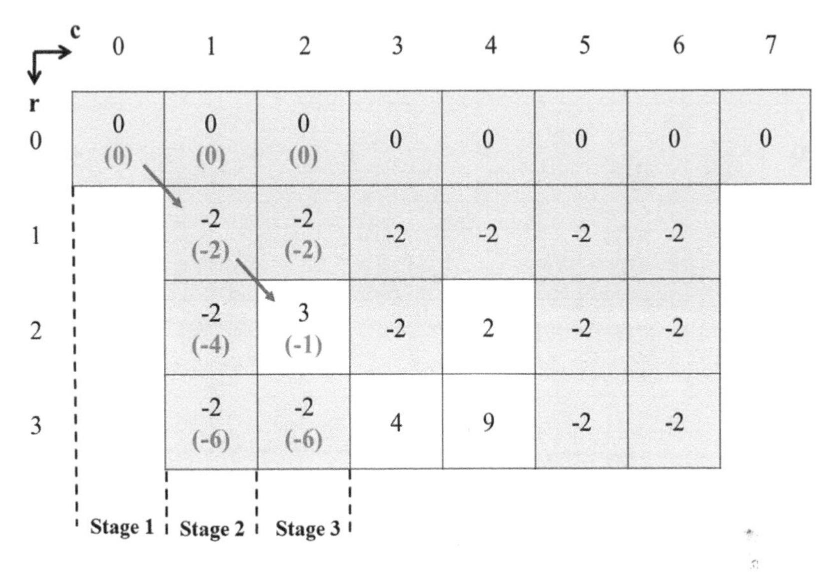

FIGURE 6.11 Pit values (P_{rc}) of blocks (represented in the orange color in parenthesis) determined in Stage 3.

Computation of Stage 4

Similar to Stage 3, the economic block value of a block in this stage (third column) is positive. The pit values (P_{r3}) for Stage 4 blocks are as follows:

$$P_{03} = \sum_{r=0} B_{r3} + max \begin{cases} P_{02} \\ P_{12} \end{cases} = B_{03} + P_{02} = 0 + 0 = 0$$

$$P_{13} = \sum_{r=1} B_{r3} + max \begin{cases} P_{02} \\ P_{12} \\ P_{22} \end{cases} = \left(B_{03} + B_{13}\right) + P_{02} = \left(0 - 2\right) + 0 = -2$$

$$P_{23} = \sum_{r=2} B_{r3} + max \begin{cases} P_{12} \\ P_{22} \\ P_{32} \end{cases} = \left(B_{03} + B_{13} + B_{23}\right) + P_{2,2}$$

$$= \left(0 - 2 - 2\right) - 1 = -5$$

$$P_{33} = \sum_{r=3} B_{r3} + max \begin{cases} P_{22} \\ P_{32} \\ P_{42} \end{cases} = \left(B_{03} + B_{13} + B_{23} + B_{33}\right) + P_{22}$$

$$= \left(0 - 2 - 2 + 4\right) + \left(-1\right) = -1$$

Block $B_{4,3}$ is not defined, and thus no feasibility to move beyond block $B_{3,3}$. The pit values (P_{rc}) of blocks of Stage 4 and arrows to optimum neighbors is shown in Figure 6.12.

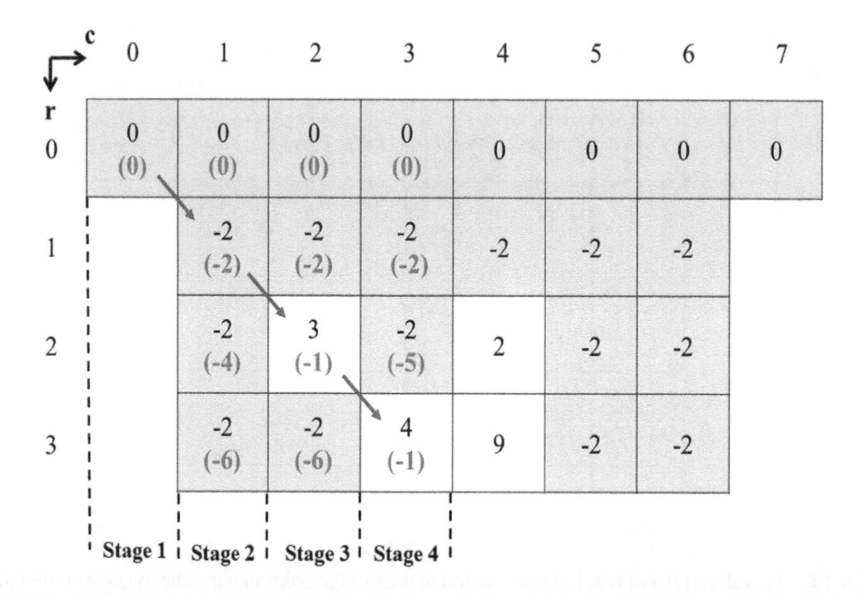

FIGURE 6.12 Pit values (P_{rc}) of blocks (represented in the orange color in parenthesis) determined in Stage 4.

Computation of Stage 5

Unlike other stages, Stage 5 has two blocks with positive economic value. The pit values (P_{r4}) for Stage 5 blocks are as follows:

$$P_{04} = \sum_{r=0} B_{r4} + max\begin{cases} P_{03} \\ P_{13} \end{cases} = B_{04} + P_{03} = 0 + 0 = 0$$

$$P_{14} = \sum_{r=1} B_{r4} + max\begin{cases} P_{03} \\ P_{13} \\ P_{23} \end{cases} = \left(B_{04} + B_{14}\right) + P_{03} = \left(0 - 2\right) + 0 = -2$$

$$P_{24} = \sum_{r=2} B_{r4} + max\begin{cases} P_{13} \\ P_{23} \\ P_{33} \end{cases} = \left(B_{04} + B_{14} + B_{24}\right) + P_{1,3}$$

$$= \left(0 - 2 + 2\right) - 1 = -1$$

$$P_{34} = \sum_{r=3} B_{r4} + max\begin{cases} P_{23} \\ P_{33} \\ P_{43} \end{cases} = \left(B_{04} + B_{14} + B_{24} + B_{34}\right) + P_{23}$$

$$= \left(0 - 2 + 2 + 9\right) + \left(-1\right) = 8$$

Block $B_{4,3}$ is not defined and thus not feasible to move towards this block. The pit values (P_{rc}) of blocks of Stage 5 and arrows to optimum neighbors is shown in Figure 6.13.

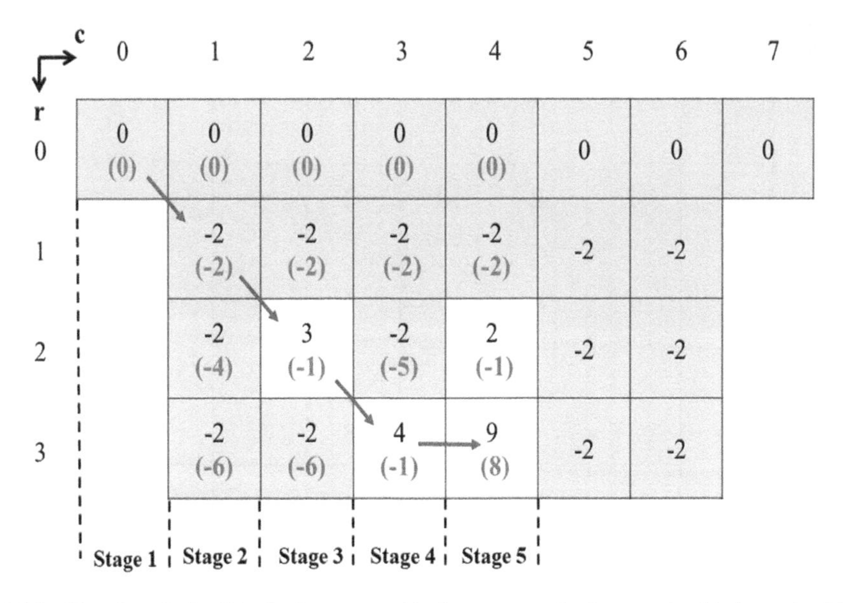

FIGURE 6.13 Pit values (P_{rc}) of blocks (represented in the orange color in parenthesis) determined in Stage 5.

Computation of Stage 6

Similar to Stage 2, the economic value of all blocks is negative. The pit values (P_{r5}) for Stage 6 blocks are as follows:

$$P_{05} = \sum_{r=0} B_{r5} + max \begin{cases} P_{04} \\ P_{14} \end{cases} = B_{05} + P_{04} = 0 + 0 = 0$$

$$P_{15} = \sum_{r=1} B_{r5} + max \begin{cases} P_{04} \\ P_{14} \\ P_{24} \end{cases} = \left(B_{05} + B_{15}\right) + P_{04} = \left(0 - 2\right) + 0 = -2$$

$$P_{25} = \sum_{r=2} B_{r5} + max \begin{cases} P_{14} \\ P_{24} \\ P_{34} \end{cases} = \left(B_{05} + B_{15} + B_{25}\right) + P_{34}$$

$$= \left(0 - 2 - 2\right) + 8 = 4$$

$$P_{35} = \sum_{r=3} B_{r5} + max \begin{cases} P_{24} \\ P_{34} \\ P_{44} \end{cases} = \left(B_{05} + B_{15} + B_{25} + B_{35}\right) + P_{34}$$

$$= \left(0 - 2 - 2 - 2\right) + \left(8\right) = 2$$

Block $B_{4,4}$ is not defined, and thus no feasibility to move towards this block. The pit values (P_{rc}) of blocks of Stage 6 and arrows to optimum neighbors is shown in Figure 6.14.

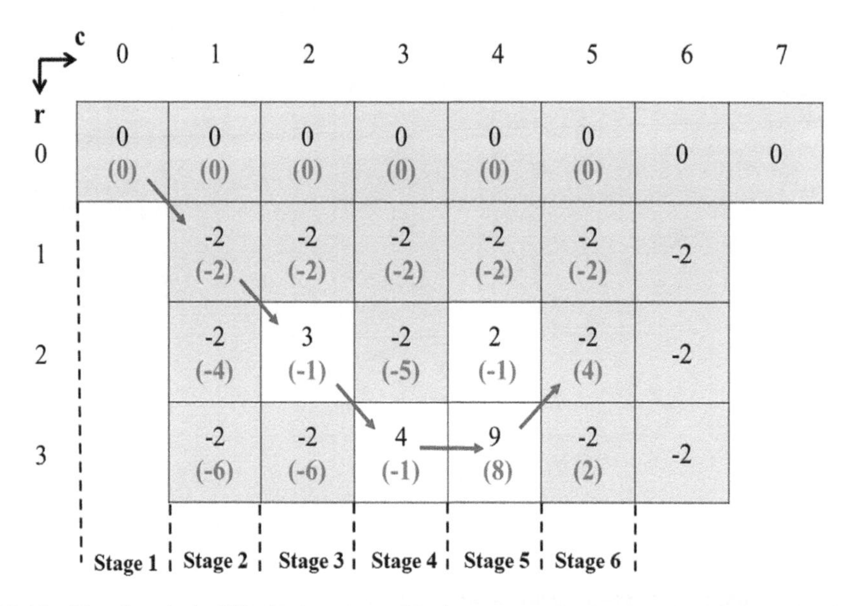

FIGURE 6.14 Pit values (P_{rc}) of blocks (represented in the orange color in parenthesis) determined in Stage 6.

Computation of Stage 7

Similar to Stage 6, the pit values (P_{r6}) for Stage 7 blocks are calculated as follows:

$$P_{06} = \sum_{r=0} B_{r6} + max \begin{cases} P_{05} \\ P_{15} \end{cases} = B_{06} + P_{25} = 0 + 0 = 0$$

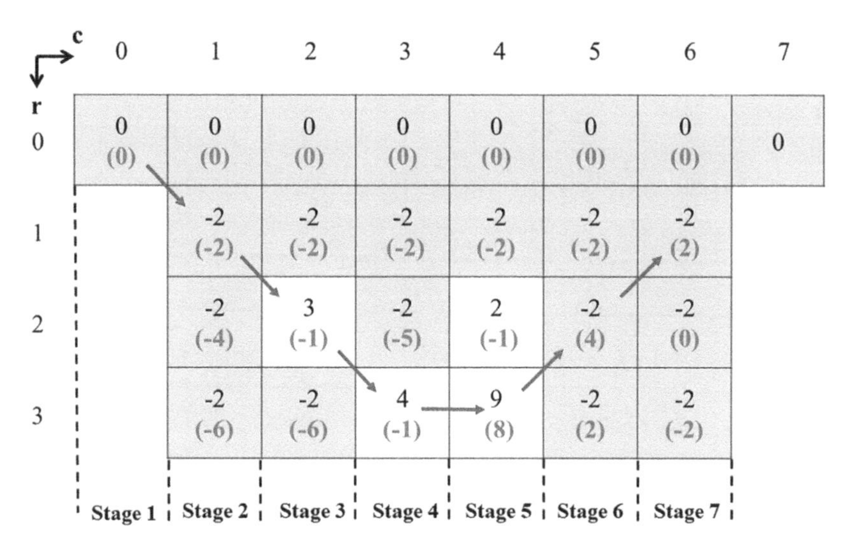

FIGURE 6.15 Pit values (P_{rc}) of blocks (represented in the orange color in parenthesis) determined in Stage 7.

$$P_{16} = \sum_{r=1} B_{r6} + max \begin{cases} P_{05} \\ P_{15} = (B_{06} + B_{16}) + P_{2,5} = (0-2) + 4 = 2 \\ P_{25} \end{cases}$$

$$P_{26} = \sum_{r=2} B_{r6} + max \begin{cases} P_{1,5} \\ P_{2,5} = (B_{06} + B_{16} + B_{26}) + P_{25} \\ P_{35} \end{cases}$$

$$= (0-2-2) + (4) = 0$$

$$P_{36} = \sum_{r=3} B_{r5} + max \begin{cases} P_{25} \\ P_{35} = (B_{05} + B_{15} + B_{25} + B_{35}) + P_{25} \\ P_{45} \end{cases}$$

$$= (0-2-2-2) + (4) = -2$$

Block $B_{4,5}$ is not defined, and thus no feasibility to move towards this block. The pit values (P_{rc}) of blocks of Stage 7 and arrows to optimum neighbors is shown in Figure 6.15.

Computation of Stage 8

Similar to Stage 0, the pit values (P_{r7}) for Stage 8 blocks are calculated as follows:

$$P_{07} = \sum_{r=0} B_{r7} + max \begin{cases} P_{0,6} \\ P_{1,6} \end{cases} = B_{06} + P_{16} = 0 + 2 = 2$$

The pit values (P_{rc}) of blocks of Stage 8 and arrows to optimum neighbours are shown in Figure 6.16.

Thus, the final optimum pit boundary is indicated by a blue thick line in Figure 6.17. The boundary offers the most economical extractions. The updated P_{rc} value of the extreme right-hand block of the arrow (this case block B_{07} in Stage 8) is 2, which is the cumulative value of the ultimate pit of the solution. If we superimpose the same pit in Figure 6.17 on the block model in Figure 6.7 and calculate the sum of the economic value within the pit from the block model,

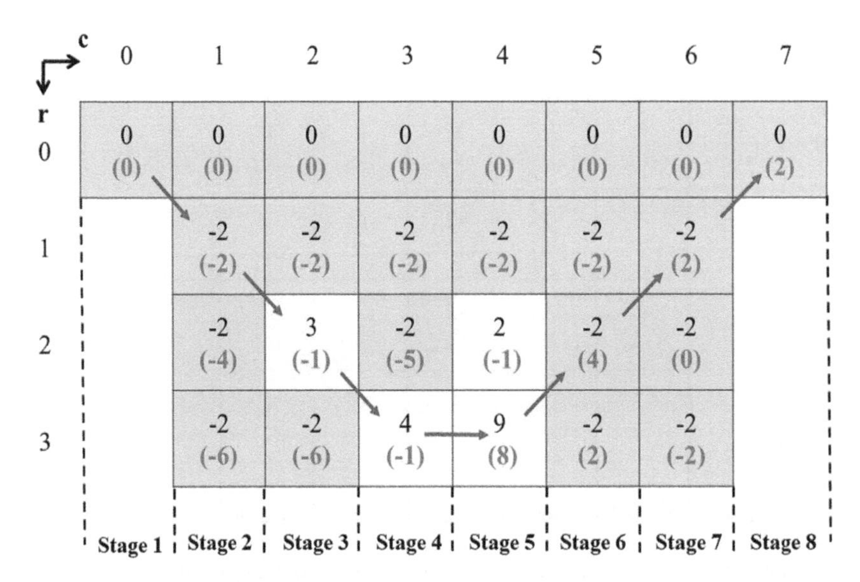

FIGURE 6.16 Pit values (P_{rc}) of blocks (represented in the orange color in parenthesis) determined in Stage 8.

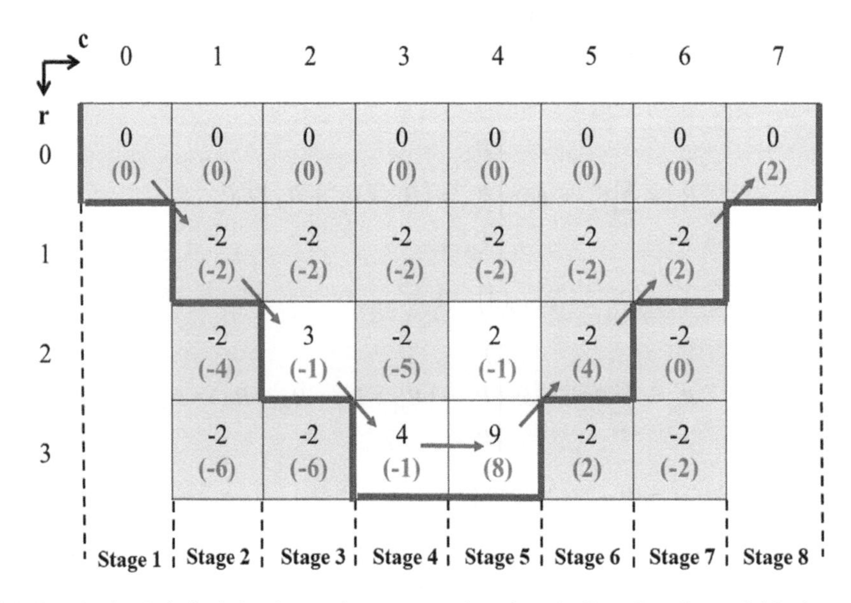

FIGURE 6.17 Optimal pit limit for the maximum return based on the P_{rc} values for each block.

we will see the cumulative value is 2, which is exactly the same as the P_{rc} value of the block where the arrow ends.

6.5 EXAMPLE 3: STOPE BOUNDARY OPTIMIZATION USING DYNAMIC PROGRAMMING

Similar to pit limit optimization, dynamic programming can be used for underground stope boundary optimization. Like other optimization problems, underground stope optimization aims to maximize profit by increasing the revenue from recovered mineral resources and decreasing the extraction costs. Stope boundary optimization is significant for underground mining projects because the optimal stope design ensures optimal production rates of ore with a minimum amount of waste, and it guides the production schedule over the life of the mine. However, developing techniques for stope optimization is a complex task because it requires the satisfaction of many mining constraints while maximizing the overall profit.

The dynamic programming provides a rigorous solution for underground stope optimization. The method is developed for a specific underground mining method, i.e., block caving method, to provide an optimum solution for a two-dimensional (2-D) slice from a three-dimensional (3-D) block economic value model. Riddle (1977) developed a FORTRAN code for stope optimization of a block caving operation using dynamic programming. The stope optimization algorithm is the modification of the pit optimization using dynamic programming. However, there are some basic differences between block cave mining and open-pit mining operations. Open-pit mining starts from the surface; whereas, the block caving operation can begin at any vertical elevation due to a vertical boundary cut-off. The entry point of the stope can vary and can start any column horizontally, not required to start at the first draw point. Figure 6.18 shows an example of block economic value and draw point locations in a block cave mining operation. Another major difference between open pit and block cave mining is that the vertical range of extraction in block cave mining depends on the mining height of the draw point under consideration, as opposed to slope constraint in open-pit mining. For example, in an open pit, in the 2D section of the deposit with a 45° slope angle, only three adjacent blocks (one above, one below, and one on the same level) are considered for calculating P_{ij} value (see Example 2 in Section 6.4). However, the number of adjacent blocks to be considered for P_{ij} calculation in block caving is determined by the row index (elevation) of the block and the percent draw control at the mine. If i is the row index, and d is the percentage of draw control, then the total number of adjacent blocks need to be considered for P_{ij} the calculation is $(2r+1)$, where r is represented as:

$$r = ceil(i * d)$$

Where, $ceil$ is the ceiling value of the product. To illustrate this adjacent block consideration, let us assume that the block is located in row 11 and draw control is 25%. Therefore, the number of adjacent blocks to consider is 7 [$2 * ceil(11 * 0.25) + 1 = 2 * 3+1$]. Figure 6.19 shows seven adjacent blocks that need to consider for P_{ij} calculation for a block in row 11 with 25% draw control in block cave mining.

The fundamental difference in implementing the DP algorithm for block caving requires allowing the entry and exit points elevation variations. The P_{ij} calculation for block caving works very similarly to the open-pit mining, except it starts from the draw point levels and goes upward, as opposed to starts at the surface and goes downward in the open-pit mine. The limit of the exit point is obtained based on the maximum P_{ij} value. If P_{mn} is the maximum P_{ij} value then exit point will be at draw point n and vertical lift of m.

In addition, to allow the horizontal variation of the starting block for the block caving, P_{ij} needs to be calculated by changing the starting block location. To handle this phenomenon of block caving in dynamic programming, an additional subscript s is introduced for P_{ij} calculation. The profit

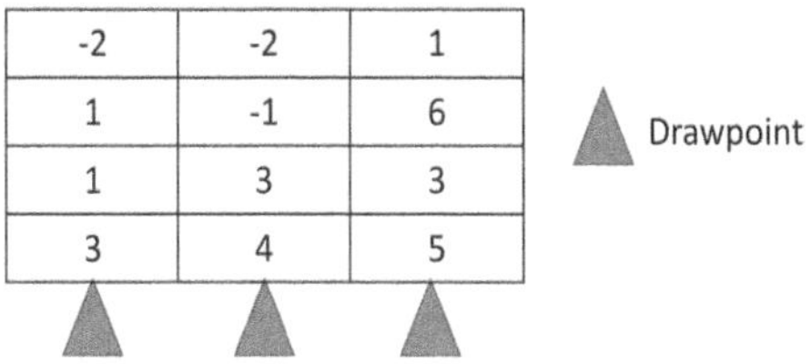

-2	-2	1
1	-1	6
1	3	3
3	4	5

▲ Drawpoint

FIGURE 6.18 Example of block caving stope with economic block value (number inside the block represents block economic value) and draw points.

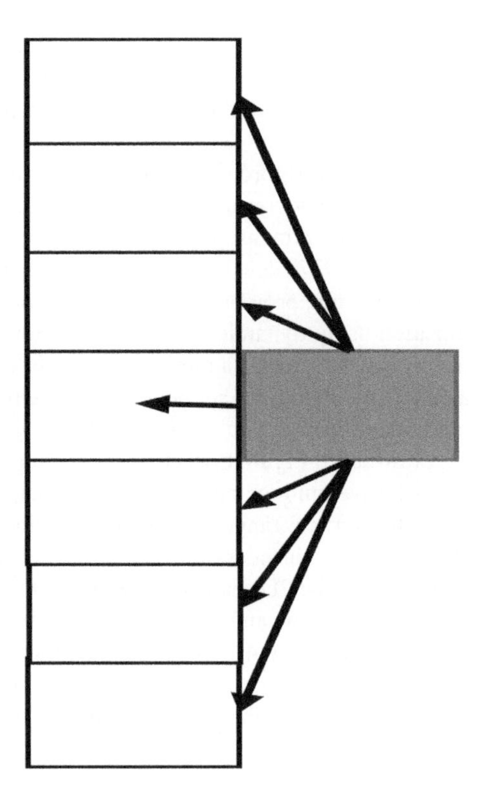

FIGURE 6.19 Seven adjacent blocks (white blocks) of a block under consideration (green block) that is located in row 11 with a draw control of 25%.

calculation for block (*i, j*) is P_{ijs} for the block in row *i* of the column (draw point) *j*, and starting at any level of the draw point *s*. The algorithm can be summarized as below:

If the 2-D section of the block economic value model has *I* rows and *J* columns (4 rows and 3 columns, for example, Figure 6.18), the dynamic programming for stope optimization has the following steps:

Step 1: Calculate P_{ij1} for all blocks of *I* rows and *J* columns within the 2-D section, considering that the mine is starting at any level of draw point 1. This is very similar to the open-pit example, as explained before.
Step 2: Eliminate all blocks from column 1, and calculate P_{ij2} for all remaining blocks assuming mine starts at any level of draw point 2.
Step 3: Repeat step b until $s=J$.
Step 4: Determine $\mathbf{max}\{P_{kmn}\}$ for $k=1,2,\ldots,I$; $m=1,2,\ldots,J$; and $n=1,2,\ldots,j$.

Figure 6.20 shows the illustration of stope optimization using example data of Figure 6.18. The percentage of draw control is used at 25%. Therefore, with the maximum level of 4, in this example, the *r* value will be 1. Figure 6.20(a) shows the block economic values with column and row numbers included. It can be seen; unlike open-pit problem, the row number starts from the bottom of the deposit. Moreover, there are no zero rows included on the problem at hand. The reason is that in open-pit mining, the entry and exit point to the pit needs to be on the surface level, which is not

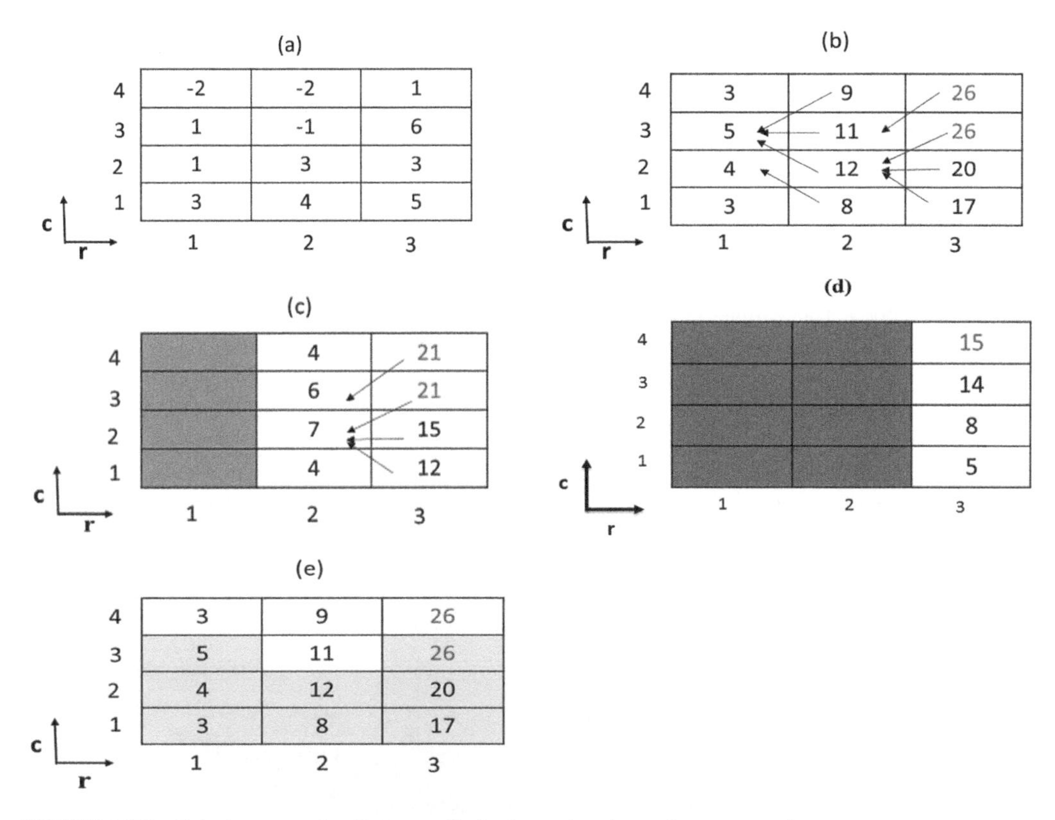

FIGURE 6.20 Solution example of stope optimization using dynamic programming.

the necessary criteria for block cave mining. Figure 6.20(b) shows the updated P_{ijs} value of all the blocks considering that the mine is starting at any level of draw point 1. Here the problem has been solved stages, similar to open-pit mine, and solved the first draw point level before solving the second draw point level, and so on. It shows the maximum cumulative block economic value is 26 when mining starts at any level of draw point level 1. It can also be observed that there are two different stope contours that provide exactly the same cumulative block economic value. It does not matter which pit stope we will choose; we will eventually get the same economic value. Figure 6.20(c) shows the updated P_{ijs} value of all the blocks considering that the mine is starting at any level of draw point 2. For solving this, all the block information at draw point 1 are eliminated, and P_{ijs} values are calculated. It is observed that the maximum cumulative P_{ijs} value is 21, and there are two stope contours that provide the same maximum cumulative values. Similar, the P_{ijs} value of all the blocks was calculated considering the mine starts at any level of draw point 3 and presented in Figure 6.20(d). It is observed the maximum cumulative value is 15. In this example, it can be seen the maximum cumulative values are 26, 21, and 15, respectively, by changing the starting location at draw points 1, 2, and 3. Out of these three maximum cumulative values, 26 is the maximum value that belongs to the solution when the starting location is at draw point 1. Therefore, the solution for the starting mining location at draw point 1 is the optimum solution for the problem. Figure 6.20(e) shows the final stope contour of the problem highlighting the blocks that are within the stope. If we sum up the block value within the stope contour from Figure 6.20(a), we will observe that some of the value is also 26.

6.6 CASE STUDY OF DYNAMIC PROGRAMMING APPLICATIONS TO DETERMINE THE ULTIMATE PIT FOR A COPPER DEPOSIT

The case study was carried out to calculate the ultimate pit limit of a copper-cobalt mine using dynamic programming. The mine is located in Central Africa. The deposit is a stratiform copper-cobalt deposit. Two principal stratiform mineralized horizons are responsible for creating this orebody. The deposit can be followed on the surface for approx. 1400 m and to a maximum vertical depth of 250 m. The deposit dips 30°–45°, but near the surface, the deposit folds on itself. The footwall consists of breccia and cuts across the orebody. The main orebody consists of Oxide, Mixed, and Sulphide body. The primary copper mineral is chalcopyrite.

A total of 462 borehole data are available from the historical records and recent exploratory drillings campaigns. The geological database was created in Surpac software using the borehole data. The geological database was used for interpreting ore sections, ore body modelling, and resource estimation. The average drill hole spacing is 50 m. The assay data were composited using the composite length of 1m for resource estimation. Figure 6.21 shows the interpreted orebody model of the deposit, which was prepared by combining the 50 m spaced vertical section lines. The ordinary block kriging algorithm was applied for estimating copper (Cu) in block size 12.5 m × 12.5 m × 5 m. Figure 6.22 shows the block kriging estimate of copper for this deposit. Table 6.2 presents total tonnage, average copper and cobalt grade of oxide, mixed, and sulfide ore zones within the deposit.

FIGURE 6.21 Orebody model of the case study deposit.

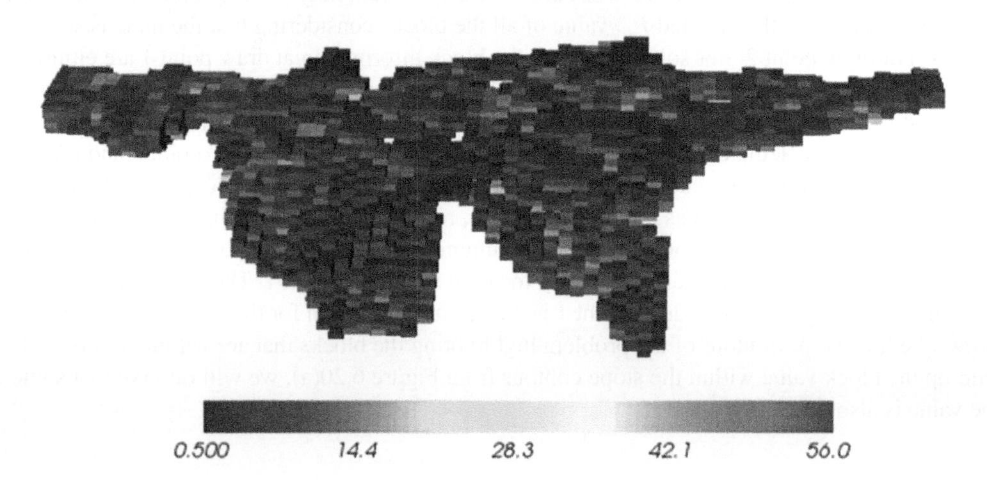

FIGURE 6.22 Block kriging estimate of case study copper deposit.

TABLE 6.2
Estimated resource of the case study copper mine in the different ore types

Ore Type	Tonnes	Average Grade (Cu %)
Oxide	9,933,319	3.37
Mixed	2,269,610	2.63
Sulphide	9,240,742	2.95

TABLE 6.3
The economic and slope parameters using in the optimization

Parameter	Sub-parameters	Units	Value
Pit Slope	–	Degree	45°
Specific gravity	–	t/m³	2.7
Mining	Mining Costs	US$/t mined	1.20
	Mining Recovery	%	100%
Process	Process Costs	$/t processed	11.1
	Recovery Cu	%	98%
Selling Price	Cu	$/lb	3.2
Selling Cost	Cu	$/lb	0.4

The pit limit optimization of the copper deposit is performed to identify the set of blocks from the deposit that will be mined over the mine life. Pit limit optimization aims to maximize profit by making a decision about several hundreds of thousands of mining blocks, whether to mine or not, subject to certain sets of constraints (Chatterjee and Dimitrakopoulos, 2020). For making the decision of which block to mine, an economic block model was calculated from the geologic grade model. The economic block value (c) of the block was calculated using the following equations:

$$c = \begin{cases} Net\,revenue - MC - PC, & \text{if } Net\,revenue > PC \\ -MC, & \text{Otherwise} \end{cases} \tag{6.2}$$

where

$$Net\ revenue = T * G * REC * (P - S)$$

MC is the mining cost, PC is the processing cost, T is the tonnage, G is the grade of Cu, REC is the recovery of Cu, P is the selling price of Cu, and S is the selling cost of Cu. To calculate the block economic value, economic parameters from Table 6.3 were used.

The ultimate pit limit problem is an integer programming (IP) formulation problem (Paithankar et al., 2020). The formulation of the problem can be written as:

$$Z = \text{Max} \sum_{i=1}^{N} c_i\, x_i \tag{6.3}$$

$$\text{subject to} \quad x_i - x_j \le 0,\ j \in \Gamma_i,\ i \in N \tag{6.4}$$

$$x_i \in \{0,1\},\ i \in N \tag{6.5}$$

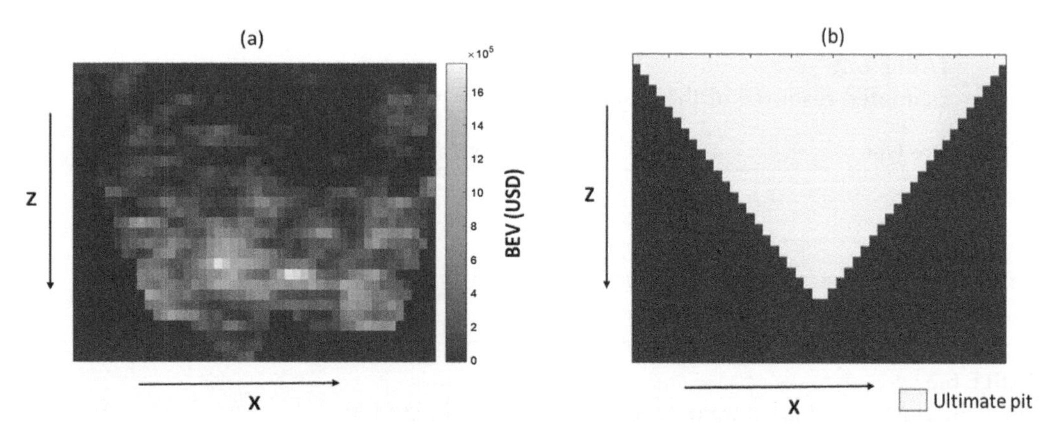

FIGURE 6.23 (a) Economic block value of the section along Easting 900 m; and (b) ultimate pit using dynamic programming for the same section.

Constraints of Eq. (6.4) are slope constraints, which ensure all overlying blocks are mined before mining an underlying block. Constraints of Eq. (6.5) are binary constraints, which ensure that blocks will be extracted or remain in the block model. Since the optimization formulation has only slope constraints; the optimization problem can be solved efficiently using a dynamic programming algorithm (Riddle, 1977). However, it is noted here that dynamic programming can provide an optimum solution for a 2-D problem. Therefore, the dynamic programming needs to apply in each vertical section of the deposit, and then finally stitch together the solutions from each section to get a 3-D ultimate pit. It is noteworthy to mention here that although the individual vertical solution using the dynamic programming is optimum, it is not necessary that 3-D solution after stitching individual solution from each section will be optimum. Moreover, the generated 3-D stitching solution from optimum 2-D solutions can violate the slope constraints for the 3-D geometry. It should also be noted here that for generating a 3-D optimum solution of the ultimate pit limit. However, the purpose of this case study is to show the application of dynamic programming for solving the real-life mining problem.

For calculating the ultimate pit using dynamic programming, the resource block model from Surpac was used. The block model was exported as a string file, which was then formatted in Excel software for using in dynamic programming. The block economic value calculation and dynamic programming implementation were performed in the Matlab software. The step-by-step process of the implementation was explained in Appendix 6. The ultimate pit limit of each section was calculated using dynamic programming, where sections are East-West direction with 50 m section interval. The economic block value and ultimate pit limit along section Easting 900 of the case study mine are presented in Figure 6.23. After combining all sections, it was observed that within the ultimate pit, the total amount of ore is 9.1 Mt at an average Cu grade of 2.03%. The average stripping ratio was calculated as 2.22 bcm: 1 tonne.

APPENDIX 6

This appendix demonstrates how to use Matlab for ultimate pit limit calculation using dynamic programming. The copper data that were used in the case study section of the dynamic programming were used for the demonstration. The data were saved in CSV files. The exported block model from Surpac is saved in CSV file 'block_cu_output0.csv'. The screenshot of the data file is shown in Figure A6.1. The first three columns represent X, Y, Z of the block centroid. Fourth column

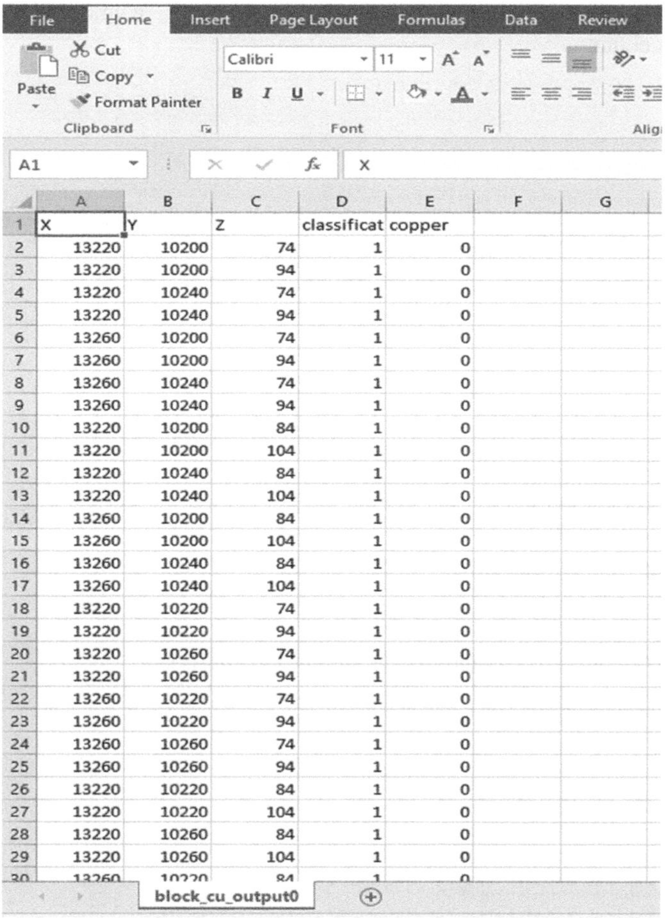

X	Y	Z	classificat	copper
13220	10200	74	1	0
13220	10200	94	1	0
13220	10240	74	1	0
13220	10240	94	1	0
13260	10200	74	1	0
13260	10200	94	1	0
13260	10240	74	1	0
13260	10240	94	1	0
13220	10200	84	1	0
13220	10200	104	1	0
13220	10240	84	1	0
13220	10240	104	1	0
13260	10200	84	1	0
13260	10200	104	1	0
13260	10240	84	1	0
13260	10240	104	1	0
13220	10220	74	1	0
13220	10220	94	1	0
13220	10260	74	1	0
13220	10260	94	1	0
13260	10220	74	1	0
13260	10220	94	1	0
13260	10260	74	1	0
13260	10260	94	1	0
13220	10220	84	1	0
13220	10220	104	1	0
13220	10260	84	1	0
13220	10260	104	1	0

block_cu_output0

FIGURE A6.1 A file structure for preparing block model data for dynamic programming model.

represents the classification of the mineralized zone (0 represents outside the main orebody, and 1 represents within the orebody), and the fifth column represents copper grade. Due to the confidentiality and Memorandum of Understanding (MoU), the name, locations, and actual data were not disclosed. However, anyone can prepare the data file from the resource model and saved it in the CSV file format.

Loading data

Please use the data set block_cu_output0.csv. Matlab has a function to read the CSV data, but it can only read the numeric value. If the data has a header row, it is not recommended to use the readcsv function; instead, we can use 'readtable' function.

This code will bring the CSV data to Matlab

```
T = readtable('block_cu_output0.csv', 'HeaderLines',1);
```

Now, we can convert the table to the matrix in Matlab by using table2array function.

```
Cu = table2array(T);
```

Column 4 does not have useful information for optimization using dynamic programming. Therefore, we can delete that column.

```
Cu(:,4)=[];
```

Block economic value calculation

In this section, we will calculate the economic block value of the Copper data. For calculating the block economic value, we need to provide a set of inputs.

The following inputs are needed for block economic value calculation:

Specific gravity (tonne per meter cube)
Recovery (between 0 and 1)
Selling price (dollar per pound)
Selling cost (dollar per pound)
mining cost (dollar per ton)
processing cost (dollar per ton)
Block size along X-axis (in m)
Block size along Y-axis (in m)
Block size along X-axis (in m)

We can pass these information to Matlab using the following codes:

```
spgr=2.7; %specific gravity of rock
rec=0.98;    %recovery
slprc=3.9;    % selling price of ore
slcst=0.4;  % selling cost of ore
prcst=11.1;    %processing cost of ore
mncst=1.2;    % mining cost of rock
nx=20;  %size of block in x direction
ny=20;  %size of block in y direction
nz=10;  %size of block in z direction
```

Now we are ready to calculate the block economic value. The below code will calculate the economic block value of the entire deposit.

```
for i=1:length(Cu(:,1))
        y(i)=(((((((nx*ny*nz*spgr)*Cu(i,4)*rec)/100)*2000*
(slprc-slcst))-(nx*ny*nz*spgr)*(prcst+mncst))));
   if y(i)<=0
        Cu(i,5)=-nx*ny*nz*spgr*mncst;
   else
        Cu(i,5)=y(i);
   end
end
```

Preparing data for dynamic programming

After block economic value calculations, we need to sort data so that we can generate sections along the Z direction. The below code will sort the data:

```
P=sortrows(Cu,[2 1 3],{'ascend' 'ascend' 'descend'});
```

Before generating vertical sections, another piece of information we need is the number of blocks along X, Y, and Z directions in the block model. This information can be obtained from the block model summary of the original block model data in the mining software. For the copper data, we know that it has the following number of blocks along X, Y, and Z directions:

Along X: 46
Along Y: 16
Along Z: 29

So, we will be creating the column vector of block economic value to the 3D block model and then take a slice for generating 2D sections. The individual 2D section will be used for dynamic programming.

The following code is used for creating the 3D blocks for section generation from the column vector input data:

```
nx=46;
ny=16;
nz=29;
Mat = reshape(P(:,5),nz,nx,ny);
```

Running dynamic programming in 2D

Well, now our block economic value data are ready for dynamic programming. Let us visualize the block economic value of one section (Figure A6.2).

```
% We are looking at Section 5
imagesc(Mat(:,:,5))
```

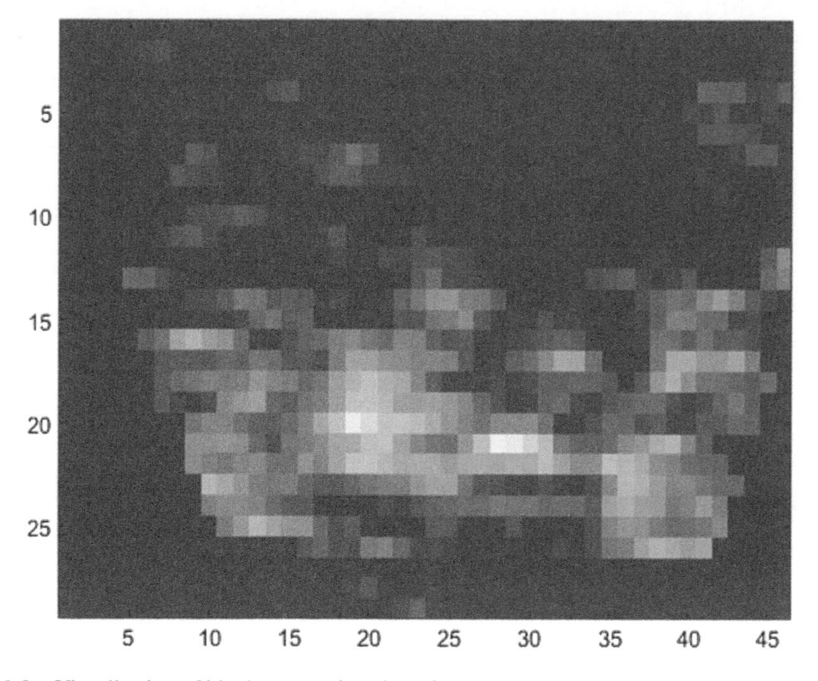

FIGURE A6.2 Visualisation of block economic value of one section.

Now let us run the dynamic programming for one section. However, before running, we need to make sure the block, which cannot be mined, should artificially be set to waste block. The following code will do that job for section 5 of the data set:

```
% Let's do it for section 5
A=Mat(:,:,5);
min_val=min(A(:));
[m,n]=size(A);
ceil_val=ceil(n/2);
floor_val=floor(n/2);
for i=1:floor_val
     A(i+1:end,i)=min_val;
     A(i+1:end,n-i+1)=min_val;
end
```

Now the data is ready, and we can get the ultimate pit for that section. The below code will be solving the ultimate pit using dynamic programming and will display the pit outline.

```
B=[zeros(1,n);A];
C=cumsum(B);
D=C;
E=[];
for i=2:n
     for j=2:m
     k=[D(j-1,i-1) D(j,i-1) D(j+1,i-1)];
     D(j,i)=D(j,i)+max(k);
     id=find(k==max(k));
     if id==1
        E=[E;i j i-1 j-1];
     elseif id==2
        E=[E;i j i-1 j];
     else
        E=[E;i j i-1 j+1];
     end

   end
   if j==m+1
       k=[D(j-1,i-1) D(j,i-1)];
       D(j,i)=D(j,i)+max(k);
       id=find(k==max(k));
         if id==1
           E=[E;i j i-1 j-1];
         else
           E=[E;i j i-1 j];
         end
   end
end
Pitvalue=max(D(2,:));
id=find(D(2,:)==Pitvalue);
f=[id 2];
F=f;
ind=100;
while (ind>2)
     k=find(E(:,1)==f(1) & E(:,2)==f(2));
     F=[F;E(k,3:4)];
```

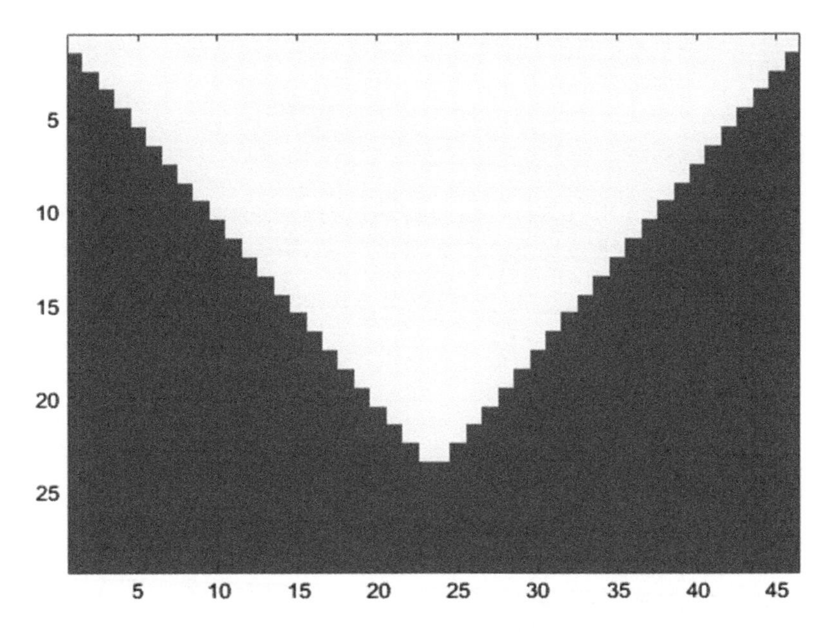

FIGURE A6.3 Ultimate pit for 2D model derived using dynamic programming.

```
        ind=E(k,4);
        f=E(k,3:4);
    end
    pit=zeros(m+1,n);
    for j=1:length(F)
        pit(1:F(j,2),F(j))=1;
    end
    pit(1,:)=[];
    imagesc(pit)
```

Running dynamic programming in 3D

Now, we can put all these codes together and run for all sections one after another within a single code. We can create a separate Matlab function 'dynamic_prog' to do that job. The function looks like this:

```
function [U_Pit]=dynamic_prog(Mat)
[a,b,c]=size(Mat);
U_Pit=[];
for ik=1:c
A=Mat(:,:,ik);
min_val=min(A(:));
[m,n]=size(A);
if min_val == max(A(:))
    pit=zeros(m,n);
else
floor_val=floor(n/2);
for i=1:floor_val
    A(i+1:end,i)=min_val;
    A(i+1:end,n-i+1)=min_val;
end
```

```
B=[zeros(1,n);A];
C=cumsum(B);
D=C;
E=[];
for i=2:n
     for j=2:m
        k=[D(j-1,i-1) D(j,i-1) D(j+1,i-1)];
        D(j,i)=D(j,i)+max(k);
        id=find(k==max(k));
        if id==1
           E=[E;i j i-1 j-1];
        elseif id==2
           E=[E;i j i-1 j];
        else
           E=[E;i j i-1 j+1];
        end

     end
     if j==m+1
        k=[D(j-1,i-1) D(j,i-1)];
        D(j,i)=D(j,i)+max(k);
        id=find(k==max(k));
          if id==1
           E=[E;i j i-1 j-1];
          else
           E=[E;i j i-1 j];
          end
        end
end
Pitvalue=max(D(2,:));
if Pitvalue < 0
      pit=zeros(m,n);
else
id=find(D(2,:)==Pitvalue);
f=[id 2];
F=f;
ind=100;
while (ind>2)
      k=find(E(:,1)==f(1) & E(:,2)==f(2));
      F=[F;E(k,3:4)];
      ind=E(k,4);
      f=E(k,3:4);
end
pit=zeros(m+1,n);

for j=1:length(F)
     pit(1:F(j,2),F(j))=1;
end
pit(1,:)=[];
end
end
U_Pit(:,:,ik)=pit;
end
```

The function needs to be saved in the working directory or within the Matlab path. Once we have the function saved, we can call that function and run the code.

```
[U_Pit]=dynamic_prog(Mat);
```

Generating a file for Surpac or any mine planning software

Now we have a solution from the 3D deposit using dynamic programming. We can export it to Surpac or other mine planning software and visualize it there. However, before doing that, we need to prepare the data.

```
Final_sol=[P(:,1:4) reshape(U_Pit,numel(U_Pit),1)];
```

Once the data is prepared, we can explore the data in the CSV file.

```
csvwrite('final_pit.csv', Final_sol)
```

The CSV file now can easily be imported to the block model in the mining software for visualization and post-processing.

7 Network analysis for mining project planning

7.1 INTRODUCTION

Project management through network analysis involves sequencing tasks or activities and identifying their predecessors appropriately for executing the project in the most efficient way. The network analysis starts with constructing a network diagram, which represents the interdependencies and time requirements of the individual tasks of the project. The individual tasks have two fundamental properties:

- An activity only occurs if and only if all the activities leading into it are complete.
- No activity can start until its tail activity has occurred.

For example, in the network diagram (Figure 7.1), 1, 2, 3, 4, and 5 represents the nodes, A, B, C, D, and E represents the five activities or tasks, and a, b, c, d, and e represents the time to complete the tasks A, B, C, D, and E, respectively. Thus, activity E cannot start until both the activities (C and D) leading into it are completed.

The definitions of a few related terms are given below:

- **Critical path**: The longest path, which takes the longest time to complete, is termed the critical path.
- **Critical tasks or activities**: All the tasks that reside on the critical path are critical tasks of the project.
- **Non-critical path**: All the paths that are not the critical paths are termed non-critical paths.
- **Non-critical tasks or activities:** All the activities that reside on the non-critical paths are non-critical tasks.

7.2 REPRESENTATION OF NETWORK DIAGRAM

In the project network diagram, each activity is represented by an arc connected by two nodes pointing in the direction of the progress of the project. The first node represents the start of the activity, and the subsequent node represents the end of the activity. Thus, the nodes of the network represent the precedence relationship among various activities.

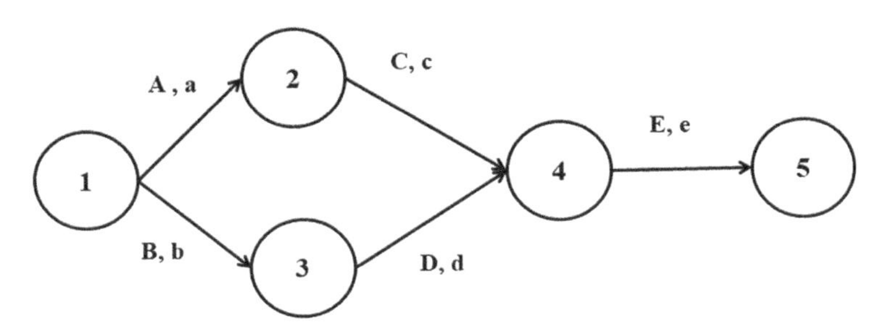

FIGURE 7.1 Sample network diagram.

DOI: 10.1201/9781003200703-7

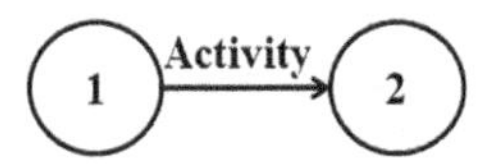

FIGURE 7.2 Activity in a network diagram.

Activity can be defined as any operation which utilizes resources, except dummy, to complete and connect two nodes, as shown in Figure 7.2. Any activity can be classified into four types as follows.

- **Predecessor activity** – Activities that must be completed immediately before the start of another activity are called predecessor activities.
- **Successor activity** – A successor activity is a task that occurs following a predecessor activity. Any successor activity cannot start until all the immediate predecessor activities are completed.
- **Concurrent activity** – Activities that can be accomplished parallel are known as concurrent activities.
- **Dummy activity** – Any activity which does not consume any resource but merely depicts the technological dependence is called a dummy activity.

Once the list of activities and predecessors are identified, the following rules should be followed to construct a project network:

a. The first node (Node 1) represents the start node of the project. Any arc lead from the start node represents activities with no predecessors.
b. The last node represents the completion of the project and should be included with unique value in every project network.
c. Assign the numbers to the nodes in ascending order. That is, the node representing completion of an activity always has a higher number than the node representing start of the activity.
d. An activity should not be represented by more than one arc or two nodes should not be connected by more than one arc.
e. Each activity must be identified by two distinct nodes.

Figure 7.3(a)–(d) shows the network construction with different predecessor and successor relationships. To avoid violation of rules (d)–(f), a dummy activity with zero duration (represented by a dotted arc in Figure 7.3(d)) may be introduced. The dummy activity can be introduced in different ways to represent the concurrent activities, as shown in Figure 7.3(d).

7.3 METHODS OF DETERMINING THE DURATION OF A PROJECT

7.3.1 Critical Path Method (CPM)

CPM-based project network analysis was first introduced by Du Pont (1957) to optimize the cost of the project and its overall completion time. The proposed method was applied first time in 1958 in the construction of a new chemical plant. CPM is used to assist the project manager in scheduling the activities involved in a typical project.

CPM-based network analysis is conducted to determine the total duration required to complete a project and identification of the critical and non-critical activities involved in the project. In a complete network of the project, the path with the maximum elapsed time from the start to the end node is called the critical path, and the activities on it are called critical activities. It is possible to have multiple critical paths if there is more than one path with the longest duration. CPM is a deterministic approach, where the durations of the various activities are assumed to be known with certainty.

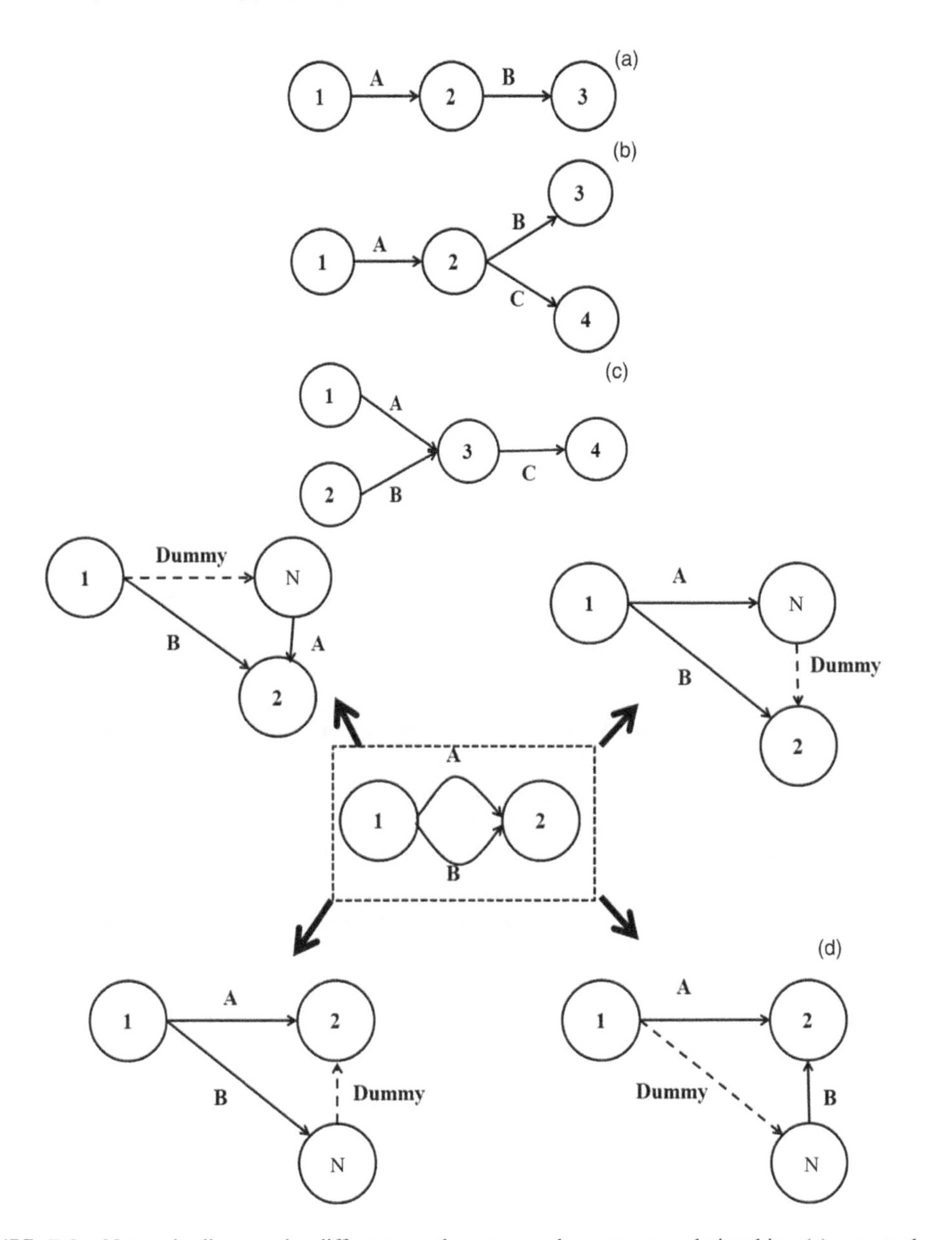

FIGURE 7.3 Network diagram in different predecessor and successor relationship: (a) one task is a predecessor of another task, (b) one task is a predecessor of two other tasks, (c) two tasks are the predecessor of another task, and (d) representation of two concurrent activities.

An activity is critical if there is no flexibility in estimating the start and finish time. On the other hand, a non-critical task always facilitates some flexibility or slack in scheduling the start and finish time without delaying the overall completion of the project.

The methodology of the CPM-based network analysis is explained with the following example.

Let,

ES_{A_k} – Earliest start time of task/activity A_k

LS_{A_k} – Latest start time of task/activity A_k

t_k – Duration of activity A_k

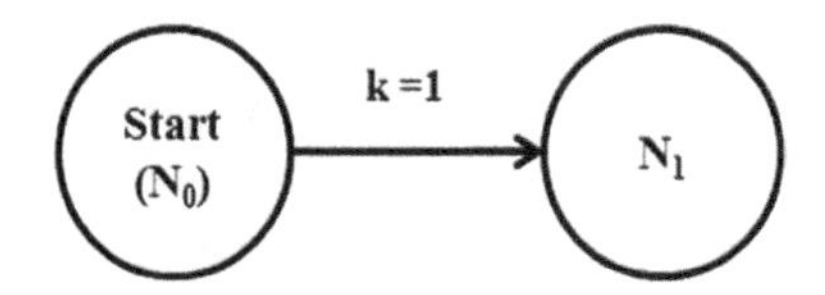

FIGURE 7.4 Path represents the first activity by connecting start node to subsequent node.

The earliest and latest occurrences of task A_k are defined with respect to the start and completion time of the project. The critical path calculation involves two passes, viz. forward pass, and backward pass. The earliest start times of the activities are calculated using the forward pass, whereas the latest start times of the activities are estimated in the backward pass.

Computation of earliest start (ES) time earliest finish (EF) time during the forward pass

Let the project start time of the starting node, N_0 (shown in Figure 7.4) is equal to zero. Then, the earliest start time of the first activity, $\theta_{k=1}$ is equal to 0.

Let r- numbers of activities are the immediate predecessor of activity A_k (shown in Figure 7.5), The earliest start time of activity, A_k, can be determined as

$$ES_{Ak} = Max\left[\left(EF_{A_1} + t_1\right), \left(EF_{A_2} + t_2\right), \cdots \left(EF_{A_r} + t_r\right)\right]$$

where EF_{A_1}, EF_{A_2}, …, EF_{A_r} are early finish time of activities A_1, A_2, …, A_r, respectively, and t_1, t_2, …,t_r are the duration of activities A_1, A_2, …, A_r, respectively.

The forward pass is completed when the *ES* time for the end node is determined.

After completion of the forward pass, computations of the latest start (LS) time and latest finish (LF) time of each activity are carried out through the backward pass.

Computation of LS time and LF time during backward pass

In the backward pass, the computation starts from the end node, N_e, and finishes at the start node, N_0. Let the end task is A_n. The backward pass start by equating the value of EF time of the end task with the LF time of the end task.

$$LF_{A_n} = EF_{A_n}$$

Let r- numbers of immediate successor activities of A_k(shown in Figure 7.6).

The LS time of activity, A_k, can be determined as

$$LS_{A_k} = Min\left[\left(LS_{A_1} - t_1\right), \left(LS_{A_2} - t_2\right), \cdots \left(LS_{A_r} - t_r\right)\right]$$

The backward pass is finished when the *LS* value for tasks start with node 0 is computed. For the first task, $ES_{A_0} = LS_{A_0}$

Any activity, A_k, connected by two nodes, *i* and *j*, will be treated as a critical activity if it satisfies the following conditions:

$$ES_{A_k} = LS_{A_k}$$

$$ES_{A_k} - ES_{A_{k-1}} = LS_{A_k} - LS_{A_{k-1}} = t_k$$

The above conditions state that the earliest start and latest start times exactly differ by the duration of the activity. All other activities are non-critical activities.

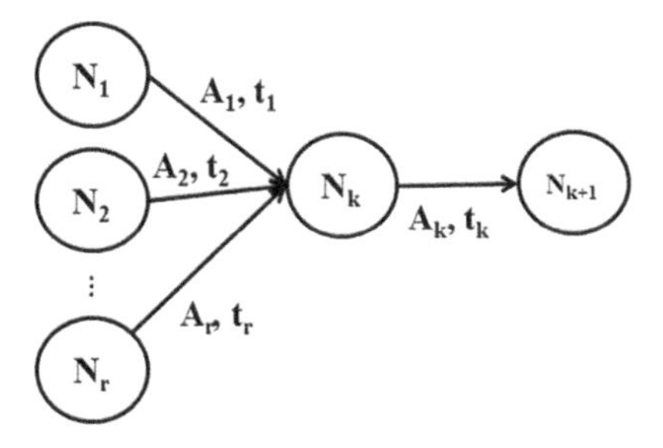

FIGURE 7.5 Earliest start time of activity, A_k in the case of multiple predecessors.

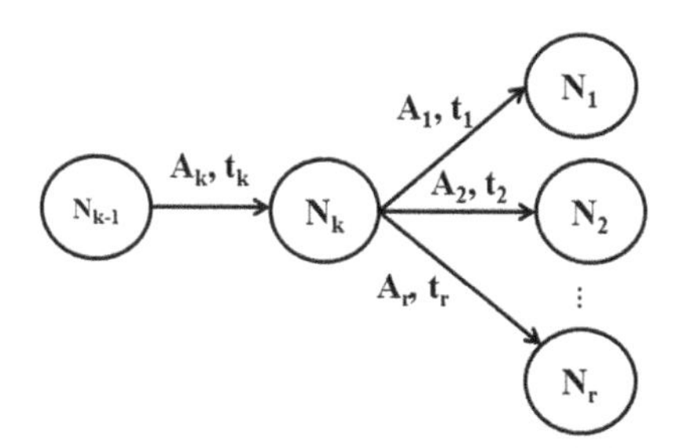

FIGURE 7.6 Latest start time of activity, A_k in the case of multiple successor activities.

Construction of the time schedule

For an activity, A_k, ES_{A_k} represents ES time and LS_{A_k} represents the latest completion time. This indicates that the interval (ES_{A_k}, LS_{A_k}) delineates the maximum span during which activity, A_k may be scheduled without delaying the project.

Determination of floats

Floats are slack times available within the allotted span of the non-critical activities. These are mainly of two types, total float (TF) and free float (FF).

The total float is the amount of time span of the earliest start of any activity, A_k, and the latest start of either of the immediately following activity, A_{k+1}, over the duration, t_k. This also indicates the time an activity can be delayed in completion without affecting the overall completion time of the project.

Mathematically, it can be defined as:

$$\text{Total Float}\left(TF_{A_k}\right) = LS_{A_{k+1}} - ES_{A_k} - t_k = LS_{A_k} - ES_{A_k}$$

$$\Rightarrow \left(TF_{A_k}\right) = LS_{A_{k+1}} - ES_{A_k} - t_k = LS_{A_k} - ES_{A_k}$$

On the other hand, the free float is the excess of the time span defined from the earliest start of an activity, A_k, and the earliest start of any of the immediate following activity, $(k+1)$, over the duration t_k. This indicates that the completion of an activity can be delayed beyond the earliest finish time without affecting the earliest start of subsequent activity.

Mathematically, it can be defined as:

$$\text{Free Float}\left(FF_{A_k}\right) = ES_{A_{k+1}} - ES_{A_k} - t_k$$

$$\Rightarrow \left(FF_{A_k}\right) = ES_{A_{k+1}} - ES_{A_k} - t_k$$

The free float of any task is always less or equal to the total float. That is, $FF_{A_k} \leq TF_{A_k}$

For a non-critical activity, If $FF_{A_k} = TF_{A_k}$, then the activity, A_k, can be scheduled any time in the span of (ES_{A_k}, LS_{A_k}) without any conflict in the time schedule.

Furthermore, if $FF_{A_k} < TF_{A_k}$, then the start of the activity, A_k, can be delayed by at most, FF_{A_k}, relative to its earliest start time, ES_{A_k} without any conflict in the time schedule.

Example 7.1

A mining company, X, has just made the winning bid to start a new mine for extraction of iron ore deposit. The company needs the mine to go into production operation within a year. For this, the company assigned a project manager to complete the construction of a mine office building within a period of 40 weeks. The project manager needs to finish the construction work within the schedule time of 40 weeks. Due to some uncertainty in the availability of labour and raw materials, the manager planned to finish within 47 weeks to save the excessive costs required in expediting the project.

The project manager has to arrange different crews to perform the various activities at different times. The following table listed the activities involved in the construction of the mine office. The duration and predecessor for each activity are listed in Table 7.1. Analyse the project using CPM for determining the critical path, total floats, and free floats.

Determine the following:

(a) Draw the network diagram of the project to visualize the flow of the activities.
(b) Determine earliest start and earliest finish time for each activity for completion of the project without any delay.
(c) Determine the latest start and latest finish of each activity to meet this project completion time.
(d) Determine the project completion time without any delays.
(e) Identify the critical activities which need to be strictly finished within the scheduled time to avoid delaying project completion.
(f) For noncritical activities, how much delay can be allowed without delaying project completion?

TABLE 7.1
Activity list for the opencast project

Activity	Activity Description	Predecessors	Duration (week)
A	Excavation work for the foundation		2
B	Placing the foundation	A	5
C	Fixing the side walls	B	12
D	Fixing the roof	C	2
E	Exterior plumbing work	C	5
F	Interior plumbing work	E	6
G	Put up the exterior siding	D	7
H	Exterior painting	E, G	6
I	Electrical and pipeline work	C	4
J	Fitting the wallboard	F, I	6
K	Fixing the floor tiles	J	8
L	Interior painting work	K	6
M	Fixing the interior /exterior fixtures	H, L	7

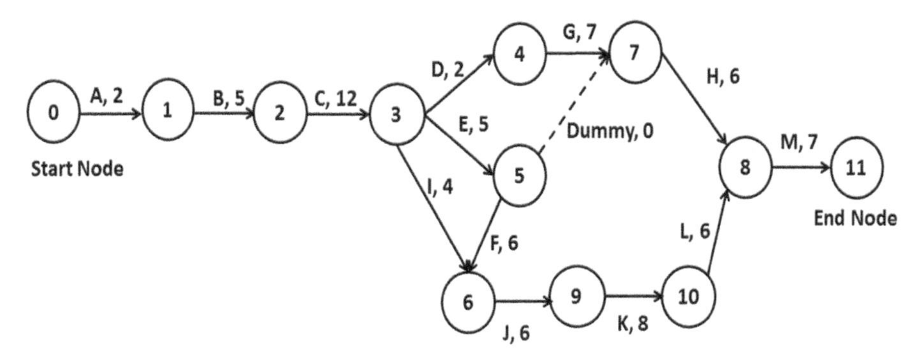

FIGURE 7.7 Network diagram of the project.

Solution

If each activity is completed one by one, the project will be completed in 76 weeks (i.e., summation of each activity time). It is much higher than the deadline of the project. Fortunately, few of the activities can be carried out in parallel, which substantially reduces the project completion time.

(a) The network diagram of the defined project is shown in Figure 7.7. As activity, A, does have any predecessor, it is considered as a starting activity, which is connected by nodes 0 and 1. Activity, A has only one successor activity, B, which is connected by nodes 1 and 2. Similarly, Activity, C, connected by nodes 2 and 3, has only one predecessor task, B. Task, C, has three successor activities (D, E, and I), as shown in Figure 7.7. Moreover, activity F has only one predecessor activity, E, but activity H has two predecessor tasks, E and G. Thus, to comply with this, a dummy activity with zero duration is assigned between nodes 5 and 7. Activity, J has two predecessor tasks, F and I. Activity, J is followed by K, which task L further follows. The last activity, M, is predeceased by two activities, H and L.

(b) The ES and EF for each activity are determined through the forward pass, as follows:

Forward pass to determine the Early Start (ES) and Early Finish (EF) time of each activity involved in the project.

$ES_A = 0;$ $\qquad EF_A = ES_A + t_A = 0 + 2 = 2$

$ES_B = Max\left(EF_A\right) = 2;$ $\qquad EF_B = ES_B + t_B = 2 + 5 = 7$

$ES_C = Max\left(EF_B\right) = 7;$ $\qquad EF_C = ES_C + t_C = 7 + 12 = 19$

$ES_D = Max\left(EF_C\right) = 19;$ $\qquad EF_D = ES_D + t_D = 19 + 2 = 21$

$ES_E = Max\left(EF_C\right) = 19;$ $\qquad EF_E = ES_E + t_E = 19 + 5 = 24$

$ES_F = Max\left(EF_E\right) = 24;$ $\qquad EF_F = ES_F + t_F = 24 + 6 = 30$

$ES_G = Max\left(EF_D\right) = 21;$ $\qquad EF_G = ES_G + t_G = 21 + 7 = 28$

$ES_H = Max\left(EF_E, EF_G\right) = Max(24, 28) = 28;$ $\qquad EF_H = ES_H + t_H = 28 + 6 = 34$

$ES_I = Max\left(EF_C\right) = 19;$ $\qquad EF_I = ES_I + t_I = 19 + 4 = 23$

$ES_J = Max(EF_F, EF_I) = Max(30, 23) = 30;$ $EF_J = ES_J + t_J = 30 + 6 = 36$

$ES_K = Max(EF_J) = 36;$ $EF_K = ES_K + t_K = 36 + 8 = 44$

$ES_L = Max(EF_K) = 44;$ $EF_L = ES_L + t_L = 44 + 6 = 50$

$ES_M = Max(EF_H, EF_L) = Max(34, 50) = 50;$ $EF_M = ES_M + t_M = 50 + 7 = 57$

(c) The LS and LF for each activity are determined through the backward pass, as follows:

Backward pass to determine the Latest Start (LS) and Latest Finish (LF) time of each activity involved in the project.

$LF_M = EF_M = 57;$ $LS_M = LF_M - t_M = 57 - 7 = 50$

$LF_L = Min(LS_M) = 50;$ $LS_L = LF_L - t_L = 50 - 6 = 44$

$LF_K = Min(LS_L) = 44;$ $LS_K = LF_K - t_K = 44 - 8 = 36$

$LF_J = Min(LS_K) = 36;$ $LS_J = LF_J - t_J = 36 - 6 = 30$

$LF_I = Min(LS_J) = 30;$ $LS_I = LF_I - t_I = 30 - 4 = 26$

$LF_H = Min(LS_M) = 50;$ $LS_H = LF_H - t_H = 50 - 6 = 44$

$LF_G = Min(LS_H) = 44;$ $LS_G = LF_G - t_G = 44 - 7 = 37$

$LF_F = Min(LS_J) = 30;$ $LS_F = LF_F - t_F = 30 - 6 = 24$

$LF_E = Min(LS_F, LS_H) = Min(24, 45) = 24;$ $LS_E = LF_E - t_E = 24 - 5 = 19$

$LF_D = Min(LS_G) = 37;$ $LS_D = LF_D - t_D = 37 - 2 = 35$

$LF_C = Min(LS_D, LS_E, LS_I) = Min(35, 19, 26) = 19;$ $LS_C = LF_C - t_C = 19 - 12 = 7$

$LF_B = Min(LS_C) = 7;$ $LS_B = LF_B - t_B = 7 - 5 = 2$

$LF_A = Min(LS_B) = 2;$ $LS_A = LF_A - t_A = 2 - 2 = 0$

(d) The summarized values of ES and EF, LS, LF, and slack values for each activity are shown in Table 7.2.

The EF time of the end task, M, is equal to the LF of the end task. Thus, the project completion time is 57 weeks without any delay.

(e) The critical bottleneck activities are those which exhibit a zero slack value. It can be easily inferred from Table 7.2 that activities A, B, C, E, F, J, K, L, and M have zero slack value.

Critical Path = A-B-C-E-F-J-K-L-M

TABLE 7.2
Computed values of ES, EF, LS, LF, and slack for each activity

Activity	Duration	Early Start (ES)	Early Finish (EF)	Latest Start (LS)	Latest Finish (LF)	Slack = LF-EF= LS-ES
A*	2	0	2	0	2	0
B*	5	2	7	2	7	0
C*	12	7	19	7	19	0
D	2	19	21	35	37	16
E*	5	19	24	19	24	0
F*	6	24	30	24	30	0
G	7	21	28	37	44	16
H	6	28	34	44	50	16
I	4	19	23	26	30	7
J*	6	30	36	30	36	0
K*	8	36	44	36	44	0
L*	6	44	50	44	50	0
M*	7	50	57	50	57	0

Note: *Critical activities having zero slack value

The arithmetic sum of all the critical tasks is equal to the project completed without any delay.

$$\text{Project completion time} = t_A + t_B + t_C + t_E + t_F + t_J + t_K + t_L + t_M$$
$$= 2 + 5 + 12 + 5 + 6 + 6 + 8 + 6 + 7 = 57 \text{ Weeks}$$

Therefore, the delays in the completion time of all the critical activities should be avoided to prevent delays in project completion time.

(f) The tolerable delays for non-critical activities without delaying project completion time can be determined from the free float and total floats, as explained below.

Determination of floats

We have the total float (TF) of activity, A_k is given by

$$TF_{A_k} = LS_{A_{k+1}} - ES_{A_k} - t_k = LS_{A_k} - ES_{A_k}$$

$$TF_D = LS_G - ES_D - t_D = 37 - 19 - 2 = 16$$

$$TF_G = LS_H - ES_G - t_G = 44 - 21 - 7 = 16$$

$$TF_H = LS_M - ES_H - t_H = 50 - 28 - 6 = 16$$

$$TF_I = LS_J - ES_I - t_I = 30 - 19 - 4 = 7$$

The free float (FF) of activity, $A_{k,}$ is given by

$$FF_{A_k} = ES_{A_{k+1}} - ES_{A_k} - t_k$$

$$FF_D = ES_G - ES_D - t_D = 21 - 19 - 2 = 0$$

$$FF_G = ES_H - ES_G - t_G = 28 - 21 - 7 = 0$$

$$FF_H = ES_M - ES_H - t_H = 50 - 28 - 6 = 16$$

$$FF_I = ES_J - ES_I - t_I = 30 - 19 - 4 = 7$$

The total float and free float values indicate the following:

$$FF_D < TF_D, FF_G < TF_G$$

$$TF_H = FF_H, TF_I = FF_I$$

That is, the total floats and free floats for the respective activities, H and I, are equal. However, for activities D and G, the respective free floats are less than the total floats.

The non-critical activities, H and I, can be scheduled any time in the span of ($ES_H = 28$ weeks, $LS_H = 44$ weeks) and ($ES_I = 19$ weeks, $LS_I = 26$ weeks), respectively, without any conflict in the time schedule.

On the other hand, the start of the non-critical activity, G, can be delayed by at most FF_D (= 0 weeks), relative to its earliest start time, $ES_D (= 19\,weeks)$ without any conflict in the time schedule.

Similarly, the start of the non-critical activity, G, can be delayed by at most FF_G (=16 weeks), relative to its earliest start time, $ES_D (= 19\,weeks)$ without any conflict in the time schedule.

The graphical representation of the start time and end time of each activity is represented in Figure 7.8.

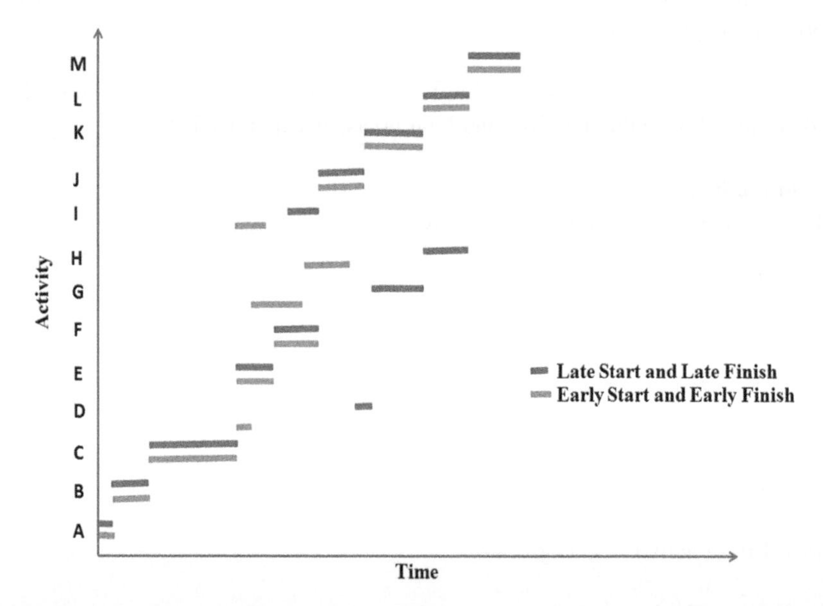

FIGURE 7.8 Schedule of the activities involves in the project.

7.3.2 Program Evaluation and Review Technique (PERT)

PERT was devised in 1958 by the US Navy to optimize the time schedule of a Polaris missile program. Additionally, the method can accommodate the uncertainty involved in the completion time of the activities. PERT is similarly used for project scheduling to that of CPM; however, the duration of the completion of activity follows a probability distribution.

In real-life projects, the duration of some of the activities are non-deterministic for a variety of reasons. It is, thus, necessary to use a probability distribution to represent the stochastic nature of the activity duration. It is assumed that this probability distribution is continuous, unimodal, with two non-negative abscissa intercepts. These intercepts denoted by 'a' and 'b' ($a < b$) are the minimum and maximum time necessary to perform the job. In the context of PERT calculation, 'a' is referred to as the optimistic time, and 'b' is known as the pessimistic time. In addition, the modal value m is commonly referred to as the most likely time used in PERT calculation.

The statistical distribution, which ideally suits these assumptions, is known as a β-distribution (shown in Figure 7.9). For an activity A_k, the expected duration (t_{ek}) and the standard deviation (σ_k) of the duration can be approximated by

$$t_{ek} = \frac{a+b+4m}{6}; \quad \sigma_k = \frac{b-a}{6}$$

where

- a: **optimistic time of completion** of an activity, when all the conditions are favourable for early completion than expected. This is the shortest possible time in which an activity can be completed in ideal conditions.
- b: **pessimistic time of completion** when all the conditions are against the normal working condition. This is the maximum possible time in which everything goes wrong, and abnormal situations prevail.
- m: **most likely time of completion** of an activity, when all the conditions are normal as expected.

7.3.2.1 PERT analysis algorithm

A typical network consists of multiple activities. It is possible to estimate the probability that an activity, A_k, in the network will start by a pre-specified time. Since the completion time of activity, A_k is a random variable and thus the earliest start time of activity, A_k, is also a random variable. It is

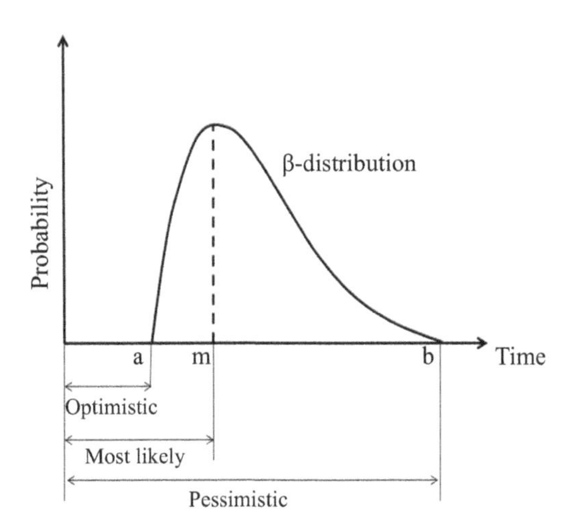

FIGURE 7.9 β-distribution showing the optimistic, pessimistic, and most likely time.

assumed that all the activities in the network are random variables, and thus, the mean and variance of the network need to be determined. If the network consists of only one longest path from the start node to the end node, then the path represents the critical path. The mean of the critical path is the sum of expected durations (t_e) of all the activities in the path, and the variance is the sum of the variances of the same activities. If the network consists of more than one path, the longest path with a wider statistical distribution of the durations is considered the critical path. Once the mean and variance of the duration of the project are obtained, the probability of completion of a project for any random time, t, can be determined as

$$P\left(t \le t_e\right) = P\left[z \le \frac{t - t_e}{\sqrt{var\left(t_e\right)}}\right] = P\left[z \le \frac{t - t_e}{\sigma}\right]$$

In the above equation, z represents a standard normal random variable with mean and standard deviations are 0 and 1, respectively. The reason for consideration of normal distribution is that the expected project completion time, t_e, is the sum of random independent variables, following the central limit theorem.

Example 7.2

A mining project has five tasks (A, B, C, D, and E). The sequence and the time required to finish the individual tasks in days are listed in Table 7.3. Determine the expected project completion time. Also, determine the probability of completing the project within 35 days.

Solution

The network diagram of the defined project with their time estimates is shown in Figure 7.10. The starting node of the network diagram is defined as 1. Two activities (A and B) do not have any predecessors and thus can be started concurrently at time zero. Activity C has only one predecessor, A, as represented in the network diagram. Activity A is also the predecessor of activity D and is thus assigned via a dummy activity. Finally, activity E is predeceased by two activities, C and D.

The expected durations along with the standard deviations of each activity are summarized in Table 7.4.

The network showing the expected duration of each activity is shown in Figure 7.11. The computation of the expected duration of completion of the project will be done using a similar approach as that of CPM.

The ES, EF, LS, LF, and slack values are summarized in Table 7.5.

Thus, the critical tasks involved in the project are A, D, and E, and the expected duration of the project is equal to the sum of the expected duration of the critical tasks. Thus, the expected duration of the project = ($t_{eA} + t_{eD} + t_{eE}$) = 29.1 days.

The critical path is **A→D→E.**

Therefore, the variance of the project completion time (V_T) = Sum of the variance of the critical activities = $\sigma_A^2 + \sigma_D^2 + \sigma_E^2$ = $1.5^2 + 1^2 + 1.5^2 = 5.5$ (days)^2

TABLE 7.3
List of activities and their predecessors and time durations

Activity	Predecessors	Optimistic time (a)	Most likely time (m)	Pessimistic time (b)
A	—	7	10	16
B	—	6	10	15
C	A	4	6	10
D	A, B	6	8	12
E	C, D	8	10	14

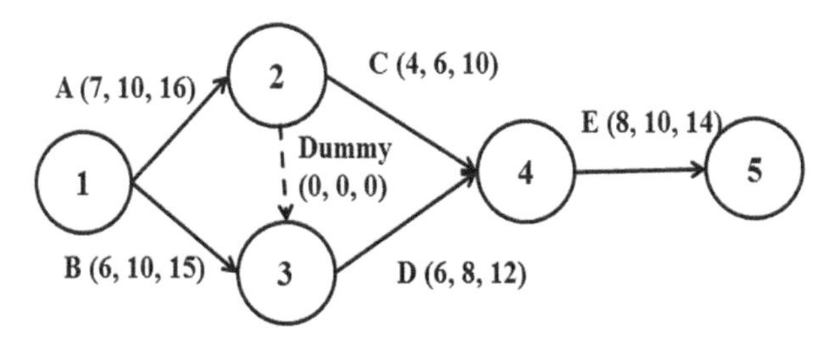

FIGURE 7.10 Network diagram of the defined project.

TABLE 7.4
Expected duration and standard deviation of individual tasks

Activity	Optimistic time (a)	Most likely time (m)	Pessimistic time (b)	$t_e = (a + 4m + b)/6$	$\sigma = (b - a)/6$
A	7	10	16	10.5	1.5
B	6	10	15	10.2	1.5
C	4	6	10	6.3	1.0
D	6	8	12	8.3	1.0
E	8	10	14	10.3	1.5

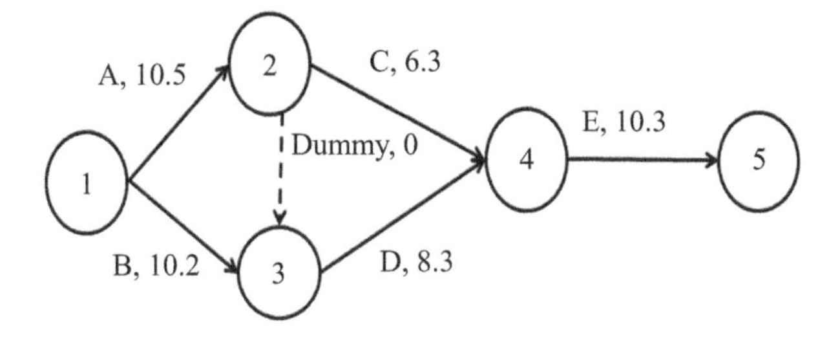

FIGURE 7.11 Network diagram showing the expected duration of each activity.

TABLE 7.5
Summarized values of ES, EF, LS, LF and slack

Activity	t_e	ES	EF	LS	LF	Slack
A*	10.5	0	10.5	0	10.5	0
B	10.2	0	10.2	0.3	10.5	0.3
C	6.3	10.5	16.8	12.5	18.8	2
D*	8.3	10.5	18.8	10.5	18.8	0
E*	10.3	18.8	29.1	18.8	29.1	0

Note: *Critical activity/task

$$\text{Standard deviation of duration of the project } (\sigma) = \sqrt{V_T} = \sqrt{5.5} = 2.345 \text{ days}$$

The area represented by the shaded portion (shown in Figure. 7.12) of the normal distribution represents the probability.

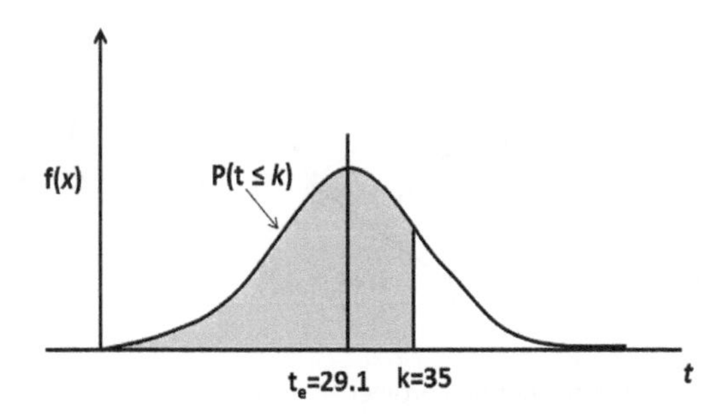

FIGURE 7.12 Distribution of expected duration of project completion.

The probability that the project will finish within 35 days ($t = 35$ days) is given by

$$P(t \le t_e) = P\left[z \le \frac{t - t_e}{\sigma}\right]$$

$$P(t \le 35) = P(t - t_e \le 35 - t_e) = P\left(\frac{t - t_e}{\sigma} \le \frac{35 - t_e}{\sigma}\right) = P\left(z \le \frac{35 - 29.1}{2.345}\right)$$

$$\Rightarrow \quad P(z \le 2.515) = 0.994$$

[Refer to the normal distribution table. The value is corresponding to 2.0 in column and 0.01 in the row.]

7.4 NETWORK CRASHING

It is often desirable to accelerate the project's progress. Thus, to reduce the project duration, the completion time of certain activities needs to be crashed. Crashing an activity refers to reduction in the completion time of the activity below to normal or estimated during through additional resources. These resources include overtime of workers, hiring additional manpower, special materials or equipment, etc. The network planning mechanism can be used to identify the activities whose duration should be crashed so that the completion time of the project can be shortened in the most economical manner. The process is known as network crashing. The crashing of duration should be done along the critical path. If there is more than one critical path, the duration of the activities should be crashed equally in both the paths.

The aim of the time-cost trade-off in the network crashing analysis is to reduce the estimated project duration through critical path analysis in order to meet the pre-defined deadline with the least cost. The reduction in the project duration often helps in the following ways:

- Major resources are released early for use in other projects.
- Working in adverse weather conditions can be avoided, which may affect the mine productivity.
- Improve project cashflow.

The network analysis of time-cost trade-off gives the solution for crash duration estimation of each activity for completion of the project in a pre-defined duration. The project duration can be reduced by adjusting overlaps between activities or by reducing activities' duration. In general, there is an inverse relationship between the time and the direct cost to complete an activity. That is, the less is the investment on the resources (labour, equipment, and material), the larger is the duration to complete an activity. Therefore, the data necessary for determining how much to crash particular

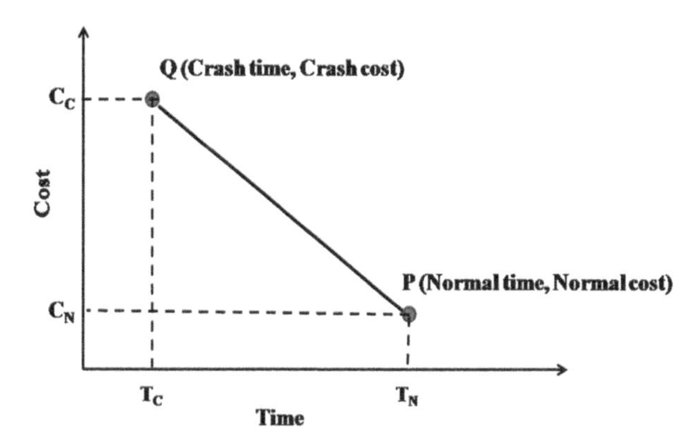

FIGURE 7.13 Typical time-cost graph showing a linear relationship.

activities can be determined from the time-cost relationship. This is demonstrated by a linear relationship, as shown in Figure 7.13.

In Figure 7.13, two points are labelled as P (T_N, C_N) and Q (T_C, C_C), which represents the normal completion time of the project and crash completion time of the project, respectively. The point, P, representing the normal cost and time on the time-cost graph for an activity indicate the time duration and cost of the activity when it is performed in the normal way. On the other hand, the point, Q, representing crash time and cost indicate the time and cost when the activity is fully crashed, i.e., it is fully expedited with no cost spared to reduce its duration as much as possible. These times and costs should be reliably predicted without significant uncertainty. The crashing of a network involves the following steps:

Step 1: Identification of critical path: In the first step, the critical path needs to be identified for the identification of the critical activities. The critical path can be identified either using CPM or PERT, which is an input for network crashing calculation.

Step 2: Determination of cost slopes for each activity in the network.

The cost slopes of all the activities need to be determined from time-cost trade-off relationships. For linear relationships, the crash cost, normal cost, crash time, and normal time data for individual activity can be used to determine the cost slope using Eq. 7.1.

$$\text{Cost slope} = \frac{\text{Crash cost} - \text{Normal cost}}{\text{Normal time} - \text{Crash time}} = \frac{C_C - C_N}{T_N - T_C} \tag{7.1}$$

The cost slope indicates the demand of additional cost per unit crashing of duration to expedite the project. For a non-linear relationship, the functional relationship between the cost and time needs to be estimated.

Step 3: Sort the cost slope in ascending order

All the activities resorted from lowest to highest cost slope. The critical activity having the minimum cost slope has to be crashed first to its minimum duration in such a way that non-critical paths should not become the critical path.

Step 4: Crashing the activity

Crash the critical activities as per the order of minimum cost slope to reduce the critical path duration. The critical activity having the lowest cost slope should be crashed first to the maximum extent possible in order to optimize the project completion cost in the desired time period. It is always desired to make the cost-time trade-off analysis by crashing the activities by one day to track the status of the non-critical paths.

Step 5: Parallel crashing of activities

In the process of crashing the activities in the critical path, other paths may become critical. In this case, activities from both paths should be crashed concurrently with an equal amount to reduce the overall completion time of the project. In the case of parallel crashing, the arithmetic sum of the cost slope of all the parallel activities should be considered for comparison.

Step 6: Determination of total project cost after crashing of activities

Crashing of activities as per steps 4 and 5 should be continued until either the desirable project duration is reached or the maximum possible crashing is achieved. For the different project durations, the total cost is determined by adding the corresponding fixed cost to the direct cost. The direct cost is calculated by adding the expediting crashing cost commutative to the normal cost.

Example 7.3
The activities along with the predecessors, normal and crash completion time, normal and crash costs for constructing canteen in the mine is given below in Table 7.6.

Draw the network diagram. Estimate the critical path and project completion time when the activities are carried out in the normal process. If the indirect cost is US$600/day, determine the time-cost trade-off for the project for reducing the project duration by 2-weeks from the expected completion time.

Solution
The network diagram of the project is shown in Figure 7.14.

The values of ES, EF, LS, LF and slack are summarized in Table 7.7.

$$\text{Critical Path} = \mathbf{A \rightarrow B \rightarrow D_1 \rightarrow E}$$

$$\text{Non-critical paths are } \mathbf{A \rightarrow C \rightarrow E} \text{ and } \mathbf{A \rightarrow D \rightarrow D_2 \rightarrow E}$$

$$\text{Normal completion time of the project} = 16 + 8 + 7 = 31 \text{ days}$$

Lengths of the other two non-critical paths are 29 days and 30 days, respectively.

TABLE 7.6
List of activities with the relevant details

		Normal		Crash	
Job description	Predecessor	Duration (Weeks)	Direct Cost in US$	Duration (Weeks)	Direct Cost (US$)
A: Lay foundation and build walls		16	18000	12	20000
B: Tile Flooring	A	8	4000	5	4900
C: Install electricity	A	6	8000	4	12000
D: Install Plumbing	A	7	4000	5	5800
E: Connect services to finish	B, C, D	7	4000	6	5000

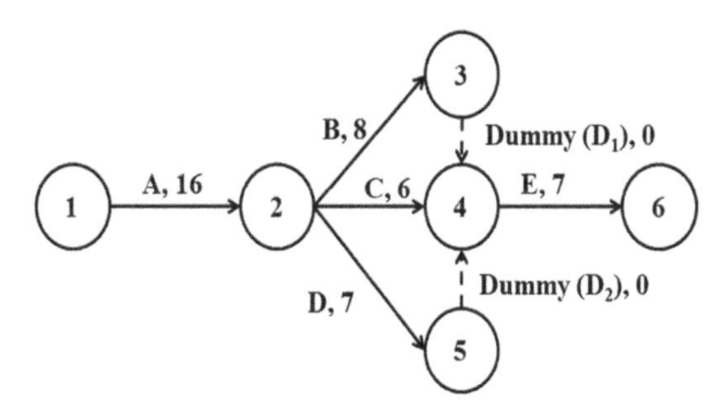

FIGURE 7.14 Network diagram of the defined project.

TABLE 7.7
Summarized values of ES, EF, LS, LF and slack

Activity	Duration (t_e)	ES	EF	LS	LF	Slack
A*	16	0	16	0	16	0
B*	8	16	24	16	24	0
C	6	16	22	18	24	2
D	7	16	23	17	24	1
E*	7	24	31	24	31	0

Note: *Critical activity/task

Normal cost of the project = Direct cost + Indirect cost

= (18000 + 4000 + 8000 + 4000 + 4000) + 31*600

= 38000 + 18600 = US$56600

The cost slope of each activity is determined using the relationship, as demonstrated in Eq. (7.1).

$$\text{Cost slope for activity A } (CS_A) = \frac{C_C - C_N}{T_N - T_C} = \frac{20000 - 18000}{16 - 12} = \frac{2000}{4} = \text{US\$500}$$

Similarly, the cost slopes for all other activities are determined and listed in Table 7.8.

The order of the cost slope is

$$B < A < D < E < C$$

Crashing the project by 1-day

The crashing is done on the critical activities till no other non-critical path becomes critical. If, after crashing, any non-critical path becomes a critical path, then a concurrent crashing should be done to reduce the project cost. The minimum cost slope is on, among critical activities, activity B. Thus, the first target for crashing is activity B. After crashing activity B by 1-day, the duration of the project becomes 30 days, and the revised network of the project is shown in Figure 7.15.

The crash cost of the project is

= Normal cost + Crash cost + Indirect cost

= (18000 + 4000 + 8000 + 4000 + 4000) + 1*300 + 30*600

= US$56300

TABLE 7.8
Cost slopes of each activity

Activity	Normal Time	Crash Time	Normal Direct Cost	Crash Direct Cost	Cost Slope
A*	16	12	18000	20000	500
B*	8	5	4000	4900	300
C	6	4	8000	12000	2000
D	7	5	4000	5200	600
E*	7	6	4000	5000	1000

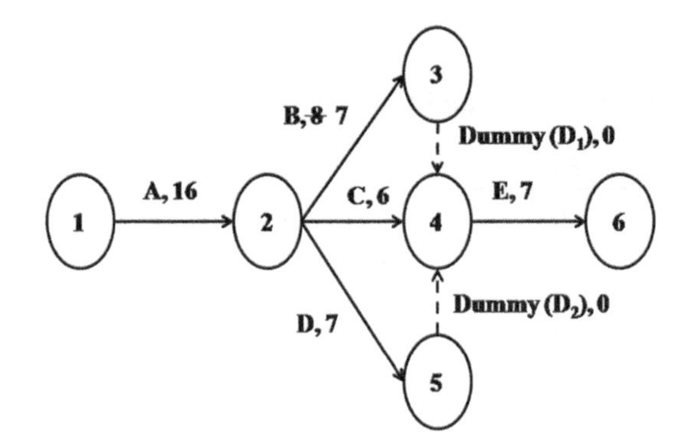

FIGURE 7.15 Resultant network diagram after 1-day crash.

The crash cost for 1-day for activity B is US$300, and the indirect cost for 1-day is US$600. Therefore, the indirect cost for 30 days is US$18000 (30*600).

After 1-day crashing of activity B, one non-critical path (**A→D→D$_2$→E**) becomes a critical path along with the existing critical path (**A→B→D$_1$→E**). Thus, the number of critical paths in the network becomes two. Therefore, a parallel crashing in both the critical paths is required to reduce the project completion time.

Thus, to further reducing the duration of the project, activities in both the critical paths need to be crushed simultaneously.

Crashing the project by 1-additional day

There are three options to reduce the duration of the project by an additional 1-day. Activities A and E, lay in both the paths and thus crashing either activity by 1-day reduces the project duration by 1-day. In another way, the crashing of the network can be done by crashing activities B and D by 1-day each. To make the cost-time trade-off, the activity with the minimum crashing cost based on the cost slope should be chosen. The crashing costs for three options to reduce the project cost by an additional 1-day are as follows:

Option 1: 1-day crashing cost of A = US$500
Option 2: 1-day crashing cost of E = US$1000
Option 3: 1-day crashing cost of B + 1-day crashing cost of D = 300 + 600 = US$900

Out of three, the crash cost of Option 1 (1-day crashing of activity A) is the lowest. Therefore, activity A should be crashed to reduce the duration of the project by one additional day. After crashing the

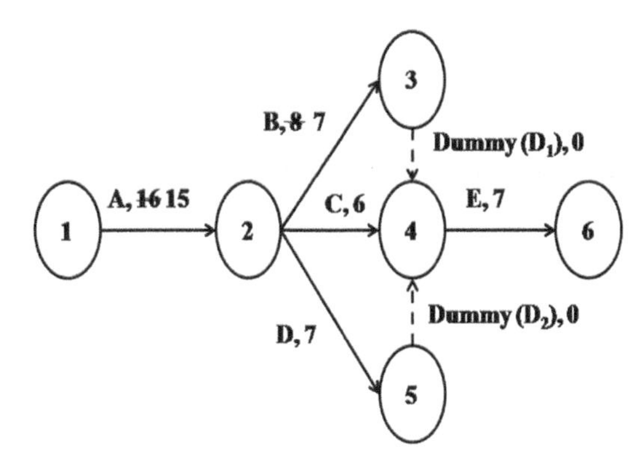

FIGURE 7.16 Resultant network diagram after 2-days crash.

activity, A by 1-day, the duration of the project becomes 29 days, and the revised network of the project is shown in Figure 7.16.

Crash cost of the project is

= Normal cost + Crash cost + Indirect cost
= (18000 + 4000 + 8000 + 4000 + 4000) + (1*300 + 1*500) + 29*600
= US$56200

The crash cost for 1-day for activity B is US$300, and that of 1-day for activity A is US$500. The indirect cost for 29 days is US$17400 (29*600).

In a similar way, the activities can be further crashed up to the maximum possible to complete the project in the desired period.

Example 7.4

The project details are given in Example 7.1 for the construction of a mine office; the project manager gathered additional information to speed up the project. The additional information like normal cost, crash cost, and maximum crash time for different activities are listed in Table 7.9.

Determine the followings by using the information given in Example 7.1 along with the additional information.

(a) If uncertainty exist in estimating durations of activity, determine the probability of completing the project within the deadline of 54 weeks.
(b) Determine the least expensive way of attempting to meet the target completion time within 55 weeks.
(c) A penalty of US$300,000 if the company has not started production activities by the deadline of 54 weeks from now. Determine the probability that the company has to pay the penalty charges.
(d) To provide additional incentive for speedy completion of the project, a bonus of US$150,000 will be paid to the company if the project completes within 52 weeks. Analyse the justification for completion of the project within 52 weeks by crashing the tasks.

Solution

(a) The critical activities of the project were identified in Example 7.1 as A, B, C, E, F, J, K, L, and M (indicated by red mark arrows in Figure 7.17). Therefore, the critical path is **A→B→C→E→F→J→K→L→M**

TABLE 7.9
Additional information of various activities of the project

Activity	Duration (week)	SD of activity duration	Crash time	Normal Direct cost (in thousand US$)	Crash Direct cost (in thousand US$)
A	2	0.25	1	4	5
B	5	1	3	8	9
C	12	2	9	12	16
D	2	0.5	2	6	6
E	5	1	4	6	8
F	6	1	5	8	10
G	7	1.5	5	8	9
H	6	1	4	4	4.5
I	4	1.2	4	10	10
J	6	1.5	4	4	5
K	8	1	7	8	10
L	6	1.2	5	3	3.5
M	7	0.75	7	6	6

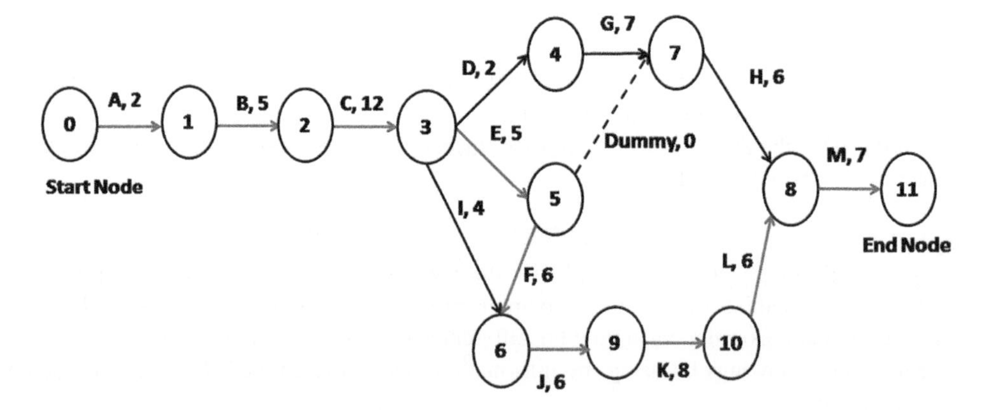

FIGURE 7.17 Network diagram of the project.

Therefore, the mean and standard deviation of the project can be determined using mean time of completion of each critical activity.

Thus, the mean duration of project completion (t_e) can be determined as

$$t_e = t_A + t_B + t_C + t_E + t_F + t_J + t_K + t_L + t_M$$
$$= 2 + 5 + 12 + 5 + 6 + 6 + 8 + 6 + 7 = 57 \text{ weeks}$$

The standard deviation of project completion (σ_P) can be determined as

$$\sigma_P = \sqrt{\text{Sum of the variance of critical tasks}}$$
$$= \sqrt{\sigma_A^2 + \sigma_B^2 + \sigma_C^2 + \sigma_E^2 + \sigma_F^2 + \sigma_J^2 + \sigma_K^2 + \sigma_L^2 + \sigma_M^2}$$
$$= \sqrt{0.25^2 + 1^2 + 2^2 + 1^2 + 1^2 + 1.5^2 + 1^2 + 1.2^2 + 0.75^2} = 3.50 \text{ weeks}$$

The probability that the project will finish within the deadline of 54 weeks ($t = 54$ weeks) is given by

$$P(t \le 54) = P(t - t_e \le 54 - t_e) = P\left(\frac{t - t_e}{\sigma} \le \frac{54 - t_e}{\sigma}\right) = P\left(z \le \frac{54 - 57}{3.5}\right)$$

$$\Rightarrow P(z \le -0.85) = 0.20$$

[Refer to the normal distribution table. The value is corresponding to .05 in column and 0.8 in the row]

Therefore, the probability that the project will complete within the deadline is equal to 0.20. There is only a 20 percent chance that the project will complete within the deadline.

(b) The mean completion time of the project is 57 weeks, and thus crashing of 2 weeks is required. The cost slope of each activity is determined using Eq. (7.1) and listed in Table 7.10.

The ascending order of the cost slope is

$$\mathbf{J^* = H \le B^* = G = L^* \le A^* \le C^* = E^* = F^* = K^*}$$

Among all the critical activities, the cost slope of activity J is the lowest (=0.25). Therefore, crashing of activity J should be done first. Before crashing of any critical activity, the length of the non-critical paths should be determined.

The lengths of the non-critical paths are as follows:

Non-critical Path 1: **A→B→C→D→G→H→M**
Mean Length (t_{e1}) = 2 + 5 + 12 + 2 + 7 + 6 + 7 = 41 weeks
Non-critical Path 2: **A→B→C→E→H→M**
Mean Length (t_{e2}) = 2 + 5 + 12 + 5 + 6 + 7 = 37 weeks
Non-critical Path 3: **A→B→C→I→J→K→L →M**
Mean Length (t_{e3}) = 2 + 5 + 12 + 4 + 6 + 8 + 6 + 7 = 50 weeks

TABLE 7.10
Cost slopes of each activity

Activity	Normal Duration (week)	Crash time	Normal cost (in thousand US$)	Crash cost (in thousand US$)	Cost Slope (thousand US$)/ week)
A*	2	1	4	5	1
B*	5	3	8	9	0.5
C*	12	9	12	16	4/3 = 1.333
D#	2	2	6	6	--
E*	5	4	6	8	2
F*	6	5	8	10	2
G	7	5	8	9	0.5
H	6	4	4	4.5	0.25
I#	4	4	10	10	--
J*	6	4	4	4.5	0.25
K*	8	7	8	10	2
L*	6	5	3	3.5	0.5
M*#	7	7	6	6	--

*Critical tasks, #Crashing of the task is not possible or allowed

As the lengths of three non-critical paths are 41 weeks, 37 weeks, and 50 weeks, crashing of 2 weeks of any critical tasks will not lead to form any non-critical to be a critical path. Thus, activity J should be crashed by two weeks to complete the project within 55 weeks.

The cost of completion of the project before crash = arithmetic sum of normal cost of completion of all the activities = 4 + 8 + 12 + 6 + 6 + 8 + 8 + 4 + 10 + 4 + 8 + 3 + 6 = US$87 thousands

On the other hand, the cost of completion of the project after crash = arithmetic sum of normal cost of completion of all the activities + Crash cost of activity J by 2 Weeks = (4 + 8 + 12 + 6 + 6 + 8 + 8 + 4 + 10 + 4 + 8 + 3 +6) + (2*0.25) = US$87.5 thousands

That is, there is an increase in the project cost by US$0.5 thousand (=87.5-87) for reducing the project completion time by two weeks.

(c) The probability of completion of the project within 54 weeks is determined as

$$P(t \leq 54) = 0.20$$

The probability is very less, and thus there is a maximum chance that the company has to pay the penalty charges of US$3,00,000 with the normal work pace. It is recommended that the company should speed up the work by crashing the activities in an optimal way to reduce the risk of paying the penalty. In solution (b), it was observed that crashing of task J by 2-weeks reduces the mean completion time to 55 weeks with the cost increment of US$500 only.

After crashing by two weeks, the probability of completion of the project within 54 weeks is given by

$$P(t \leq 54) = P(t - t_e \leq 54 - t_e) = P\left(\frac{t - t_e}{\sigma} \leq \frac{54 - t_e}{\sigma}\right) = P\left(z \leq \frac{54 - 55}{3.5}\right)$$

$$\Rightarrow P(z \leq 0.285) = 0.61$$

Thus, the chances of completion have been improved to 61% from 20% by crashing a task by two weeks. Additional cost incurred in this is US$500. If we want to further improve the chances, additional crashing of activities is needed. It should be noted that the crashing cost should not be exceeded by the penalty cost.

(d) To achieve the reward, the project needs to be completed within 52 weeks. It was estimated that the expected normal time of completion of the project is 57 weeks and thus needs to be crashed by five weeks. The maximum crashing of each activity, along with the cost slope, is shown in Table 7.10. The network diagram (shown in Figure 7.17) indicates that four paths exist, including critical paths. The lengths of each path are calculated below:

Length of the Critical Path (A-B-C-E-F-J-K-L-M) = **57 weeks**

Lengths of the other paths are as follows:

Non-critical Path 1: A-B-C-D-G-H-M = 41 Weeks

Non-critical Path 2: A-B-C-E-H-M = 37 Weeks

Non-critical Path 3: A-B-C-I-J-K-L-M = 50 Weeks

Table 7.10 clearly indicates that the minimum cost slope is observed for task J and the maximum crashing possible to task J is two weeks, as defined in the problem. After crashing by two weeks, Path 1 is still the critical path. The next minimum cost slopes among critical tasks are B and L. The maximum crashing possible for tasks B and L is two weeks and one week, respectively. Thus, activities B and L are crashed by two weeks and one week, respectively. It should be kept in mind that

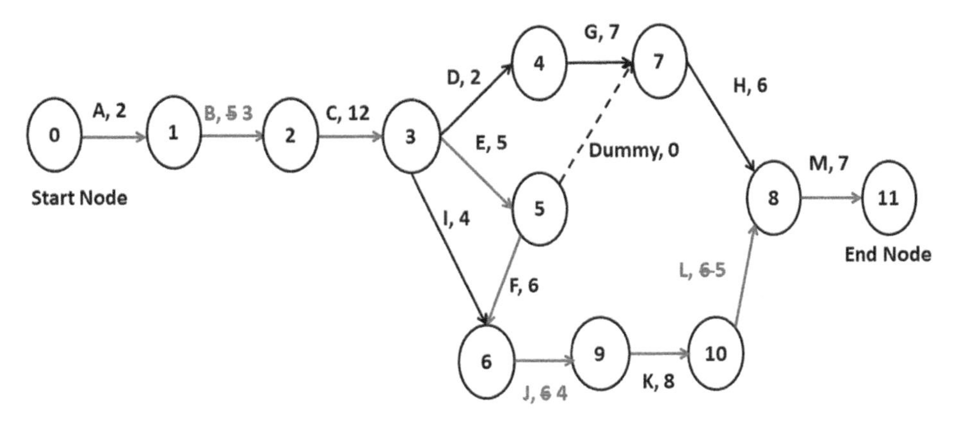

FIGURE 7.18 Network diagram showing the crashing of critical activities.

after crashing of the critical tasks, neither of the non-critical paths becomes a critical path. In this case, the length of the critical path after crashing of five weeks is 52 weeks, which is still higher than the length of each non-critical path.

The revised network diagram after crashing by five weeks (2 weeks for activity J, two weeks for activity B, and one week for activity L) is shown in Figure 7.18. The total crashing cost of the project is given by

$$\text{Total crashing cost} = 0.25*2 + 0.5*2 + 0.5*1 = \text{US\$2 thousand}$$

Thus, the expected project completion time after crashing is

$$t_e = 52 \text{ weeks}$$

For a normal distribution, the probability that the project will finish within the deadline of 52 weeks ($t = 52$ weeks) is given by

$$P(t \leq 52) = 0.5$$

That is, the chances of project completion are 50%. It is up to the project manager that he will take the risk of crashing to achieve the reward.

Exercise 7

Q1. Draw a project network diagram consists of activities A to L with the following precedence relationships:

Activity	Predecessors
A	--
B	--
C	--
D	A, B
E	B
F	B
G	F, C
H	B
I	E, H
J	E, H
K	C, F, J
L	K

Q2. Explain the method for determining the maximum possible delay allowed in the start time in respect to the earliest start time of all the immediately succeeding activities to be scheduled anywhere between their earliest and latest completion time.

Q3. The activities and their predecessors of a project are identified as follows:

Activity	A	B	C	D	E	F	G	H	I	J	K
Predecessor	--	--	--	A	B	B	C	D	E, G	F	H, J
Time in Weeks	3	3	2	5	9	6	4	2	6	5	4

Draw a project network diagram and compute the total float for each activity, critical path and project completion time.

[**Ans.** Critical Paths: B→E→I and B→F→J→K; Project Completion Time = 18 weeks]

Q4. The project manager of a mine identified eight tasks (A to H) in a typical mine project. The interdependencies of the tasks and estimated completion time are as follows:

Tasks	**Predecessor**	**Time duration (weeks)**		
		Optimistic	**Most Likely**	**Pessimistic**
P	--	4	4	10
Q	--	12	12	24
R	P	8	10	12
S	P	12	14	16
T	P	6	7	14
U	Q, R	8	8	8
V	S	4	6	8
W	T, U, V	5	5	5

a. Determine the expected completion time and variance of each activity.
b. Construct the project network diagram and estimate the expected completion time and variance of the project.
c. Determine the probability of completing the project within 32 weeks.

[**Ans.** b. Expected completion time of the project = 30 weeks and variance of the project = 1.88 week^2; c. 0.9265]

Q5. A project manager in mines gathers the following information with respect to a typical project.

Activity	**Predecessors**	**Completion time (in days)**	**Cost of expedite (in $ day)**
P	--	3	100
Q	--	5	80
R	--	6	70
S	P, Q	3	100
T	S	6	150
U	T	2	50
V	R	11	100
W	U, V	1	100

Based on the information, determine the followings:

a. Determine the earliest completion time of the project.
b. Is it economical to reduce the project duration if the indirect cost per day is US$ 100?

[**Ans.** a. Earliest completion time = 18 days; **b.** It is economical to crash the project duration]

Q6. In a power transmission line project, the normal and the crash estimates are as follows:

Activity	Predecessor	Normal Estimate		Crash Estimate	
		Time (Weeks)	Direct cost for the activity (US$)	Time (weeks)	Direct cost for the activity (US$)
A	--	12	10000	9	25000
B	A	4	–	3	4000
C	A	20	–	20	–
D	B	20	50000	14	65000
E	B	8	–	4	2000
F	B	8	–	4	2000
G	C	8	5000	4	10000
H	E	8	4000	5	7000
I	F, G, H	12	30000	9	39000
J	I	4	1000	1	4000

The indirect cost of the project is US$2000 per week.

a. Determine the critical path and project completion time.
b. Determine the project completion cost for the expected completion time.
c. Crash the project to 49 weeks and calculate the total cost for completion in reduced time.

[**Ans.** a. Critical Path is A →C→G→I→J and Project Length = 56 weeks; b. Project completion cost = US$ 212000; c. Total cost of the project after crashing = US$ 206000]

Q7. A small project has seven activities (A to G). The relevant data about these activities is given below:

Activity	Dependence	Duration (Days)		Direct Cost (US$)	
		Normal	Crash	Normal	Crash
A1	--	8	6	500	900
A2	A1	5	3	400	600
A3	A1	6	6	500	500
A4	A1	6	4	800	1000
A5	A2, A3	8	5	700	1000
A6	A3, A4	6	3	800	1400
A7	A5, A6	7	5	800	1600

a. Construct the project network diagram and estimate project completion time.
b. Determine the percentage change in cost to finish the project within 24 days.

[**Ans.** a. Project length = 29 days; b. Percentage change in cost = 20%]

8 Reliability analysis of mining systems

8.1 DEFINITION

The reliability of a mining system is the probability that it will work at the specified level without failure for a specified time duration under a given environmental condition. In other words, it can be defined as the probability that a system performs adequately with a designated service level for the intended period under the specified operating conditions.

Mathematically, the reliability, $R(t)$, of a mining system working successfully for the duration, T, can be represented as:

$$R(t) = P(t > T) \qquad T \geq 0 \tag{8.1}$$

where, t is a random variable denoting the time-to-failure or failure time, and $P(t > T)$ is the probability of successful operation up to time T.

On the other hand, the failure probability, $F(t)$, of a system measures the probability of failure within the time T, can be represented as

$$F(t) = P(t \leq T) \qquad T \geq 0 \tag{8.2}$$

8.2 STATISTICAL CONCEPTS OF RELIABILITY

Let the life of a machine be denoted by the random variable t with cumulative distribution function $F(t)$ and probability density function $f(t)$. In the context of reliability, $F(t)$ and $f(t)$ are also referred to as failure distribution. The cumulative failure distribution, $F(t)$, is a function of time and can be expressed in terms of a probability density function, $f(t)$ as:

$$F(t) = \int_0^T f(t)\,dt \tag{8.3}$$

The function $F(t)$ represents the probability that the system will fail anywhere in the time interval of 0 and T.

Thus, the reliability, $R(t)$ is given by

$$R(t) = 1 - F(t) \tag{8.4}$$

$R(t)$ indicates the probability that the system will work at least up to time, thereby adding the time dimension to the definition. For any given time, t, the sum of failure probability and reliability value will always be constant.

8.3 HAZARD FUNCTION

The hazard function, $\lambda(t)$, also referred to as the conditional failure rate function, is defined as the probability that a system fails after it has been in use for a given time. In other words, the limit of

DOI: 10.1201/9781003200703-8

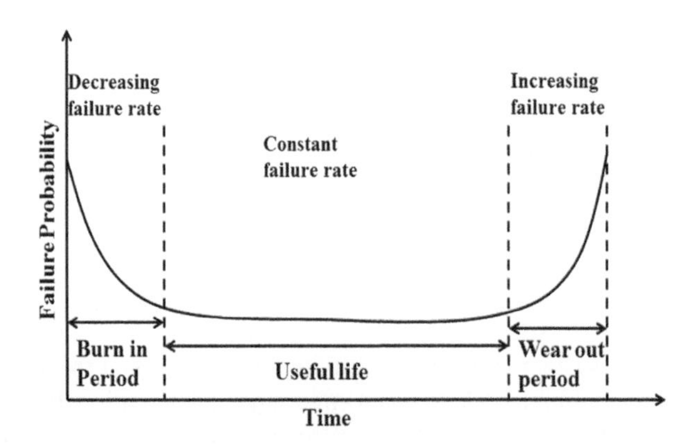

FIGURE 8.1 Bathtub curve representing the failure probability.

the failure rate tends to zero for a specified time interval. Thus, $\lambda(t)$ is the instantaneous failure rate and can be represented as

$$\lambda(t) = \lim_{\Delta t \to 0} \left(\frac{P(\text{failure of the system in the interval of } (t, t+\Delta t)) \big/ \Delta t}{P(\text{successful operation of the system up to time } t)} \right)$$

$$= \frac{1}{R(t)} \lim_{\Delta t \to 0} \left(\frac{F(t+\Delta t) - F(t)}{\Delta t} \right) = \frac{1}{R(t)} \frac{d}{dt} F(t) = \frac{f(t)}{R(t)} = \frac{f(t)}{1 - F(t)}$$

Therefore,

$$\lambda(t) = \frac{f(t)}{R(t)} = \frac{f(t)}{1 - F(t)} \tag{8.5}$$

The above derivation indicates that the hazard function, $\lambda(t)$, is the ratio of failure probability density function to reliability function. The failure rates of all the systems are changed with time. In general, the rate of failure per unit time in a mining system follows a bathtub curve shape (Figure 8.1). It can be inferred from Figure 8.1 that the failure rate is higher at the beginning, followed by a period of constant failure rate, followed by a period of increasingly high failure rates. These three periods are known as the burn-in period, useful life, and wear-out period, respectively.

8.4 CUMULATIVE HAZARD RATE

The cumulative hazard rate for a continuous hazard rate distribution function represents the area under the hazard rate function. The cumulative hazard rate function ($\Lambda(t)$) is useful to determine the average failure rates using the following equation.

$$\Lambda(t) = \int_{-\infty}^{t} \lambda(t) dt \tag{8.6}$$

Example 8.1

The reliability function of a drill machine in an opencast mine is given by

$$R(t)=\begin{cases}\left(1-\dfrac{t}{100}\right), & 0\le t\le 100\\ 0, & t>100\end{cases}$$

Find the failure rate function and failure rate trend of the drill machine.

Solution

We know that the hazard function can be presented as

$$\lambda(t)=\frac{f(t)}{R(t)}=\frac{\frac{d}{dt}F(t)}{R(t)}=\frac{\frac{d}{dt}\left[1-R(t)\right]}{R(t)}$$

$$\Rightarrow \lambda(t)=\begin{cases}\dfrac{\frac{d}{dt}\left[1-\left(1-\frac{t}{100}\right)\right]}{\left(1-\frac{t}{100}\right)}=\dfrac{\frac{1}{100}}{\left(1-\frac{t}{100}\right)}=\dfrac{100}{100-t} & \text{for } 0\le t\le 100\\ 0 & \text{for } t>100\end{cases}$$

The failure rate increases from 1 at $t=0$ to ∞ at $t=100$

8.5 RELIABILITY FUNCTIONS

The reliability function is the statistical distribution function, which is used to model the reliability of a mining system theoretically. Depending on the type of mining system, continuous and discrete statistical distributions are applied to model the reliability. The common continuous distributions are exponential, normal, Weibull, and discrete distributions are binomial and Poisson distribution to model reliability.

8.5.1 Reliability calculation with an exponential distribution function

The probability density function (pdf) for an exponential distribution with a constant failure rate, λ, can be represented as

$$f(t)=\lambda e^{-\lambda t}=\frac{1}{\mu}e^{-t/\mu},\qquad t\ge 0 \tag{8.7}$$

In the above equation, λ represents the mean number of failures per unit time, and μ represents the mean time to failure.

The probability of failure within time T, $F(t\le T)$ can be determined as

$$F(t\le T)=\int_0^T f(t)\,dt=\int_0^T \frac{1}{\mu}e^{-\frac{t}{\mu}}\,dt$$

$$\Rightarrow F(t\le T)=\frac{1}{\mu}\left[-\mu e^{-t/\mu}\right]_0^T=1-e^{-T/\mu}=1-e^{-\lambda T} \tag{8.8}$$

Hence, the reliability within time T is presented as

$$R(t>T)=1-F(t\leq T)=e^{-T/\mu}=e^{-\lambda T} \tag{8.9}$$

The hazard function, $\lambda(t)$, can be determined as

$$\lambda(t)=\frac{f(t)}{R(t)}=\frac{\frac{1}{\mu}e^{-t/\mu}}{e^{-t/\mu}}=\frac{1}{\mu}=\lambda$$

The above hazard function is constant and independent of time t. Thus, it can be inferred that the hazard function of an exponential distribution is constant and independent of time.

The other property of an exponential distribution function is the memory-less property. This is proved as follows.

Assume that a mining system (e.g., mining machine) successfully runs for time, T without any failure. Therefore, the conditional probability that the machine will run for $(T + \Delta T)$ time after a successful run of T time is given by

$$P(t>(T+\Delta T)|t>T)=\frac{P((t>(T+\Delta T),t>T)}{P(t>T)}$$

$$=\left(\frac{\text{Reliability of the machine successfully run for}(T+\Delta T)\text{time}}{\text{Reliability of the machine successfully run for } T \text{ time })}\right)$$

$$=\frac{\left[e^{-\lambda(T+\Delta T)}\right]}{e^{-\lambda T}}=e^{-\lambda\Delta T}=P(t>\Delta T)$$

The above relationship indicates that the probability of running the machine successfully for $(T+\Delta T)$ time after a successful run of time T is equal to the probability of running the machine successfully for ΔT time.

Example 8.2

The hazard function of a machine can be defined linearly as $\lambda(t)=5*10^{-6}t$, where t indicates the operating hours of the machine. For desirable reliability of 0.98, determine the life of the machine assuming the reliability function is exponentially distributed.

Solution

We have,

$$R(t)=e^{-\int_0^t \lambda(t)dt}$$

$$\Rightarrow 0.98=e^{-\int_0^t 5\text{x}10^{-6}\,\text{t}\,dt}$$

$$\Rightarrow 0.98=e^{-\left[\frac{5\text{x}10^{-6}*t^2}{2}\right]_0^t}=e^{-2.5*10^{-6}t^2}$$

$$\Rightarrow t=-\frac{ln0.98}{2.5*10^{-6}}=\frac{0.0202}{2.5*10^{-6}}=808\text{ hrs.}$$

Example 8.3
The failure probability of a mining machine is an exponential distribution with a constant failure rate of $\lambda = 0.02$ hours.

a. Determine the probability that the machine will fail within the first 10 hours
b. Suppose that the machine has successfully operated for 100 hours. What is the probability that the machine will fail during the next 10 hours of operations?

Solution

a. The pdf for exponential distribution is given by

$$f(t) = \lambda e^{-\lambda t} dt$$

The probability of failing the machine within 10 hours is given by

$$F(t \leq 10) = \int_{0}^{10} \lambda e^{-\lambda t} dt = \left[-e^{-\lambda t}\right]_{0}^{100} = 1 - e^{-0.02*10} = 1 - 0.82 = 0.18$$

Therefore, the probability of failure is 0.18.

b. Again, the probability that the machine will run another 10 hours after 100 hours of successful operation is given by

$$P(t > 100 + 10)|t > 100) = \frac{R\{(100+10) \cap 100\}}{R(100)} = \frac{1 - \int_{0}^{100+10} f(t)dt}{1 - \int_{0}^{100} f(t)dt}$$

Therefore,

$$P(t > 100 + 10)|t > 100) = \frac{1 - \int_{0}^{100+10} \lambda e^{-\lambda t}\, dt}{1 - \int_{0}^{100} \lambda e^{-\lambda t} dt} = \frac{1 - \left[-e^{-\lambda t}\right]_{0}^{110}}{1 - \left[-e^{-\lambda t}\right]_{0}^{100}}$$

$$= \frac{1 + 0.11 - 1}{1 + 0.134 - 1} = 0.82$$

Hence, the probability that the machine will fail during next 10 hours after 100 hours of successful operation is given by

$$P(t \leq 100 + 10)|t > 100) = 1 - P(t > 100 + 10)|t > 100) = 1 - 0.82 = 0.18$$

Alternative Approach
The probability that the machine will fail within 10 hours after 100 hours of successful operation is given by

$$P(t < 100 + 10)|t > 100) = \frac{F\{(100+10) \cap 100\}}{R(100)} = \frac{\int_{100}^{100+10} f(t)dt}{1 - \int_{0}^{100} f(t)dt}$$

Therefore,

$$P(t<100+10)|t>100)=\frac{\int_{100}^{100+10}\lambda e^{-\lambda t}\,dt}{1-\int_{0}^{100}\lambda e^{-\lambda t}dt}=\frac{\left[-e^{-\lambda t}\right]_{100}^{110}}{1+\left[e^{-\lambda t}\right]_{0}^{100}}=\frac{-0.11+0.134}{1+(0.134-1)}=0.18$$

Example 8.4

The lifetime of a mining cap lamp follows an exponential distribution with a mean time between failures is 8000 hours. Determine the probability for the failure of a mining cap lamp before 7000 working hours.

Solution

Given that, mean time between failure (μ) = 8000 hours

The probability that the cap lamp will fail within 7000 hours is given by

$$F(t\le 7000)=\int_{0}^{7000}\frac{1}{8000}e^{-t/8000}dt$$

$$\Rightarrow F(t\le 7000)=\frac{1}{8000}\left[-8000*e^{-t/8000}\right]_{0}^{7000}=1-e^{-7000/8000}=0.4908$$

Thus, the chance that the mining cap lamp will fail within 7000 working hours is 49%.

Example 8.5

A longwall coal mine system operates using double-ended ranging drum Shearer for cutting operation the coal and Armoured Face Conveyor (AFC) for transportation of coal. The lifetime of both the machines (shearer and AFC) follows an exponential distribution with a mean time between failures are 600 hours and 700 hours, respectively. Determine the probability that the longwall coal mine system is expected to fail before 800 hours of operation.

Solution

For Shearer, $R_1(t>800)=e^{-\frac{t}{\mu_1}}=e^{-\frac{t}{600}}$

$$R_1(t>800)=e^{-\frac{800}{600}}=e^{-4/3}=0.2636$$

For the AFC, $R_2(t>800)=e^{-\frac{t}{\mu_2}}=e^{-\frac{t}{700}}$

$$R_2(t>800)=e^{-\frac{800}{700}}=e^{-\frac{8}{7}}=0.3189$$

It is assumed that the two machines operate independently. Thus, the probability that the system will continue to function after 800 hours of operation is given by

$$R(t>800)=R_1(t>800)*R_2(t>800)=0.2636*0.3189=0.0841$$

Hence, the probability that the device will fail within 8000 hours of operation is given by

$$F(t\le 800)=1-R(t>800)=1-0.0841=0.916$$

Thus, the chance that the system will fail before 800 hours of operation is 91.6%.

8.5.2 Reliability calculation with a normal probability density function

The pdf of a normal distribution can be represented as

$$f(t,\ \mu,\sigma)=\frac{1}{\sqrt{2\pi\sigma}}e^{-\frac{(t-\mu)^2}{2\sigma^2}} \qquad t\epsilon(-\infty,\infty) \tag{8.10}$$

where
μ = mean time to failure
σ^2 = variance of the time to failure

A normal distribution is symmetrical about the mean, and the spread is measured by variance.

The probability of failure for a time T is given by

$$F(t\le T)=\int_0^T f(t)dt=\int_0^T \frac{1}{\sqrt{2\pi\sigma}}e^{-\frac{(t-\mu)^2}{2\sigma^2}}dt$$

The reliability function for a normal distribution with time T is given by

$$R(t>T)=1-F(t\le T)=1-\int_0^T \frac{1}{\sigma\sqrt{2\pi}}e^{-\frac{(t-\mu)^2}{2\sigma^2}}dt$$

A normal distribution function can be converted into a standardized normal distribution function by equating, $\frac{(t-\mu)}{\sigma}=z$

On differentiating both the sides of the above equation, we have

$$dz = dt/\sigma$$

Therefore,

$$f(z)=\frac{1}{\sqrt{2\pi}}e^{-\frac{Z^2}{2}} \qquad -\infty\le z\le\infty$$

In the standardized normal distribution function, $\mu=0$ and $\sigma=1$. The failure probability for time T is given by:

$$F(t\le T)=\varphi(z\le Z)=\int_{-\infty}^{Z} f(z)dz=\frac{1}{\sqrt{2\pi}}\int_{-\infty}^{Z} e^{-\frac{z^2}{2}}dz=\frac{1}{2}erf\left(\frac{Z}{\sqrt{2}}\right) \tag{8.11}$$

In the above equation, $\Phi(z\le Z)$ is a cumulative distribution of standard normal distribution function, and the *erf* is an error function.

For a normal distribution function, with mean, μ, and standard deviation, σ, the probability of failure of the machine within time, *T*, is given by

$$P(t\le T)=\Phi(z\le Z)$$

In the above equation, Φ represents the area under the curve between 0 and T. This can be determined from the standard statistical table of normal distribution.

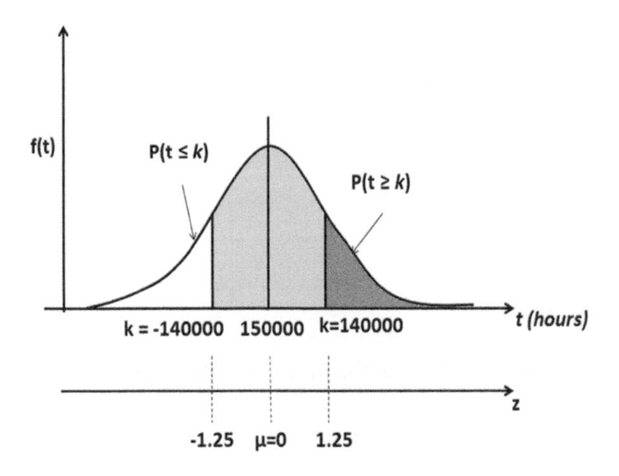

FIGURE 8.2 Shaded region of normal distribution represents the desired probability.

Example 8.6

The average life of a winding (or hoisting) rope fitted in a hoisting system in an underground coal mine is 1,50,000 hours with a standard deviation of 8000 hours. Determine the probability of failure of rope in the first 1,40,000 working hours. It is given that the failure density function of the rope follows the normal distribution.

Solution

Given data:

Mean life = μ = 1,50,000 hours
Standard deviation of the life of rope = σ = 8,000 hours
Time period for which the probability has to be determined = T = 1,40,000 hours

$$\text{Therefore, } Z = \frac{T-\mu}{\sigma} = \frac{140000-150000}{8000} = -1.25$$

The desired probability is represented by the area under the curve of normal distribution function, as shown in Figure 8.2.

For a standardized normal distribution function, the reliability for a mission of 1,40,000 hours can be derived as

$$R(t>T) = P(z>Z) = P(z>-1.25) = 1-\phi(-1.25) = 1-0.1056 = 0.8944$$

Therefore, the probability that the rope will fail within 140000 hours = 1 – 0.8944 = 0.1056

Example 8.7

The failure of a repairable mine machine follows a normal distribution with mean life (μ) of 2000 hours and standard deviation (σ) of 200 hours. Determine the reliability and the hazard function of the machine after 1800 hours of successful operation.

Solution

Given data: Mean life = μ = 2000 hours
The standard deviation of the life of machine = σ = 200 hours

The time period for which the probability has to be determined = t = 1800 hours

Therefore, $z = \frac{t-\mu}{\sigma} = \frac{1800-2000}{200} = -1$

The reliability function for a standard normal deviate, z, can be determined as

$$R(1800) = P(z > -1) = 1 - \phi(-1) = 1 - 0.1587 = 0.8413$$

Therefore, the failure probability of the machine is (1 – 0.8413) = 0.1587

The hazard function for standard normal distribution failure function can be presented as

$$\lambda(t) = \frac{f(t)}{R(t)} = \frac{\Phi(z)}{\sigma R(t)}$$

where ϕ is a cumulative distribution function of standard normal density.

Here

$$\lambda(1800) = \frac{f(t)}{R(t)} = \frac{\Phi(-1)}{100 * R(1800)} = \frac{0.1587}{100 * 0.8413} = 0.0019 \text{ failures/cycle}$$

Example 8.8

The failure rate of a shovel deployed in an opencast mine is 0.001 failures per week, and the mean life is 500 weeks with a standard deviation of 20 weeks. The mean time between failures of shovels follows a normal distribution. Determine the reliability of the shovel for a successful operation of 70 weeks starting from 450 weeks from now.

Solution

Given data: Mean life = μ = 500 weeks.

Standard deviation of the life of rope = σ = 20 weeks.

Time period for which the probability has to be determined = $450 \leq t \leq (450+70)$ weeks.

The desired probability is represented by the area under the curve of normal distribution function, as shown in Figure 8.3.

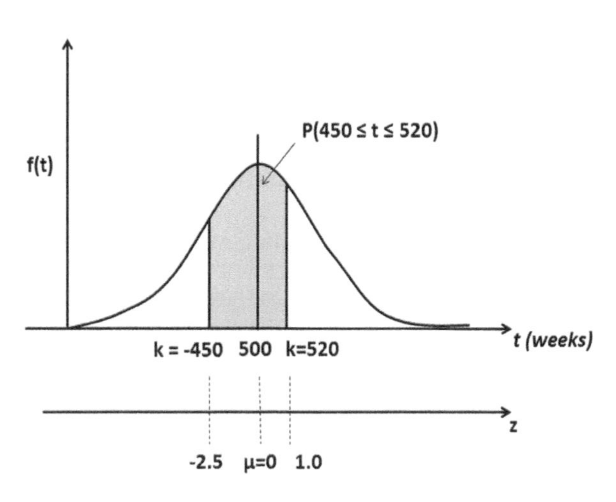

FIGURE 8.3 Shaded region of normal distribution represents the desired probability.

The reliability for a standardized normal distribution, z, can be determined as

$$R(t) = P\left(\frac{t_1 - \mu}{\sigma} \le z \le \frac{t_2 - \mu}{\sigma}\right) = P\left(\frac{450 - 500}{20} \le z \le \frac{520 - 500}{20}\right)$$

$$\Rightarrow P(-2.5 \le z \le 1) = \phi(1) - \phi(-2.5) = \{1 - \phi(-1)\} - \phi(-2.5)$$

$$\Rightarrow P(-2.5 \le z \le 1) = (1 - 0.3413) - 0.4938 = 0.1649$$

8.5.3 Reliability calculation with a Weibull distribution probability density function

The exponential function of the probability density function has memory-less property, and thus the function has limitations in wide varieties of reliability analysis. A more generalized form of the exponential distribution can be represented by the Weibull distribution (Weibull, 1951). The Weibull distribution has higher flexibility and is appropriate for modelling lifetimes of systems (or machines) for diversified engineering applications. The Weibull three-parameter probability density function is given by

$$f(t) = \frac{\beta(t - \gamma)^{\beta - 1}}{\theta^\beta} e^{-\left(\frac{t-\gamma}{\theta}\right)^\beta} \qquad t \ge \gamma \ge 0 \tag{8.12}$$

In the above function, θ is known as the scale parameter, β is the shape parameter, and γ is the location parameter. All these parameters are positive, and based on these parameter values; a Weibull distribution can be converted to the exponential distribution, the Rayleigh distribution, the normal distribution, and so on.

For $\beta = 1$ and $\gamma = 0$, a Weibull distribution function reduces to exponential distribution function as follows:

$$f(t) = \frac{1(t - 0)^{1-1}}{\theta^1} e^{-\left(\frac{t-0}{\theta}\right)^1} = \frac{1}{\theta} e^{-\frac{t}{\theta}} \tag{8.13}$$

The reliability and hazard function for a Weibull distribution is given by

$$R(t) = e^{-\left(\frac{t-\gamma}{\theta}\right)^\beta} \qquad \text{for} \quad t > \gamma > 0,\ \beta > 0,\ \theta > 0 \tag{8.14}$$

$$\lambda(t) = \frac{\beta(t - \gamma)^{\beta - 1}}{\theta^\beta} \qquad \text{for} \quad t > \gamma > 0,\ \beta > 0,\ \theta > 0 \tag{8.15}$$

The hazard function for a Weibull distribution function has decreasing trend for $\beta < 1$, an increasing trend for $\beta > 1$, and constant when $\beta = 1$.

As the exponential distribution is a typical case of the Weibull distribution at $\beta = 1, \gamma = 0$, the reliability and hazard function of the same is reduced to

$$R(t) = e^{-\frac{t}{\theta}} \tag{8.16}$$

$$\lambda(t) = \frac{1}{\theta} \tag{8.17}$$

The two-parameter Weibull distribution is represented as:

$$f(t) = \frac{\beta}{\theta^{\beta}}(t)^{\beta-1} e^{-\left(\frac{t}{\theta}\right)^{\beta}} \tag{8.18}$$

For β=1, the two parameters Weibull distribution function reduces to exponential distribution function as

$$f(t) = \frac{1}{\theta} e^{-\frac{t}{\theta}} \tag{8.19}$$

The above function is a typical form of the Weibull distribution function and follows an exponential distribution.

The mean and variance of a Weibull distribution can be determined in terms of the Gamma function, $\Gamma(t)$. The probability density function of a Weibull distribution can be defined as:

$$f(t) = \frac{\beta}{\theta^{\beta}}(t)^{\beta-1} e^{-\left(\frac{t}{\theta}\right)^{\beta}} \tag{8.20}$$

The mean of the Weibull distribution is given by

$$E(t) = \theta\Gamma\left(1+\frac{1}{\beta}\right) \tag{8.21}$$

And, the variance of the distribution is given by

$$V(t) = \theta\left\{\Gamma\left(1+\frac{2}{\beta}\right) - \left(\Gamma\left(1+\frac{1}{\beta}\right)\right)^2\right\} \tag{8.22}$$

In the above equations, Γ represents a gamma function. The value of Γ (n) can be defined as

$$\Gamma(n) = \int_0^{\infty} t^{n-1} e^{-t} dt$$

The Gamma function, Γ (n), can be expressed as a simple factorial, as explained below. Applying the integration by parts, the following relationship can be derived

$$\Gamma(n) = (n-1) * \Gamma(r-1)$$

From the above relationship, we have,

$$\Gamma(n) = (n-1) * (n-2) * \ldots * \Gamma(1)$$

But,

$$\Gamma(1) = \int_0^{\infty} t^{1-1} e^{-t} dt = \int_0^{\infty} e^{-t} dt = 1$$

Therefore,

$$\Gamma(n) = (n-1) * (n-2) * \ldots (1) = (n-1)!$$

Example 8.9

The failure time of heavy earth-moving machine (HEMM) in a mine follows Weibull distribution with scale parameter (θ) = 2500, the shape parameter (β) = 2, and the location parameter (γ) = 1200. Determine the reliability and hazard rate of the machine for an operating time of 1600 hours.

Solution

Reliability of the machine is given by

$$R(t) = e^{-\left(\frac{t-\gamma}{\theta}\right)^{\beta}} \qquad \text{for} \quad t > \gamma > 0,\ \beta > 0,\ \theta > 0$$

$$\Rightarrow R(1600) = e^{-\left(\frac{1600-1200}{2500}\right)^{2}} = 0.974$$

Again, the hazard function is given by

$$\lambda(t) = \frac{\beta(t-\gamma)^{\beta-1}}{\theta^{\beta}} \qquad \text{for} \quad t > \gamma > 0,\ \beta > 0,\ \theta > 0$$

$$\Rightarrow \lambda(1600) = \frac{2(1600-1200)^{2-1}}{2500^{2}} = 1.25*10^{-4} \text{ failures per hour}$$

Example 8.10

The lifetime (represented by random variable t) of a hoist drum in an underground mine hoist system follows a Weibull distribution. The scale and shape parameters of the Weibull distribution are, $\theta = 2000$ and $\beta = 0.5$, respectively.

(a) Determine the probability of successful running time of the hoist drum before failure.
(b) Determine the probability for a minimum operating hour of 6000 of the hoist drum without failure.

Solution

(a) The expected time that the hoist drum runs before failure is given by

$$E(t) = \theta\Gamma\left(1+\frac{1}{\beta}\right) = 2000*\Gamma\left(1+\frac{1}{0.5}\right) = 2000*\Gamma(1+2) = 2000*2$$

$$= 4000 \text{ hours}$$

(b) The probability that the hoist drum run for a minimum of 6000 hours without failure is given by

$$P(t > 6000) = 1 - P(t \le 6000) = 1 - F(6000) = 1 - \left[1 - e^{-(t/\theta)^{\beta}}\right]$$

$$= 1 - \left[1 - e^{-(6000/2000)^{0.5}}\right] = e^{-(6000/2000)^{0.5}} = 0.176$$

8.5.4 Reliability calculation with a Poisson distribution probability mass function

We have discussed different continuous distribution functions to model reliability by calculating failure probability within a given time window. However, sometimes, we might be interested to know the number of failures, which is a discrete outcome, in time t. To model reliability based on the discrete outcome, we can use discrete probability distribution. The Poisson distribution is one of the commonly used probability distribution functions in reliability analysis.

The probability mass function of a Poisson distribution with a discrete random variable, x, is represented as

$$f(x, \lambda, t) = \frac{(\lambda t)^x e^{\lambda t}}{x!} \quad \text{for } x = 0, 1, 2, \ldots \tag{8.23}$$

In the above probability mass function, λ represents the mean number of failures in a specified time period t, also called shape parameter, and x represents the number of failures in time t.

For a unit time, $t = 1$, the above probability mass function can be reduced to

$$f(x, \lambda) = \frac{(\lambda)^x e^{-\lambda}}{x!} \quad \text{for } x = 0, 1, 2, \ldots \tag{8.24}$$

The function is generally used to model the number of failure occurrences in a specified time frame. Thus, the probability of n number of failures in time t is given by

$$F(n, t) = \frac{(\lambda t)^n e^{-\lambda t}}{n!} \tag{8.25}$$

For a unit time, $t = 1$, the above equation can be reduced to

$$F(n) = \frac{(\lambda)^n e^{-\lambda}}{n!} \tag{8.26}$$

Therefore, the reliability using a Poisson distribution function, *R(n)*, for n or less failures in time t, is given by

$$R(n) = \sum_{x=0}^{n} \frac{(\lambda t)^x e^{-\lambda t}}{x!} \tag{8.27}$$

Example 8.11

The mean number of failures of a dumper deployed in an opencast mine is 3 per month. If the number of failures follows a Poisson distribution function, determine the probability that exactly four failures will occur in the next month.

Solution

Given data:

Mean number of failures in a month $(\lambda) = 3$

The probability of n failures in the next month is given by

$$F(n) = \frac{(\lambda)^n e^{-\lambda}}{n!}$$

For n = 4,

$$F(4) = \frac{(3)^4 (2.718)^{-3}}{4!} = 0.168$$

Example 8.12

The control panel of a crushing plant has a mean failure rate (λ) of 0.002 failures per hour. If the number of failures follows a Poisson distribution function, determine the reliability for a 600-hour mission with a maximum of 2 failures.

Solution

Given data: Mean number of failure per hour (λ) = 0.002
Time (t) = 600 hours
Maximum number of failure (n) = 2

Therefore, the reliability of a Poisson distribution function, R(n), for n or less failures in time t, is given by

$$R(n) = \sum_{x=0}^{n} \frac{(\lambda t)^x e^{-\lambda t}}{x!}$$

$$\Rightarrow R(3) = \frac{(0.002 * 600)^0 e^{-0.002*600}}{0!} + \frac{(0.002 * 600)^1 e^{-0.002*600}}{1!} + \frac{(0.002 * 600)^2 e^{-0.002*600}}{2!}$$

$$\Rightarrow R(3) = 0.301 + 0.361 + 0.2168 = 0.8788$$

8.5.5 Reliability calculation for a binomial distribution

Similar to Poisson distribution, the binomial distribution is used for reliability calculation when the outcomes are discrete. The binomial distribution is used in reliability analysis when there are only two outcomes. It is one of the widely used discrete random variable distributions in reliability and quality monitoring.

The probability mass function of a binomial distribution is given by

$$f(x, n, p) = \binom{n}{x} p^x (1-p)^{n-x} \qquad \text{for } x = 0, 1, 2, \ldots, n \tag{8.28}$$

In the above equation, x indicates the number of successes out of n trials, and p represents the probability of success.

The probability for k successes out of n trials is given by

$$R(k; n, p) = \frac{n!}{k!(n-k)!} p^k (1-p)^{n-k} \tag{8.29}$$

The reliability for at least k successes out of n trials is given by

$$R(x \geq k) = \sum_{x=k}^{n} \binom{n}{x} p^x (1-p)^{n-x} \tag{8.30}$$

Example 8.13

It is given that 90% of workers deployed in an underground mine successfully complete the assigned work. Determine the probability of selecting at least eight workers who successfully completed the assigned job out of a random sample of ten workers.

Solution

The probability of selection of at least eight workers who successfully completed the assigned job out of a random sample of ten workers is given by

$$R(x \geq 8) = \sum_{x=8}^{10} \binom{10}{x} 0.9^x (1-0.9)^{20-x}$$

$$\Rightarrow R(8) = \binom{10}{8} 0.9^8 (1-0.9)^{10-8} + \binom{10}{9} 0.9^9 (1-0.9)^{10-9} + \binom{10}{10} 0.9^{10} (1-0.9)^{10-10}$$
$$= 0.193 + 0.387 + 0.348 = 0.928$$

8.6 MEAN TIME BETWEEN FAILURE (MTBF) AND MEAN TIME TO FAILURE (MTTF)

MTBF measures the predicted time that a system (or machine) runs before the next unplanned breakdown happens. The MTBF is generally used for a repairable system without routine scheduled maintenance or routine preventive sub-systems replacement. The MTBF can be determined as the ratio of up time of the machine to a total number of failures in the scheduled time. Mathematically, it can be represented as

$$\text{MTBF} = \frac{\text{system (or machine) up time}}{\text{Number of failures}}$$

For example, the performance of the machine is monitored for time T, as shown in Figure 8.4. During this time, the machine failed twice, first at time t_1 and second at time t_2, where $T > t_2 > t_1$. The machine was down for Δt_1 time and Δt_2 the time during the first and second breakdown, respectively. The mean time between failures (MTBF), which is the average time between successive failures, can be determined as

$$\text{MTBF} = \frac{\text{System up time}}{\text{Number of failures}} = \frac{T - \Delta t_1 - \Delta t_2}{2}$$

MTBF can be calculated directly from the reliability function $R(t)$ by calculating the arithmetic mean using integration operation.

$$MTBF = m = \int_0^\infty R(t)\,dt = \int_0^\infty t f(t)\,dt$$

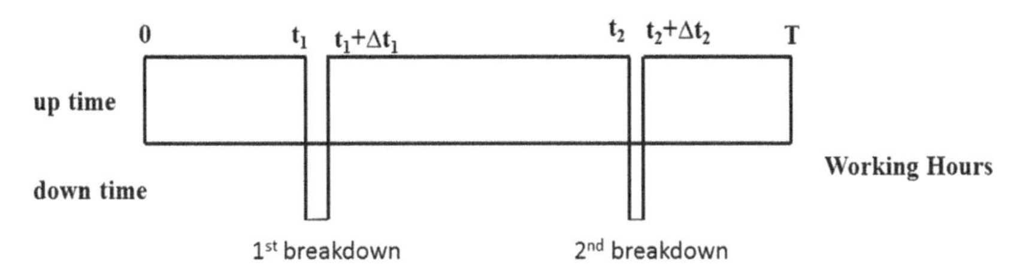

FIGURE 8.4 Break-down hours in a specified monitoring period.

Putting the value of *R(t)*, for constant failure time and exponential distribution, we have

$$\left\{\text{for } m < \infty, \left[te^{-\lambda t}\right]_0^\infty = 0\right\}$$

Thus, for the exponential distribution, the reliability or the probability that the equipment run for time *t* is equal to $e^{-t/m}$

On the other hand, MTTF is a measure of reliability used for non-repairable systems. It represents the length of time that a system is expected to last in operation until it fails. The MTTF can be determined as the ratio of total operational hours to the total number of units used in the systems. Mathematically, it can be represented as

$$\text{MTTF} = \frac{\text{Operational hours}}{\text{Number of units}}$$

In the mine systems, MTTF helps in improving the maintenance and inventory management strategy.

- MTTF is used in schedule maintenance on non-repairable items to extend the life of the items. For example, lubricating bearings on a larger machine.
- MTTF is also used to make decisions about purchasing parts and equipment in the inventory system of the mines. Good quality and durable parts lead to a higher MTTF, which requires fewer resources for replacing old parts.

Example 8.14

The performance of a shovel is monitored for 24 hours. The down time and up time of shovel are as follows (Figure 8.5).

Determine the MTBF.

Solution

The shovel was down two times for 10 and 5 minutes, respectively.

The MTBF is given by

$$\text{MTBF} = \frac{\text{Up} - \text{time}}{\text{Number of failures}} = \frac{24 - \frac{10}{60} - \frac{5}{60}}{2} = \frac{24 - 0.25}{2} = \frac{23.75}{2} = 11.875 \text{ hours}$$

Example 8.15

The life of five identical drill bits was studied in the mine operation. The failure times of five bits are respectively 30 hours, 36 hours, 28 hours, 42 hours, and 40 hours. Determine the MTTF.

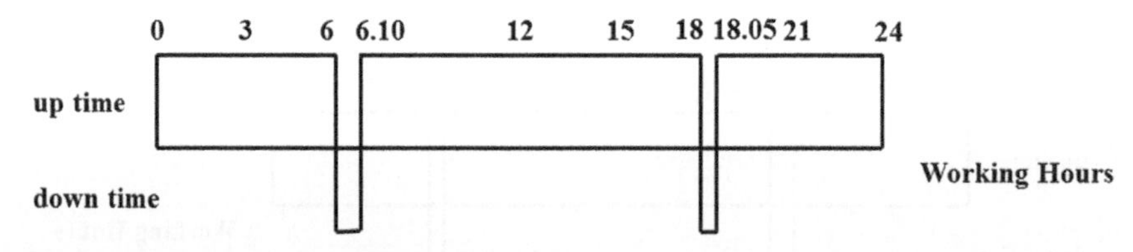

FIGURE 8.5 Up time and down time of shovel.

Solution
The MTTF of the drill bit is given by

$$MTTF = \frac{30+36+28+42+40}{5} = \frac{176}{5} = 35.2 \text{ hours}$$

This indicates that this particular type of drill bit will need to be replaced, on average, every 35.2 hours.

Example 8.16
The reliability function of a drill machine in an open-pit mine is

$$R(t) = \begin{cases} \left(1 - \frac{t}{100}\right), & 0 \le t \le 100 \\ 0, & t \ge 100 \end{cases}$$

Determine the MTBF of the drill machine.

Solution
The mean time to failures (MTBF) is given by

$$MTBF = E(t) = \int_0^{\infty} R(t)\,dt$$

$$\Rightarrow MTBF = \int_0^{10} R(t)\,dt + \int_{10}^{\infty} R(t)\,dt$$

$$\Rightarrow MTBF = \int_0^{10} \left(1 - \frac{t}{100}\right) dt + \int_{10}^{\infty} 0\,dt$$

$$\Rightarrow MTBF = \left[t - \frac{t^2}{200}\right]_0^{100} + 0 = 50 \text{ hours}$$

8.7 MAINTAINABILITY AND MEAN TIME TO REPAIR (MTTR)

The probability of restoring a failed system to a specified condition within a scheduled time by repairing the system per the prescribed guidelines is termed maintainability. It is the process of isolating and repairing the faulty items in a system at a scheduled time. The repair time of any system involves two types of time intervals, viz. passive repair time and active repair time. Passive repair time is the time taken by a service engineers to reach the machine location and time needed to get the spare parts needed to repair the machine, whereas the active repair time is the actual time the service engineer takes to repair the faulty system. The active repair time depends on the time needed to know the failure occurrence in a machine, fault detection time in the machine, replace or repair time of the faulty components, and checking time of the repaired machine.

For example, the time to repair or the total down time is denoted by a random variable, t, which follows a probability density function $f(t)$. It is assumed that the failed system is restored in time T. Then the maintainability, $M(t)$, of the system can be determined as

$$M(t) = P(t \le T) = \int_0^{T} f(t)\,dt$$

If the repair time follows an exponential probability distribution function, this can be defined as

$$f(t) = \mu e^{-\mu t}$$

where, μ (> 0) represents a constant repair rate.

Then the maintainability is given by

$$M(t) = \int_0^T \mu e^{-\mu t} dt = \left(1 - e^{-\mu T}\right) \tag{8.31}$$

The above equation represents the exponential form of the maintainability function. In maintenance studies, one important parameter often used to measure maintainability, is mean time to repair (MTTR) or the mean down time. It can be defined as the expected value of the down time random variable, *t*. For an exponential distribution, it can be determined as

$$MTTR = E(t) = \int_0^\infty t f(t) dt = \int_0^\infty t \mu e^{-\mu t} dt = \frac{1}{\mu}$$

The MTTR can also be determined as the ratio of down time of the machine to the total number of failures in the scheduled time. Mathematically, it can be represented as

$$\text{MTTR} = \frac{\text{System down time}}{\text{Number of failures}}$$

MTTR measures availability, whereas the MTBF measures availability and reliability. The higher the value of the MTBF, the longer the system will likely run before failing.

Example 8.17

A device has a decreasing failure rate characterized by two parameters Weibull distribution with a wear-out linear hazard function

$$\lambda(t) = \frac{2}{1000}\left(\frac{t}{1000}\right) = 2 * 10^{-6} t$$

In the above hazard function, *t* is in hours. The shape parameter $\beta = 2$ and the scale parameter, $\theta = 1000$. The device is required to have a design life reliability of 0.99. Determine the design life and MTTF.

Solution

Let the design life of the device is T.

The reliability *R(t)* for two parameters Weibull distribution is given by

$$R(t) = e^{-\left(\frac{t}{\theta}\right)^\beta}$$

$$\Rightarrow 0.99 = e^{-\left(\frac{T}{1000}\right)^2}$$

$$\Rightarrow \left(\frac{T}{1000}\right)^2 = \ln\left(\frac{1}{0.99}\right)$$

$$\Rightarrow \frac{T}{1000} = 0.1007$$

$\Rightarrow T = 100.7$ hours

$$MTTF = E(t) = \theta\Gamma\left(1+\frac{1}{\beta}\right) = 1000 * \Gamma\left(1+\frac{1}{2}\right) = 1000 * \Gamma(1.5) = 1000 * 0.886 = 886 \text{ hours}$$

8.8 RELIABILITY OF A SYSTEM

In general, a system consists of multiple components or subsystems configured in different types of networks, and thus the reliability of the entire system depends on the reliability of its components along with the network configuration. For example, the network configurations may be series, parallel, a combination of series-parallel, bridge, and so on. To calculate the reliability of a complex system, we need to know each sub-systems' reliability and how the subsystems are configured in the complex system.

8.8.1 System reliability on a series configuration

In a series configuration, the system will operate successfully if all the sub-systems (or units) in the system operate normally. Figure 8.6 shows a system consists of n number of components configured in series with individual's reliability $R_1, R_2 \ldots R_n$, respectively.

If x_j denotes the jth component's event in the system, then the reliability of the system (R_s) with 'n' independent components arranged in a series configuration, is given by

$$R_s = P(x_1 x_2 \ldots x_n)$$

where, P is a multivariate probability distribution function with n events. Since the components are independents to each other, we have

$$R_s = P(x_1)P(x_2)\ldots P(x_n))$$

In the above equation, $P(x_j)$ represents the probability of occurrence of a successful event x_j, for $j = 1, 2, 3, \ldots, n$.

Let $R_j = P(x_j)$ for $j = 1, 2, 3, \ldots, n$, the above equation becomes

$$R_s = R_1 R_2 R_3 \ldots R_n = \prod_{j=1}^{n} R_j \tag{8.32}$$

where R_j is the reliability of jth unit.

For exponential distribution and constant failure rate (λ_j) for j^{th} unit, we have

$$R_j = e^{-\lambda_j t}$$

Putting the values of R_j, the reliability of the system becomes

$$R_s = e^{-\sum_{j=1}^{n} \lambda_j t} = e^{-(\lambda_1 + \lambda_2 \ldots \lambda_n)t} \tag{8.33}$$

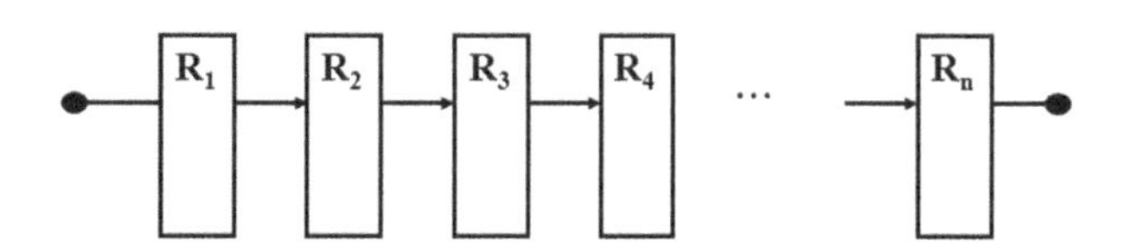

FIGURE 8.6 Series configured system.

The MTBF of the series system with exponential distribution can be determined as

$$MTBF_S = \int_0^{\infty} e^{-(\lambda_1+\lambda_2\ldots\lambda_n)t}\,dt = \frac{1}{\lambda_1+\lambda_2\ldots\lambda_n} \tag{8.34}$$

Example 8.18

A conveyor transportation system transport ore from the mine to processing plants. The system consists of five conveyor belts working in series to transport the ores. The MTBF of five conveyor belts are 720 hours, 600 hours, 480 hours, 600 hours, and 720 hours, respectively. Determine the reliability and MTBF of the system for a 120-hour mission. The failure rates of all the conveyor belts follow an exponential distribution function.

Solution

The system can be represented as shown in Figure 8.7.

The reliability of the individual components for exponential failure density function is given by

$$R_j = e^{-\lambda_j t} = e^{-t/m_j}$$

where λ_j and m_j represent constant failure rate and mean time between failures of jth component.

$$R_1 = e^{-t/m_1} = e^{-120/720} = 0.846$$

$$R_2 = e^{-t/m_2} = e^{-120/600} = 0.818$$

$$R_3 = e^{-t/m_3} = e^{-120/480} = 0.778$$

$$R_4 = e^{-t/m_4} = e^{-120/600} = 0.818$$

$$R_5 = e^{-t/m_5} = e^{-120/720} = 0.846$$

The reliability of the system is given by

$$R_s = R_1R_2R_3R_4R_5 = 0.846*0.818*0.778*0.818*0.846 = 0.372$$

Mean time between failures (MTBF) of the system is given by

$$\text{MTBF} = \frac{1}{\lambda_1+\lambda_2+\lambda_3+\lambda_4+\lambda_5} = \frac{1}{\left(\frac{1}{720}+\frac{1}{600}+\frac{1}{480}+\frac{1}{600}+\frac{1}{720}\right)} = 122 \text{ hours}$$

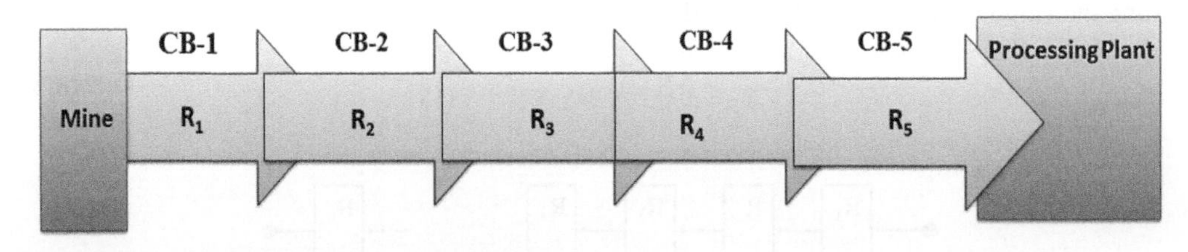

FIGURE 8.7 Conveyor belt transportation system.

8.8.2 System reliability on parallel configuration

If either of the components operates normally in a parallel configured system, the system will work. Suppose a system consists of n number of components configured in parallel, as depicted in the block diagram given in Figure 8.8. In that case, at least one of these components must operate normally for the successful operation of the system.

If $\overline{x}_j$ denotes the event that the jth component in the above-defined system is unsuccessful, then the probability of failure of the system (R_f) with 'n' independent components arranged in parallel configuration is given by

$$R_f = P\left(\overline{x}_1\overline{x}_2\ldots\overline{x}_n\right)$$

Since the components are independents to each other, we have

$$R_f = P\left(\overline{x}_1\right)P\left(\overline{x}_2\right)\ldots P(\overline{x}_n) = \left[1-P\left(x_1\right)\right]\left[1-P\left(x_2\right)\right]\ldots\left[1-P\left(x_n\right)\right]$$

Let the reliability of jth component is $R_j\left[=P\left(x_j\right)\right]$, for $j = 1,2,3,\ldots,n$, the above equation becomes

$$R_f = \left(1-R_1\right)(1-R_2)\left(1-R_3\right)\ldots(1-R_n) = \prod_{j=1}^{n}\left(1-R_j\right)$$

Therefore, the reliability of the successful operation of the system is given by

$$R_s = 1-R_f = 1-\prod_{j=1}^{n}\left(1-R_j\right) \tag{8.35}$$

For exponential distribution and constant failure rate (λ_j) for j[th] component, we have

$$R_j = e^{-\lambda_j t}$$

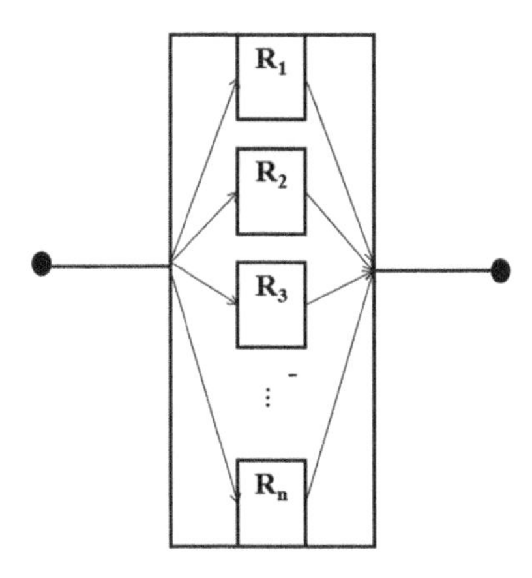

FIGURE 8.8 Parallel configured system.

Therefore, the reliability of the system is

$$R_s = 1 - \prod_{j=1}^{n}(1 - R_j) = 1 - \prod_{j=1}^{n}\left(1 - e^{-\lambda_j t}\right)$$

$$\Rightarrow R_s = 1 - \left[\left(1 - e^{-\lambda_1 t}\right) * \left(1 - e^{-\lambda_2 t}\right) * \ldots \left(1 - e^{-\lambda_n t}\right)\right] \tag{8.36}$$

The MTBF of the parallel system can be determined as

$$MTBF = \int_0^{\infty}\left[1 - \left\{\left(1 - e^{-\lambda_1 t}\right) * \left(1 - e^{-\lambda_2 t}\right) * \ldots \left(1 - e^{-\lambda_n t}\right)\right\}\right] dt$$

For identical failure rate, $\lambda_1 = \lambda_2 = \lambda_3 = \ldots.\ \lambda_n = \lambda$,
The above equation reduces to

$$MTBF_P = \int_0^{\infty}\left[1 - \left\{\left(1 - e^{-\lambda t}\right) * \left(1 - e^{-\lambda t}\right) * \ldots \left(1 - e^{-\lambda t}\right)\right\}\right] dt$$

$$\Rightarrow MTBF_P = \int_0^{\infty}\left[1 - \left(1 - e^{-\lambda t}\right)^n\right] dt = \frac{1}{\lambda}\sum_{j=1}^{n}\frac{1}{j} \tag{8.37}$$

Example 8.19

Three pumps are installed in an open-pit mine to dewater the pit. It was estimated that either of the pumps is enough for successful dewatering of the pit. The failure rates of the three pumps are 0.001 per hour, 0.0005 per hour, and 0.00075 per hour, respectively. Determine the reliability and MTBF of the system for a mission of 500 hours. The failure rates of all the pumps follow an exponential distribution.

Solution

The pump system of the given problem can be represented using a block diagram, as shown in Figure 8.9.

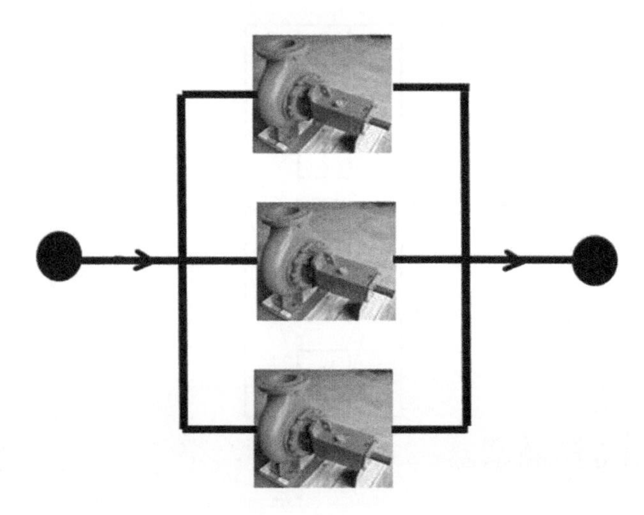

FIGURE 8.9 Block diagram of the defined system.

The reliability of this system is given by

$$R_s = 1 - \prod_{j=1}^{3}\left(1 - e^{-\lambda_j t}\right)$$

$$\Rightarrow R_s = 1 - \left[\left(1 - e^{-\lambda_1 t}\right) * \left(1 - e^{-\lambda_2 t}\right) * \left(1 - e^{-\lambda_3 t}\right)\right]$$

$$\Rightarrow R_s = 1 - \left[\left(1 - e^{-0.001*500}\right) * \left(1 - e^{-0.002*500}\right) * \left(1 - e^{-0.00075*500}\right)\right]$$

$$\Rightarrow R_s = 1 - \left[(1 - 0.60) * (1 - 0.78) * (1 - 0.68)\right]$$

$$\Rightarrow R_s = 1 - (0.4 * 0.22 * 0.32)$$

$$\Rightarrow R_s = 1 - 0.028 = 0.972$$

The MTBF of the system is given by

$$MTBF = \int_0^\infty \left[1 - \left\{\left(1 - e^{-\lambda_1 t}\right) * \left(1 - e^{-\lambda_2 t}\right) * \left(1 - e^{-\lambda_3 t}\right)\right\}\right] dt$$

$$\Rightarrow MTBF = \int_0^\infty \left[1 - (1 - e^{-\lambda_1 t} - e^{-\lambda_2 t} - e^{-\lambda_3 t} + e^{-(\lambda_1 + \lambda_2)t} + e^{-(\lambda_1 + \lambda_3)t} + e^{-(\lambda_3 + \lambda_2)t} - e^{-(\lambda_1 + \lambda_2 + \lambda_3)t}\right] dt$$

$$\Rightarrow MTBF = \int_0^\infty \left[(e^{-\lambda_1 t} + e^{-\lambda_2 t} + e^{-\lambda_3 t} - e^{-(\lambda_1 + \lambda_2)t} - e^{-(\lambda_1 + \lambda_3)t} - e^{-(\lambda_3 + \lambda_2)t} + e^{-(\lambda_1 + \lambda_2 + \lambda_3)t}\right] dt$$

$$\Rightarrow MTBF = \frac{1}{\lambda_1} + \frac{1}{\lambda_2} + \frac{1}{\lambda_3} - \frac{1}{\lambda_1 + \lambda_2} - \frac{1}{\lambda_1 + \lambda_3} - \frac{1}{\lambda_2 + \lambda_3} + \frac{1}{\lambda_1 + \lambda_2 + \lambda_3}$$

$$\Rightarrow MTBF = \frac{1}{0.001} + \frac{1}{0.002} + \frac{1}{0.00075} - \frac{1}{0.001 + 0.002} - \frac{1}{0.001 + 0.00075} - \frac{1}{0.002 + 0.00075} + \frac{1}{0.001 + 0.002 + 0.00075}$$

$$\Rightarrow MTBF = 631.6 \text{ hrs.}$$

8.8.3 System reliability of a combination of series and parallel system

In this type of system, few components are configured in series, and few are parallel, as depicted in Figure 8.10. In a system, there are two sub-systems, A and B. There are m components in sub-system A and n components in sub-system B, whose individual reliability is shown in Figure 8.10.

$$\text{Reliability of sub-system } A = 1 - \left(1 - P_1\right) * \left(1 - P_2\right) * \ldots \left(1 - P_m\right)$$

$$\text{Reliability of sub-system } B = 1 - \left(1 - S_1\right) * \left(1 - S_2\right) * \ldots \left(1 - S_n\right)$$

Reliability of the system is given by

$$R_s = \left[1 - \left(1 - P_1\right) * \left(1 - P_2\right) * \ldots \left(1 - P_m\right)\right] * \left[1 - \left(1 - S_1\right) * \left(1 - S_2\right) * \ldots \left(1 - S_n\right)\right] \qquad (8.38)$$

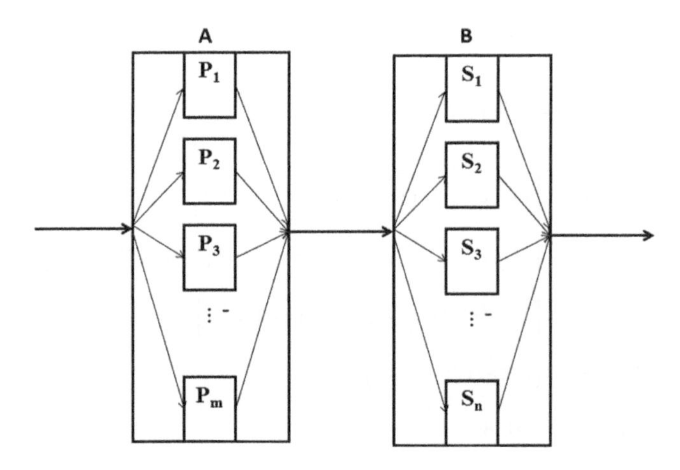

FIGURE 8.10 Combination of series and parallel configured system.

Example 8.20

In an open-pit mine, ores are transported from the production area to the processing plant in three phases, as shown below in the following block diagram (Figure 8.11). The materials are loaded to trucks from the production area by a loader, and then dump trucks transport the materials to a conveyor belt system, which transport the materials to the process plant. The reliability of each component is shown in Figure 8.11. Determine the reliability of the transportation system.

In the given system, five trucks are working in parallel, and the subsystem of dump trucks is working in series with loader and conveyor belt. That is, if any one of the dump trucks works, the system will work, provided both the shovel and conveyor belt are in operation.

Thus, the reliability of the transportation system is given by

$$R_{System} = R_S * \left[1-(1-R_d)^5\right] * R_C$$

$$\Rightarrow R_{System} = 0.95*\left[1-(1-0.9)^5\right]*0.85 = 0.95*0.999*0.85 = 0.80$$

8.8.4 System reliability of k-out-of-n configuration

In this type of system, n components are active, but at least k units must work at a time for a successful operation of the system. This system is a special case of parallel and series configured systems for $k = 1$ and $k = $ n.

Consider a system consists of 'n' identical and independent components. The reliability of each component is R. The system will work provided 'k' components in the system work at a time. Thus, the system can be refereed as the k-out-of-n configured system, and the reliability of the system can be determined using binomial distribution as:

$$\text{Reliability of } k\text{-out-of-}n \text{ configured system} = R_{k/n} = \sum_{i=k}^{n} nC_i R^i (1-R)^{n-i} \tag{8.39}$$

If each component has a constant failure rate, λ, with exponential distribution failure function, the above equation can be reduced to

$$R_{k/n} = \sum_{i=k}^{n} nC_i e^{-i\lambda t} \left(1-e^{-\lambda t}\right)^{n-i} \tag{8.40}$$

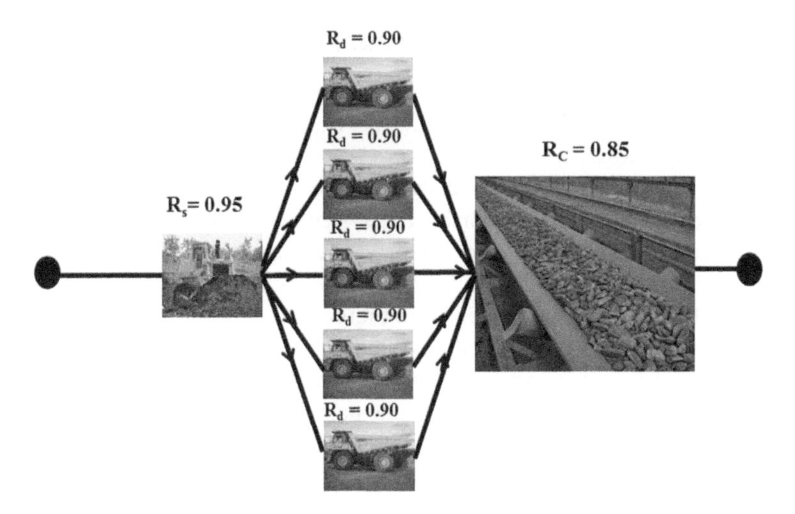

FIGURE 8.11 Transportation system.

The MTBF of the k out of n system can be determined as

$$MTBF_{k/n} = \int_0^\infty \left[\sum_{j=k}^{n} nC_j e^{-j\lambda t} \left(1 - e^{-\lambda t}\right)^{n-j} \right] dt = \frac{1}{\lambda} \sum_{j=k}^{n} \frac{1}{j} \tag{8.41}$$

Example 8.21

Recall **Example 8.19**. If it was found that at least two pumps need to be in operation for successful dewatering of the pit, determine the reliability and MTBF of the system for a mission of 500 hours.

Solution

There are four possibilities that the system will work. Any two pumps will work, and the third will not work in three ways. There is another possibility that the system will work when all three pumps work.

Therefore, the reliability of the system can be calculated as,

$$R_{k/n} = R_{2/3} = R_1 R_2 \overline{R_3} + R_1 \overline{R_2} R_3 + \overline{R_1} R_2 R_3 + R_1 R_2 R_3$$

The reliabilities of the three pump for 500-hour mission was determined as

$$R_1 = 0.60, \ R_2 = 0.78, \ R_3 = 0.68$$

Therefore, the failure probabilities of the three pumps are

$$\overline{R_1} = 1 - R_1 = 0.4$$

$$\overline{R_2} = 1 - R_2 = 0.22$$

$$\overline{R_3} = 1 - R_3 = 0.32$$

So, the system reliability can be calculated by

$$R_{2/3} = 0.60 * 0.78 * 0.32 + 0.60 * 0.22 * 0.68 + 0.40 * 0.78 * 0.68 + 0.60 * 0.78 * 0.68$$

$$\Rightarrow R_{2/3} = 0.769$$

8.8.5 System reliability of bridge configuration

In a system, five independent components are configured in a bridge network, as shown in Figure 8.12. The reliabilities of the five components are R_1, R_2, R_3, R_4, and R_5, respectively, and the direction of arrows shows the flow process.

Any bridge configuration system can be converted into a logic diagram through multiple simple parallel paths between IN and OUT terminals. The system can successfully work, provided all the components in a specified path should work. In the above diagram, no specified path on R_3 is given and thus flow to R_3 can be occurred in both ways like $R_1 \rightarrow R_3$ and $R_2 \rightarrow R_3$. The logical diagram of the bridge network (Figure 8.12) is represented in Figure 8.13.

The logic diagram is a plain series-parallel system whose reliability can be estimated by combining the series-parallel model as explained before but with the consideration interdependency of paths. Since some subsystems exist in more than one path, it is assumed that the failure of each subsystem is independent of the failure of the paths, if the same does not exist in the path. This can be implemented using the following rule:

If $P(x) = P_1 * P_2$ and $P(y) = P_2 * P_3$, then

$$P(x) * P(y) = P_1 * P_2 * P_3$$

That is, $P_i * P_i = P_i$ for any value of i.

The system reliability of a bridge network configuration is given by

$$R_S = 1 - (1 - R_1R_4) * (1 - R_2R_5) * (1 - R_1R_3R_5) * (1 - R_2R_3R_4)$$

$$\Rightarrow R_S = 1 - (1 - R_1R_4 - R_2R_5 - R_1R_3R_5 - R_2R_3R_4 + R_1R_2R_4R_5 + R_1R_3R_4R_5 + R_1R_2R_3R_4 + R_1R_2R_3R_5 + R_2R_3R_4R_5 + R_1R_2R_3R_4R_5 + 4R_1R_2R_3R_4R_5 + R_1R_2R_3R_4R_5$$

$$\Rightarrow R_S = 2R_1R_2R_3R_4R_5 - R_1R_2R_4R_5 - R_1R_3R_4R_5 - R_1R_2R_3R_4 - R_1R_2R_3R_5 - R_2R_3R_4R_5 + R_1R_4 + R_2R_5 + R_1R_3R_5 + R_2R_3R_4$$

For components with equal reliability value, R, the reliability of the system can be re-written as

$$R_S = 2R^5 - 5R^4 + 2R^3 + 2R^2 \tag{8.42}$$

For a constant failure rate, λ, with exponential failure probability, the reliability for time, t is given by

$$R_S(t) = 2e^{-5\lambda t} - 5e^{-4\lambda t} + 2e^{-3\lambda t} + 2e^{-2\lambda t} \tag{8.43}$$

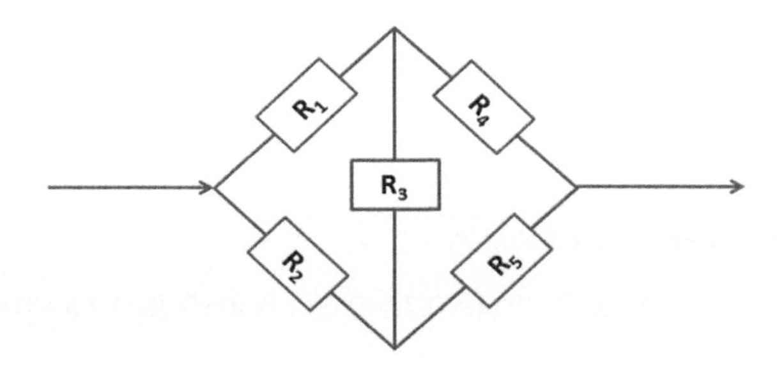

FIGURE 8.12 Bridge network configuration.

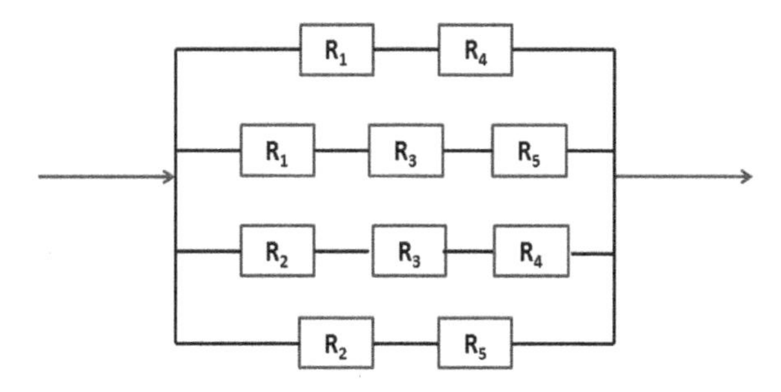

FIGURE 8.13 Logical diagram of the bridge network.

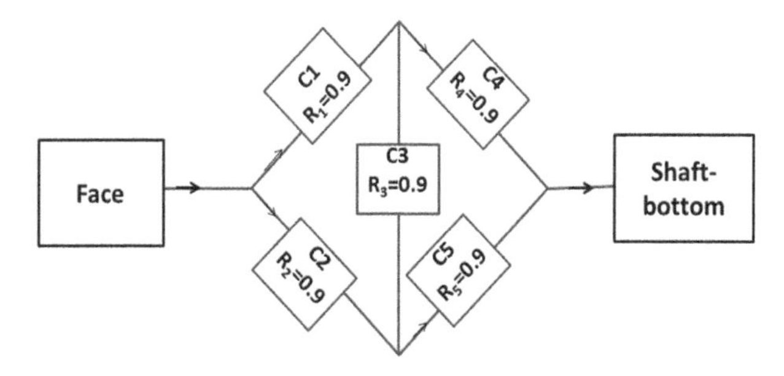

FIGURE 8.14 Configuration of the transportation system.

MTTF of system with exponentially distributed subsystems with constant failure rates can be determined as

$$MTTF = \int_0^\infty R_S(t)\,dt$$

$$\Rightarrow MTTF = \int_0^\infty \left(2e^{-5\lambda t} - 5e^{-4\lambda t} + 2e^{-3\lambda t} + 2e^{-2\lambda t}\right)dt$$

$$\Rightarrow MTTF = \left(\frac{49}{60}\right)\frac{1}{\lambda}$$

Example 8.21

The transportation system in an underground mine consists of 5 conveyor belts for transporting coal from the face to the shaft bottom transportation point. The configuration of the conveyor belts is shown in Figure 8.14. The reliability of each conveyor belt is 0.9. There are multiple paths to send coal from the face to the shaft bottom transportation point. The flows of coal in four conveyors (C1, C2, C4, and C5) are unidirectional, and conveyor C3 is bidirectional. Determine the reliability of the transportation system.

Solution

The reliability of the system can be calculated as

$$R_S = 2R^5 - 5R^4 + 2R^3 + 2R^2 = 2*0.9^5 - 5*0.9^4 + 2*0.9^3 + 2*0.9^2$$

$$= 1.18098 - 3.2805 + 1.458 + 1.62 = 0.97848$$

Example 8.22

In Example 8.21, if the flows of coal in all the five conveyors (C1, C2, C3, C4, and C5) are unidirectional in nature (shown in Figure 8.15), determine the reliability of the system.

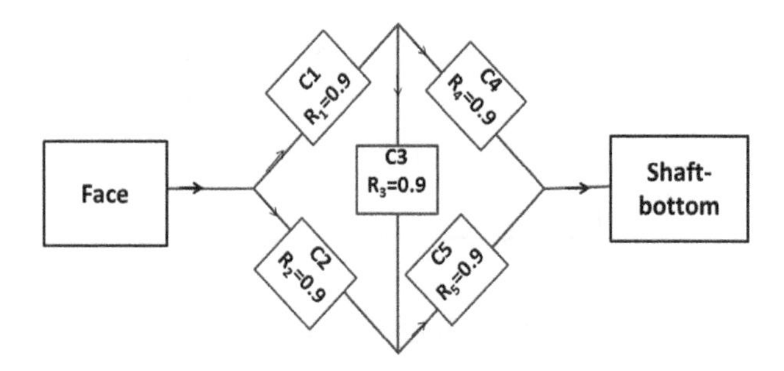

FIGURE 8.15 Configuration of the system.

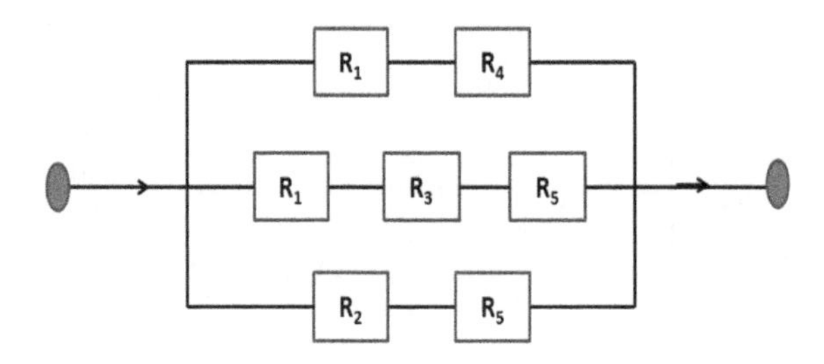

FIGURE 8.16 Logical diagram of the bridge network.

Solution

The system can successfully work provided all the elements in at least one continuous path between IN and OUT terminals successfully work. The logic diagram for the system (Figure 8.15) is represented in Figure 8.16. The subsystem R_3 is unidirectional like any other subsystem.

The reliability of the above network configuration system is given by

$$R_S = 1-(1-R_1R_4)*(1-R_2R_5)*(1-R_1R_3R_5)$$

$$\Rightarrow R_S = R_1R_4 + R_2R_5 + R_1R_3R_5 - R_1R_4R_2R_5 - R_1R_4R_3R_5 - R_1R_3R_2R_5 + R_1R_2R_3R_4R_5$$

For, $R_1 = R_2 = R_3 = R_4 = R_5 = R$

$$\Rightarrow R_S = R^2 + R^2 + R^3 - R^4 - R^4 - R^4 + R^5 = 2R^2 + R^3 - 3R^4 + R^5$$

$$\Rightarrow R_S = 2*0.9^2 + 0.9^3 - 3*0.9^4 + 0.9^5 = 1.62 + 0.729 - 1.9683 + 0.59049$$

$$\Rightarrow R_S = 0.97119$$

8.8.6 System reliability of standby redundancy

The standby redundancy represents a situation with one unit is in operation, and n other identical units are in standby mode. The configuration of this is shown in Figure 8.17. In contrast to the parallel redundancy where all units are in operation, all the components in standby redundancy are not active. The reliability of the system involves $(n+1)$ components, in which one component is operating, and the rest of the components are on standby mode till the operating component fails.

For 'n' identical units with exponential pdf, the reliability can be determined as

$$R_{St}(t) = \sum_{i=0}^{n} \frac{(\lambda t)^i e^{-\lambda t}}{i!} \tag{8.44}$$

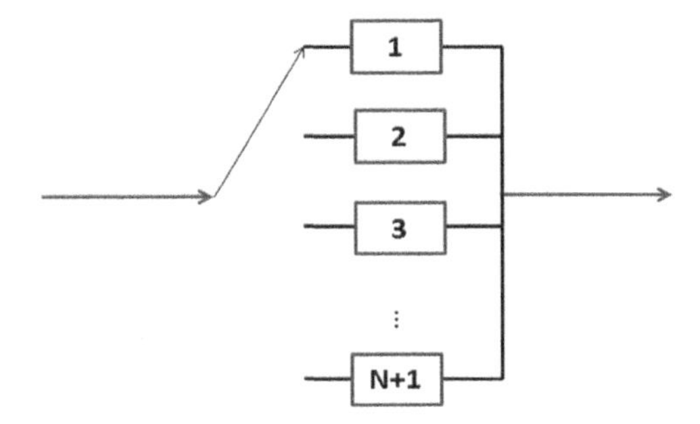

FIGURE 8.17 Standby redundancy system.

The above equation holds good in the following conditions:

- Switching arrangement is ideal.
- All the components in the system are identical, independent, and have a constant failure rate.
- Standby components are as good as new.

For two non-identical units (one active and one standby) with exponential pdf, the reliability can be determined as

$$R_{S_t}(t) = \frac{\lambda_2}{\lambda_2 - \lambda_1} e^{-\lambda_1 t} + \frac{\lambda_1}{\lambda_1 - \lambda_2} e^{-\lambda_2 t} \tag{8.45}$$

where λ_1 and λ_2 are the failure rate of active and standby unit, respectively.

For three non-identical units (one active and two standby) with exponential pdf, the reliability can be determined as:

$$R_{S_t}(t) = \frac{\lambda_2\lambda_3}{(\lambda_2 - \lambda_1)(\lambda_3 - \lambda_1)} e^{-\lambda_1 t} + \frac{\lambda_1\lambda_3}{(\lambda_1 - \lambda_2)(\lambda_3 - \lambda_2)} e^{-\lambda_2 t} + \frac{\lambda_1\lambda_2}{(\lambda_1 - \lambda_3)(\lambda_2 - \lambda_3)} e^{-\lambda_3 t} \tag{8.46}$$

Example 8.23

Two identical mine pumps working in a parallel configuration deployed in a mine for dewatering operation. The capacity of one pump is sufficient for dewatering, but the second pump was deployed as standby and will be automatically activated when the first pump fails. If the failure rate for each component is 0.001 failure/hour, determine the reliability for 300 hours of operation.

Solution

The system reliability of the pump is

$$R_{S_t}(t) = \sum_{i=0}^{1} \frac{(\lambda t)^i e^{-\lambda t}}{i!}$$

$$\Rightarrow R_{S_t}(t) = \frac{(\lambda t)^0 e^{-\lambda t}}{0!} + \frac{(\lambda t)^1 e^{-\lambda t}}{1!} = e^{-\lambda t} + (\lambda t)e^{-\lambda t}$$

$$\Rightarrow R_{S_t}(t) = e^{-0.001*300} + (0.001*300)e^{-0.001*300} = 0.74(1+0.3) = 0.963$$

Example 8.24

In Example 8.23, if the failure rate of the working pump is 0.001 failure/hour and the standby unit has a failure rate of 0.002 failures/hour. If the reliability of the switch can be considered to be 1 (perfect), determine the reliability for 200 hours.

Solution

The reliability of the system is given by

$$R_{St}(t) = \frac{\lambda_2}{\lambda_2 - \lambda_1} e^{-\lambda_1 t} + \frac{\lambda_1}{\lambda_1 - \lambda_2} e^{-\lambda_2 t}$$

$$\Rightarrow R_{St}(t) = \frac{0.002}{0.002 - 0.001} e^{-0.001*200} + \frac{0.001}{0.001 - 0.002} e^{-0.002*200} = 0.967$$

8.9 AVAILABILITY

The availability of a system (or machine) can be defined as the probability that the system is available for performing the desired job. In other words, it is the ratio of time the system available for a successful operation to the total scheduled time. The total schedule time includes both down time/breakdown time and up time of the system. Availability is generally used to measure the performance of a repairable system.

Mathematically, it can be represented as:

$$\text{Availability} = \frac{\text{System up time}}{\text{System up time} + \text{System down time}} \tag{8.47}$$

For example, the up time and down time of shovel in an open-pit mine is recorded for an 8-hour working shift, as shown in Figure 8.18.

The down time of the shovel is from 2 to 2.5 hours and from 6 to 7 hours. Therefore, the total down time of the shovel is 1.5 hours (= 0.5+1) out of an 8-hour working shift. The up time hour of the shovel is 6.5 hours (= 8 – 1.5). Therefore, the availability of the shovel for the recorded shift is given by

$$\text{Availability} = \frac{\text{System up time}}{\text{System up time} + \text{System down time}} = \frac{6.5}{8} = 0.8125 \text{ or } 81.25\%$$

Availability can also be represented as the ratio of mean time between failures (MTBF) to the sum of MTBF and mean time to repair (MTTR). Mathematically, it can be represented as

$$\text{Availability} = \frac{\text{MTBF}}{\text{MTBF} + \text{MTTR}} \tag{8.48}$$

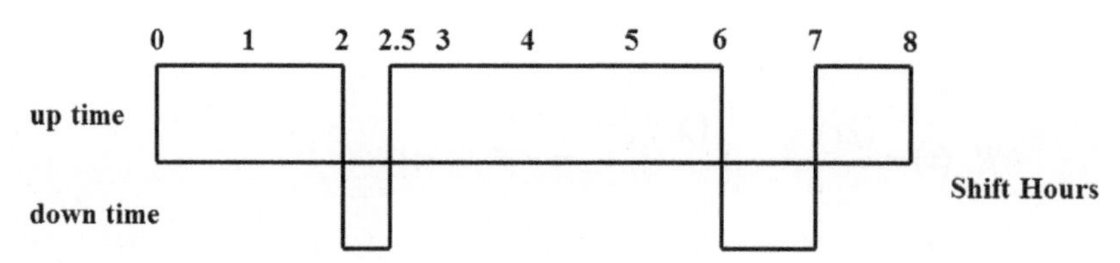

FIGURE 8.18 Performance of shovel.

In the above example, the number of failures in an 8-hour shift were observed as two, and the total up time and down time of the shovel were recorded as 6.5 hours and 1.5 hours, respectively. Therefore, the MTBF and MTTR for the above example can be determined as

$$\text{MTBF} = \frac{6.5}{2}$$

$$\text{MTTR} = \frac{1.5}{2}$$

Therefore,

$$\text{Availability} = \frac{\text{MTBF}}{\text{MTBF} + \text{MTTR}} = \frac{6.5/2}{6.5/2 + 1.5/2} = \frac{6.5}{8} = 0.8125 \text{ or } 81.25\%$$

The aim of the reliability analysis is to reduce the MTTR for increasing the availability of the systems. By reducing the value of MTTR, the performance of the system can be improved for more economical operation. It may always happen in the mines that the faults in the system or machine are difficult to diagnose, and thus, it increases the down time of the system or machine.

For a non-repairable system, the availability of the system represents the reliability or life of the system. But, for repairable systems, availability is equal to or greater than reliability.

Example 8.26

A pumping unit is installed in an underground mine for dewatering operation has a reliability of 0.95 for 800 hours of successful operation. The pump's failure probability function follows an exponential distribution. The availability of the pump during the same period is 0.98. If the hazard rates of failure and repair are constant, determine the MTBF and MTTR.

Solution

We have,

$$R(t) = e^{-\lambda t}$$

where λ is constant failure rate.

Given $R(t)$ is equal to 0.95 and t is equal to 800, we calculate

$$\Rightarrow 0.95 = e^{-\lambda * 800}$$

$$\Rightarrow \lambda = -\frac{\ln(0.95)}{800} = -\frac{-0.051}{800} = 6.41 * 10^{-5}$$

From failure rate λ, we can calculate *MTBF*

$$\text{MTBF} = 1/\lambda = 15600.62 \text{ hours}$$

Again, we know

$$\text{Availability}(A) = \frac{\text{MTTF}}{\text{MTTF} + \text{MTTR}} = \frac{15600.62}{15600.62 + \text{MTTR}}$$

$$\Rightarrow 0.98 = \frac{15600.62}{15600.62 + \text{MTTR}}$$

$$\Rightarrow \text{MTTR} = \frac{15600.62 * (1 - 0.98)}{0.98} = 318.38 \text{ hours}$$

8.10 IMPROVEMENT OF SYSTEM RELIABILITY

The improvement of reliability of a system is always a challenging task for engineers due to the decision-making process of identifying the individual component(s) that helps to improve the overall reliability of the system.

The improvement of system reliability can be made by following two approaches:

- **Fault avoidance**: The fault in the system can be avoided or minimized by using highly reliable components.
- **Fault tolerance**: This approach, known as the redundancy approach, tolerates the fault with the use of redundant components in the system. The drawbacks of the redundancy approach are increased design complexity of the system along with the costs, weight, and space requirement. In general, the approach is relatively expensive than the fault avoidance approach.

8.10.1 Redundancy optimization

The optimization is a critical step for improving the reliability of a system using the redundancy approach. The optimization technique is used for designing a system to achieve the desired reliability or maximize the reliability by using redundant components with optimized cost. In this optimization method, similar components are added in parallel at each stage to enhance the reliability of the system with optimum size or units, cost, and complexity. Thus, it is required to optimize the cost for obtaining the desired reliability of the system using the redundancy method. The cost and reliability of a simple parallel configured system can be represented as

$$\text{Optimum Cost} = f(\text{cost, number of parallel element, method of redundancy})$$

$$\text{Desired Reliability} = g(\text{reliability, number of parallel element, method of redundancy})$$

Thus, the aim of the redundancy approach is to achieve the maximum or desired reliability with an optimal number of redundant identical units (n) for a given budgetary cost or minimum cost. The method of redundancy approach is demonstrated with **Examples 8.27** and **8.28**.

Example 8.27

In a mining system, three machines must operate in series for a successful operation with a budgetary cost of US$700,000. The reliabilities and cost of each unit are shown in Table 8.1. Determine the optimum number of redundant units based on the marginal analysis. Also, compute the improved reliability of the system.

Solution

In marginal analysis, the optimum number of redundant units is decided by examining the additional benefits obtained by the addition of a unit compared to the additional costs incurred by that same activity. The results of the marginal analyses are summarized in Table 8.2.

TABLE 8.1
Reliability and cost of each unit

Machine	Reliability	Unit Cost (US$$10^3$)
1	0.80	200
2	0.90	100
3	0.95	75

TABLE 8.2
Marginal analyses results of different system configuration

Type	System Configuration	System Reliability	Cost (US$10³)	Remarks
A	R_1=0.80 R_2=0.90 R_3=0.95	0.684	375	
B	R_1=0.80 R_2=0.90 R_3=0.95 R_1=0.80	0.821	575	
C	R_1=0.80 R_2=0.90 R_3=0.95 R_2=0.90	0.752	475	
D	R_1=0.80 R_2=0.90 R_3=0.95 R_3=0.95	0.718	450	
E	R_1=0.80 R_2=0.90 R_3=0.95 R_1=0.80 R_1=0.80	0.848	775	Cost exceeded the available budget
F	R_1=0.80 R_2=0.90 R_3=0.95 R_2=0.90 R_2=0.90	0.759	575	
G	R_1=0.80 R_2=0.90 R_3=0.95 R_3=0.95 R_3=0.95	0.720	525	
H	R_1=0.80 R_2=0.90 R_3=0.95 R_1=0.80 R_2=0.90	0.903	675	
I	R_1=0.80 R_2=0.90 R_3=0.95 R_1=0.80 R_2=0.90 R_3=0.95	0.948	750	Cost exceeded the available budget

Given that, three components of the machine are connected in series. The reliabilities of three components for a certain time are R_1 (=80%), R_2 (=90%), and R_3 (=95%), respectively. The unit cost of each machine are C1 (=US$200000), C2 (=US$100000), and C3 (=US$75000), respectively. The objective of the study is to obtain the maximum reliability of the system with a budgetary cost of US$700,000.

The existing reliability of the system (Type A) is given by:

$$R_S = R_1 R_2 R_3 = 0.8 * 0.9 * 0.95 = 0.684$$

The total cost of the system having a single unit of each machine is given by

$$\text{Total cost} = \text{C1} + \text{C2} + \text{C3} = 200000 + 100000 + 75000 = \text{US\$}375{,}000$$

The first component has the lowest reliability (= 0.80) in the given system but has the highest input unit cost. In the first step of marginal analysis, the reliability of the system is examined with the addition of one redundant unit in parallel to the same one by one. The reliabilities of the systems for three cases are demonstrated by system configuration B, C, and D, respectively. The corresponding costs of the system are US$575,000, US$475,000, and US$45,000, respectively. The above analyses indicate that the reliability of the system cannot be increased desirably without adding a redundancy unit of the first component.

Since the incurred cost is still less than the budgetary cost, two redundant units can be added with additional cost. The results further indicate that two redundancy units of component 1 (condition E) increases the reliability of the system to 0.848. But, the cost of the system exceeded the budget. It can also be inferred that one redundancy unit of component 1 and component 2 each (condition H) increases the reliability of the system to 0.903 without exceeding the budget. The reliability of the system can be improved to 0.948 (Condition I) with one redundancy unit of all three components, but the cost exceeded the budget. Thus, the optimum condition is H.

Example 8.28

A mining machine consists of three components in series has the following characteristics, as shown in Table 8.3. The reliability goal is 0.90 for the system. Construct the optimum redundancy network for achieving the desired reliability by minimizing the cost.

Solution

Given that, three components of the machine are connected in series. The reliabilities of three components for a certain time are R_1 (=70%), R_2 (=80%), and R_3 (=90%), respectively. Reliability of R_G (=0.90%) is desired for the machine.

The existing reliability of the machine is given by:

$$R_S = R_1 R_2 R_3 = 0.504$$

The existing reliability (0.504) of the machine is less than the desired reliability of 0.90%. Thus, the reliability of the individual component of the system needs to be enhanced to achieve the desired goal. The change in the reliability of the machine will be examined by increasing the reliability of one component at a time. Table 8.4 shows the system configuration by adding one redundancy unit at each stage, one at a time.

TABLE 8.3
Reliability and cost of each unit

Components	Reliability	Unit Cost (US$$10^3$)
1	0.70	300
2	0.80	400
3	0.90	500

TABLE 8.4
Marginal analyses results of the machine configuration by adding one redundancy component

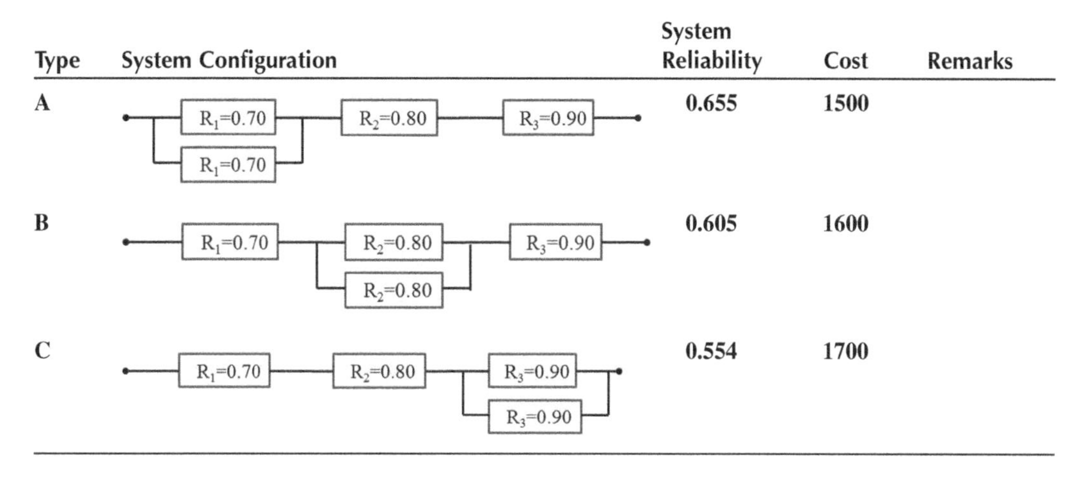

Type	System Configuration	System Reliability	Cost	Remarks
A	R_1=0.70, R_1=0.70, R_2=0.80, R_3=0.90	0.655	1500	
B	R_1=0.70, R_2=0.80, R_2=0.80, R_3=0.90	0.605	1600	
C	R_1=0.70, R_2=0.80, R_3=0.90, R_3=0.90	0.554	1700	

TABLE 8.5
Marginal analyses results of the machine configuration by adding two redundancy components

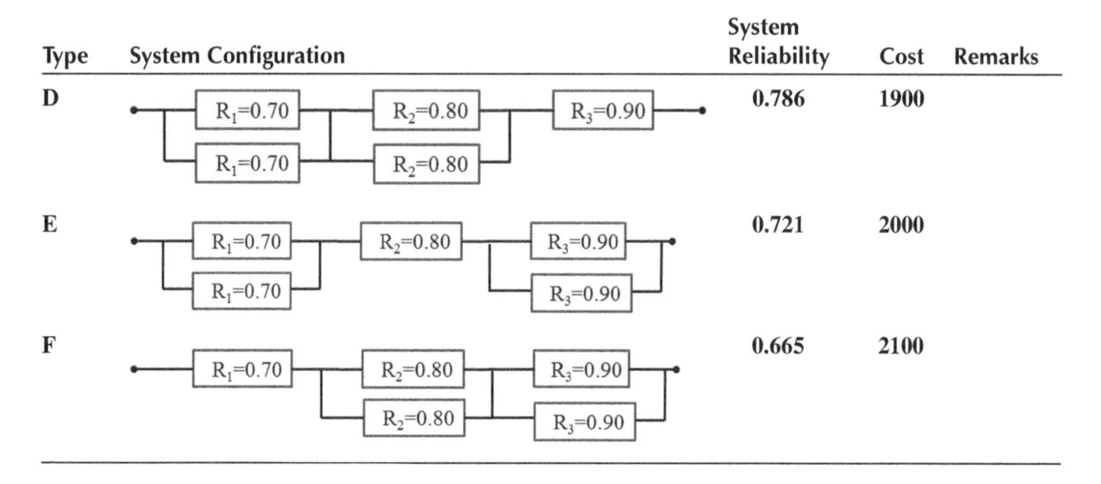

Type	System Configuration	System Reliability	Cost	Remarks
D	R_1=0.70, R_1=0.70, R_2=0.80, R_2=0.80, R_3=0.90	0.786	1900	
E	R_1=0.70, R_1=0.70, R_2=0.80, R_3=0.90, R_3=0.90	0.721	2000	
F	R_1=0.70, R_2=0.80, R_2=0.80, R_3=0.90, R_3=0.90	0.665	2100	

The results shown in Table 8.4 indicate that the redundancy of one component cannot improve the reliability of the system up to the desired level. Now, we will try increasing the reliability of two components at a time to see whether the reliability goal can be achieved. Table 8.5 shows different configuration results of two components' redundancy.

Still, the reliability of the system is far short of the desirable reliability performance. Now, we will try increasing the reliability of three components at a time to see whether the reliability goal can be achieved. The results are summarized in Table 8.6.

The redundancy of all three components cannot facilitate the desired level of reliability performance. Therefore, we will examine the system reliability with two redundant components of the first component and one each for the rest of the component. The results are summarized in Table 8.7.

TABLE 8.6
Marginal analyses results of the machine configuration by adding three redundancy components

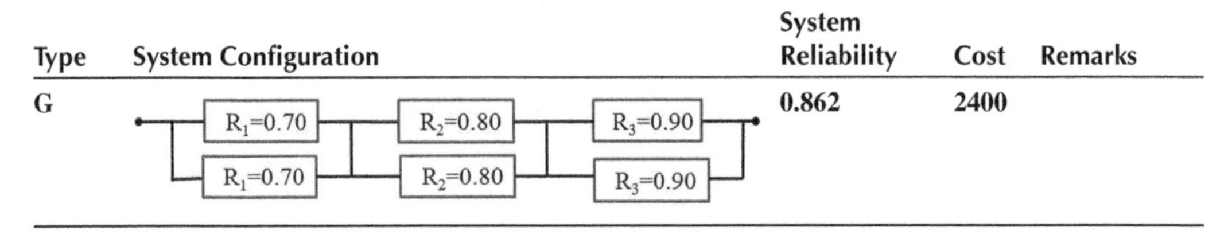

Type	System Configuration	System Reliability	Cost	Remarks
G	R_1=0.70, R_1=0.70 – R_2=0.80, R_2=0.80 – R_3=0.90, R_3=0.90	0.862	2400	

TABLE 8.7
Marginal analyses results of the machine configuration by adding two redundant components of the first component and one each for the rest of the component

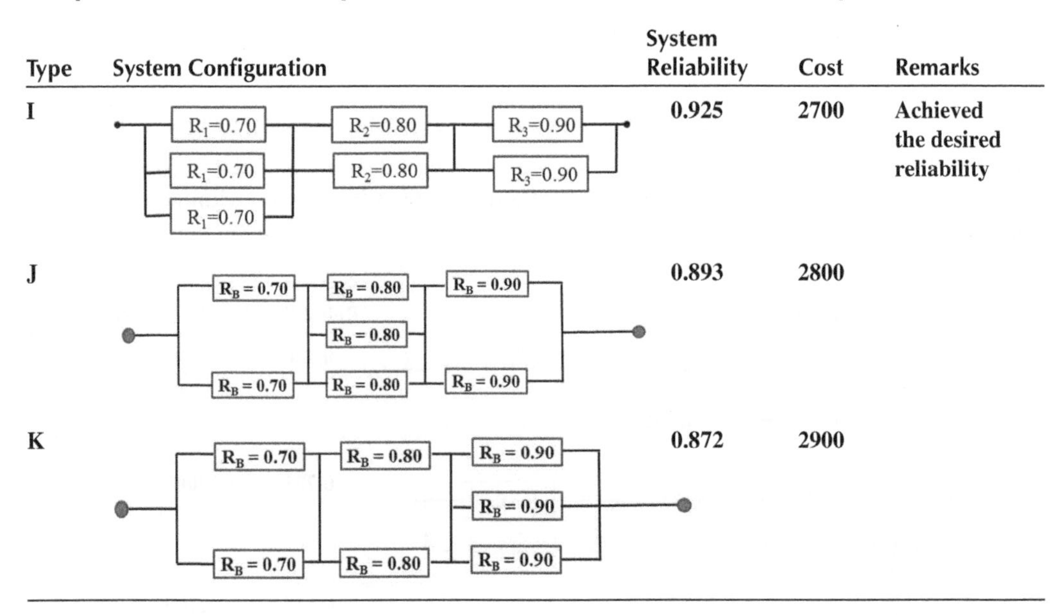

Type	System Configuration	System Reliability	Cost	Remarks
I	R_1=0.70, R_1=0.70, R_1=0.70 – R_2=0.80, R_2=0.80 – R_3=0.90, R_3=0.90	0.925	2700	Achieved the desired reliability
J	R_B = 0.70, R_B = 0.70 – R_B = 0.80, R_B = 0.80, R_B = 0.80 – R_B = 0.90, R_B = 0.90	0.893	2800	
K	R_B = 0.70, R_B = 0.70 – R_B = 0.80, R_B = 0.80 – R_B = 0.90, R_B = 0.90, R_B = 0.90	0.872	2900	

In the next step, the reliabilities of two components are enhanced at a time to improve the reliability of the machine. It is also suggested that increasing the reliability of two components at a time shall not provide the desired goal. Furthermore, the desired goal cannot be achieved with the redundancy of all three components. The condition, I, when there is two redundancy unit of the first component and single redundancy unit of each for rest of the component, reliability of the system is more than the desired performance with a lower cost than the conditions J and K. It can be inferred from the analysis that the reliability of the system should be improved by targeting the lowest reliability component.

The sensitivity analysis of the change in the reliability of the machine with the changing of individual component's reliability is shown in Figure 8.19. The results indicate that even by enhancing the reliability of individual components to a hypothetical value of 1 (R=100%), the reliability of the machine cannot facilitate the desired goal by improving the reliability of one component at a time. Thus, the reliability goal can be achieved with the addition of identical redundant components in parallel. The sensitivity analysis of the change in the reliability of the machine with the addition of identical redundant components of different reliability is shown in Figure 8.20.

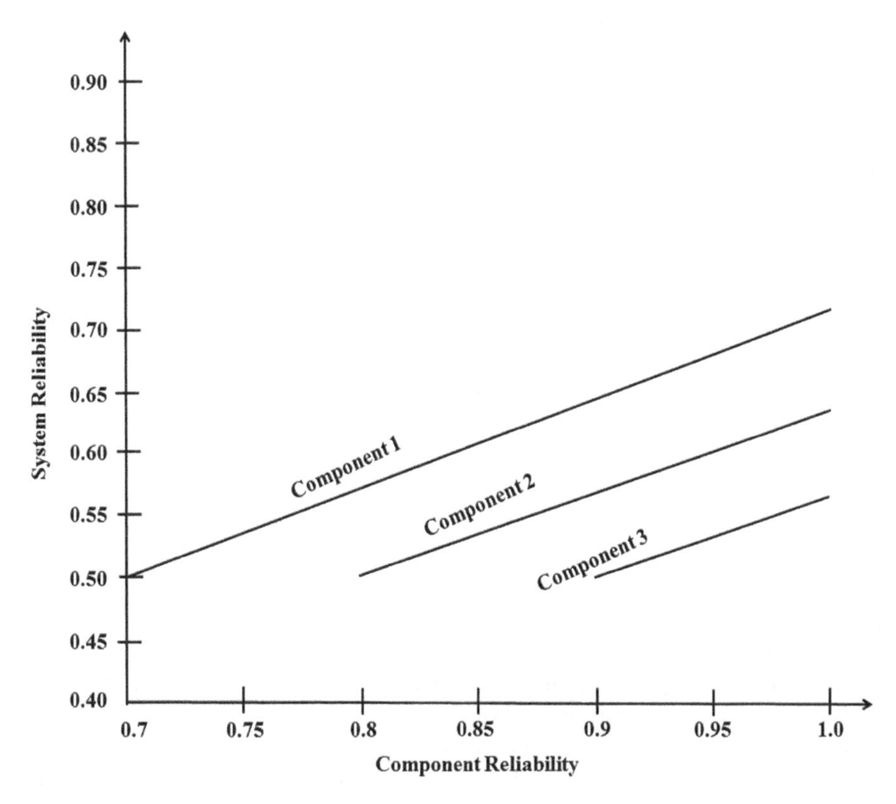

FIGURE 8.19 Changes in system's reliability with change in component's reliability.

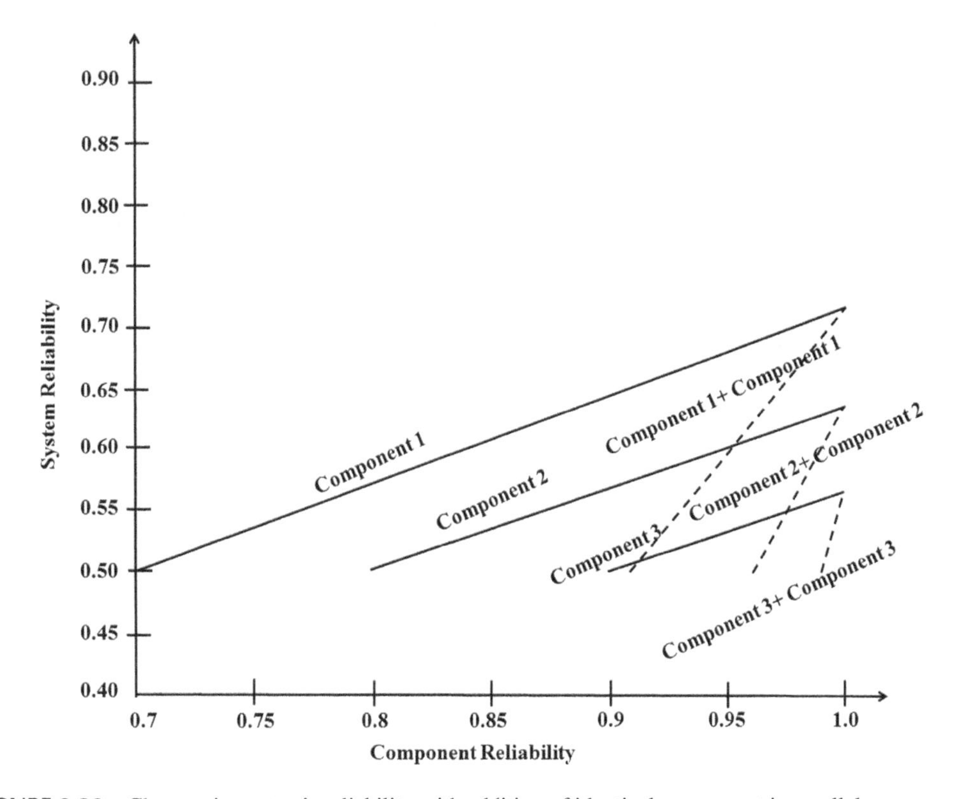

FIGURE 8.20 Changes in system's reliability with addition of identical component in parallel.

8.11 RELIABILITY ANALYSIS TO A MINE SYSTEM: A CASE STUDY

8.11.1 Introduction

The reliability analyses of the mine system are important tasks to prevent the unpredicted breakdown of the production system. In the mine, many types of HEMMs are deployed for operation in harsh environments. Thus, scheduled maintenance of the system is needed to eliminate the system's undesirable failure. The size and complexity of the HEMMs have been increased in the last few decades, and hence improvement in the reliability or availability of mine systems is a challenging task (Aven, 2006). In the past, many studies are conducted to analyse the reliability of different types of machines and mine systems for the improvement of the performances (Barabady and Kumar, 2008; Samanta et al., 2004; Vayenas et al., 2009; Gustafson et al., 2013; Roy et al., 2001; Samanta et al., 2002; Hoseinie et al., 2011a; Wang et al., 2012; Hoseinie et al., 2011b; Al-Chalabi et al., 2014; Rahimdel et al., 2013; Allahkarami et al., 2016; Morad et al., 2014; Uzgören et al., 2010; Demirel et al., 2014; Gupta et al., 2011; Gharahasanlou et al., 2014; Yunusa-Kaltungo et al., 2017). The other critical application areas of reliability analysis in the mining industry are preventive replacement decision of the equipment (Demirel and Gölbaşı, 2016; Palei et al., 2020), maintainability (Barabady and Kumar, 2008), availability (Kajal et al., 2013), total productive maintenance (Kalra et al., 2015), and impact of climate conditions on the failure rates (Furuly et al., 2013). Recently, researchers have successfully applied different forecasting algorithms to forecast failure time from historical mining equipment failure data (Chatterjee and Bandopadhyay, 2012; Chatterjee et al., 2014; Paithankar and Chatterjee, 2018).

Chatterjee and Bandopadhyay (2012) proposed a genetic algorithm (GA)-based artificial neural network model for the prediction of the reliability of a load haul dump (LHD) deployed in a coal mine. The study results indicated that the proposed model predicts the reliability of the LHD satisfactorily.

The case study, presented herein, analyses the reliability of a conveyor belt installed in a mine for coal transportation. The data for the case study mine was taken from Gorai et al. (2017).

8.11.2 Data

The production of coal in the case study mine was simultaneously done in two different working faces and subsequently transported through series of seven conveyor belts, as shown in Figure 8.21. The production line is affected by the unwanted failure of the system. The present case study demonstrates the reliability analysis of one conveyor belt. In a similar way, the reliability of other belts and subsequently for the entire conveyor belt system can be analysed.

The first step of the reliability assessment is the collection of failure and repair time data of the defined system (in this case, conveyor belt). The failure and repair time data of the conveyor belt for six months (1 July 2015 to 31 December 2015), as shown in Table 8.8, was taken from Gorai et al. (2017).

8.11.3 Exploratory data analysis

Once the failure and repair data are collected for a machine/system, the next step is to determine the time between failures (TBF) and time to repair (TTR) from the failure and repair data. TBF and TTR represent the duration between two consecutive failures and the repair of a particular failure event. The cumulative TBF (CTBF) and cumulative TTR (CTTR) were determined from the TBF and TTR data. All these calculated data (TBF, TTR, CTBF, and CTTR) are represented in Table 8.9.

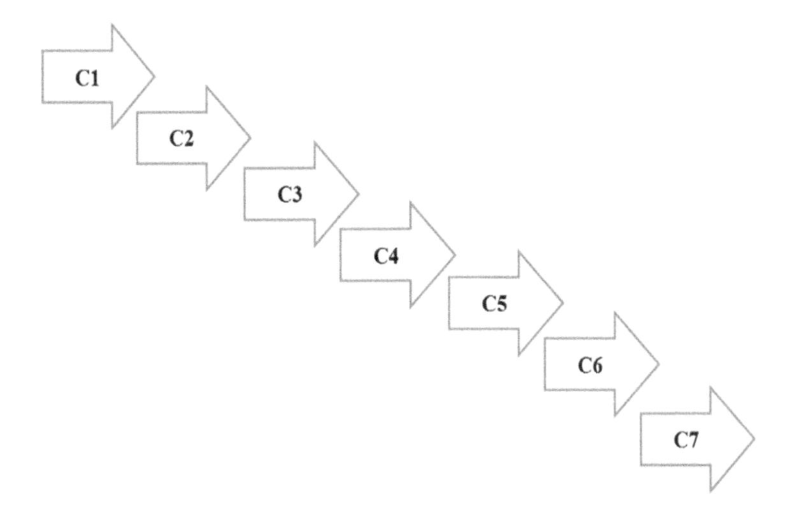

FIGURE 8.21 Transportation system from working face to pit bottom of the mine consists of seven conveyors arranged in series.

The TBF and TTR data should be independent and identical distribution (IID) for reliability analysis. The IID test can be done using trend and serial correlation tests (Kumar et al., 1989). The same can be validated using the U-statistics test (Sen 1992). In the trend test, the CTBF and CTTR data were plotted against the cumulative number of failures or the cumulative number of repairs, respectively, as shown in Figure 8.22. It is desired that the line drawn through the data points should follow an approximately straight line in order to ensure data are free from any trend, which implies that the data set is IID (Kumar et al. 1989).

In the serial correlation test, the $(i\text{-}1)^{th}$ TBF or TTR data are plotted against the ith TBF or TTR data, respectively (Figure 8.22). If the data points are randomly scattered without any clear pattern, it implies a data set free from serial correlation, which implies that the data are independent (Kumar et al. 1989).

In the existing case, the trend test and serial correlation test results clearly indicate that the TBF and TTR dataset for the conveyor belt are independent and identically distributed and thus can be used for reliability analysis.

8.11.4 Estimating the best fit probability density function (PDF) for TBF and TTR

After examining the data for IID, the best fit PDF are estimated separately for the TBF and TTR datasets. The best fit analyses can be conducted based on the modified Kolmogorov-Smirnov (K-S) test (Chakravarty, Laha, and Roy, 1967). The current study examined the five PDF (exponential 2-parameter, normal, lognormal 3-parameter, Weibull 3-parameter, and Weibull 2-parameter) for fitting the TBF and TTR datasets. The generalized form of five PDF (exponential 2-parameter, normal, lognormal 3-parameter, Weibull 3-parameter, and Weibull 2-parameter) are as follows:

- **A Weibull distribution with 3-Parameter**

$$f(t) = \alpha \beta^{-\alpha} (t-\gamma)^{\alpha-1} e^{-\left(\frac{t-\gamma}{\beta}\right)^{\alpha}} \tag{8.49}$$

TABLE 8.8
Failure and repair time data of the system

Sl. No.	Date of failure (Time)	Date of repair (Time)	Sl. No.	Date of failure (Time)	Date of repair (Time)	Sl. No.	Date of failure (Time)	Date of repair (Time)
1	03/07/15 (10:45)	03/07/15 (12:15)	19	01/09/15 (11:30)	01/09/15 (13:15)	37	05/11/15 (14:35)	05/11/15 (15:45)
2	07/07/15 (19:30)	07/07/15 (20:15)	20	04/09/15 (07:20)	04/09/15 (10:45)	38	09/11/15 (11:15)	09/11/15 (12:00)
3	09/07/15 (07:30)	09/07/15 (10:45)	21	6/09/15 (14:45)	06/09/15 (15:25)	39	15/11/15 (10:00)	15/11/15 (11:25)
4	13/07/15 (22:15)	14/7/15 (6:30)	22	10/09/15 (05:10)	10/09/15 (06:30)	40	17/11/15 (23:25)	18/11/15 (3:30)
5	18/07/15 (11:25)	18/7/15 (12:45)	23	13/09/15 (20:15)	13/09/15 (21:30)	41	21/11/15 (17:30)	21/11/15 (19:25)
6	21/07/15 (7:25)	21/7/15 (8:30)	24	18/09/15 (19:30)	18/09/15 (20:30)	42	26/11/15 (5:45)	26/11/15 (6:20)
7	22/07/15 (17:55)	22/7/15 (20:15)	25	19/09/15 (10:30)	19/09/15 (17:50)	43	27/11/15 1(8:30)	27/11/15 (19:00)
8	26/7/15 (10:25)	26/7/15 (13:35)	26	25/09/15 (19:25)	25/09/15 (20:15)	44	30/11/15 (10:10)	30/11/15 (11:00)
9	29/7/15 (6:20)	29/7/15 (7:45)	27	30/09/15 (6:30)	30/09/15 (7:30)	45	01/12/15 (13:00)	01/12/15 (14:30)
10	02/08/15 (5:20)	2/08/15 (7:15)	28	03/10/15 (14:35)	03/10/15 (15:40)	46	03/12/15 (07:15)	03/12/15 (07:45)
11	05/08/15 (10:15)	5/08/15 (11:25)	29	05/10/15 (06:30)	05/10/15 (07:15)	47	08/12/15 (10:25)	08/12/15 (14:25)
12	07/08/15 (22:20)	08/08/15 (02:45)	30	09/10/15 (07:45)	09/10/15 (08:30)	48	10/12/15 (19:10)	10/12/15 (21:40)
13	12/08/15 (6:30)	12/08/15 (07:15)	31	16/10/15 (15:25)	16/10/15 (20:40)	49	15/12/15 (14:25)	15/12/15 (15:30)
14	14/08/15 (18:20)	14/08/15 (20:30)	32	18/10/15 (18:35)	18/10/15 (19:40)	50	18/12/15 (17:00)	19/12/15 (10:30)
15	18/08/15 (12:45)	18/08/15 (13:50)	33	21/10/15 (13:30)	21/10/15 (14:00)	51	23/12/15 (11:20)	23/12/15 (11:45)
16	21/08/15 (22:35)	22/08/15 (1:30)	34	25/10/15 (20:00)	25/10/15 (22:15)	52	26/12/15 (1:25)	26/12/15 (2:00)
17	25/08/15 (7:45)	25/08/15 (10:25)	35	29/10/15 (3:15)	29/10/15 (7:30)	53	29/12/15 (17:50)	30/12/15 (12:40)
18	27/08/15 (13:20)	27/08/15 (14:50)	36	04/11/15 (06:15)	04/11/15 (07:00)			

Source: Gorai et al. 2017.

TABLE 8.9
TBF, TTR, CTBF, and CTTR for the system

Sl. No	TBF in hour	TTR in hour	CTBF in hour	CTTR in hour	Sl. No	TBF in hour	TTR in hour	CTBF in hour	CTTR in hour
1	104.75	1.50	104.75	1.50	28	39.92	1.08	2248.09	61.32
2	36.00	0.75	140.75	2.25	29	97.25	0.75	2345.34	62.07
3	110.75	3.25	251.50	5.50	30	175.66	0.75	2521	62.82
4	109.17	8.25	360.67	13.75	31	51.17	5.25	2572.17	68.07
5	68.00	1.33	428.67	15.08	32	66.92	1.08	2639.09	69.15
6	30.50	1.08	459.17	16.16	33	102.5	0.50	2741.59	69.65
7	88.50	2.33	547.67	18.49	34	79.25	2.25	2820.84	71.9
8	67.92	3.17	615.59	21.66	35	147.00	4.25	2967.84	76.15
9	95.00	1.42	710.59	23.08	36	32.33	0.75	3000.17	76.9
10	76.92	1.92	787.51	25.00	37	92.66	1.17	3092.83	78.07
11	60.08	1.17	847.59	26.17	38	142.75	0.75	3235.58	78.82
12	104.17	4.42	951.76	30.59	39	61.42	1.42	3297	80.24
13	59.83	0.75	1011.59	31.34	40	90.08	4.08	3387.08	84.32
14	90.42	2.17	1102.01	33.51	41	108.25	1.92	3495.33	86.24
15	81.83	1.08	1183.84	34.59	42	36.75	0.58	3532.08	86.82
16	81.17	2.92	1265.01	37.51	43	63.66	0.50	3595.74	87.32
17	53.58	2.66	1318.59	40.17	44	26.83	0.83	3622.57	88.15
18	118.17	1.50	1436.76	41.67	45	42.25	1.50	3664.82	89.65
19	67.83	1.75	1504.59	43.42	46	123.17	0.50	3787.99	90.15
20	55.42	3.42	1560.01	46.84	47	56.75	4.00	3844.74	94.15
21	86.75	0.66	1646.76	47.50	48	115.25	2.50	3959.99	96.65
22	87.08	1.33	1733.84	48.83	49	74.58	1.08	4034.57	97.73
23	119.25	1.25	1853.09	50.08	50	114.33	17.50	4148.9	115.23
24	15.00	1.00	1868.09	51.08	51	62.08	0.42	4210.98	115.65
25	152.92	7.33	2021.01	58.41	52	88.42	0.58	4299.4	116.23
26	107.08	0.83	2128.09	59.24	53		18.83		135.06
27	80.08	1.00	2208.17	60.24					

where α – scale parameter, β – shape parameter, γ – location parameter.

- **A Weibull distribution with 2-Parameter**

$$f(t) = \alpha\beta^{-\alpha}(t)^{\alpha-1} e^{-\left(\frac{t}{\beta}\right)^{\alpha}} \tag{8.50}$$

In this case, the location parameter (γ) is equal to zero.

- **Lognormal distribution with 3-Parameter**

$$f(t) = \frac{1}{(t-\gamma)\sigma\sqrt{2\pi}} e^{\left\{-\frac{[ln(t-\gamma)-\mu]^2}{2\sigma^2}\right\}} \tag{8.51}$$

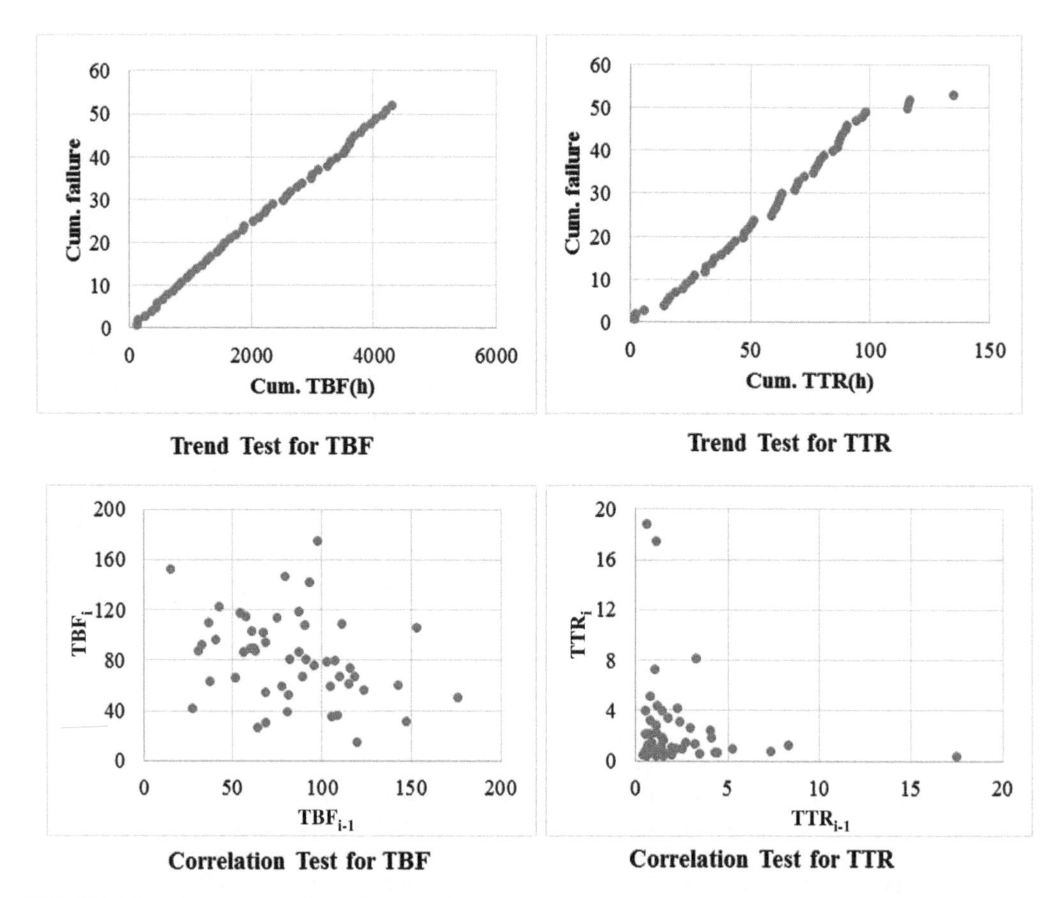

FIGURE 8.22 Trend test and serial correlation test.

where, μ – mean, σ – standard deviation, γ – location parameter.

- **Lognormal distribution with 2-parameter**

$$f(t) = \frac{1}{t\sigma\sqrt{2\pi}} e^{\left\{-\frac{[ln(t)-\mu]^2}{2\sigma^2}\right\}} \tag{8.52}$$

In this case, the location parameter (γ) is equal to zero.

- **Normal distribution**

$$f(t) = \frac{1}{\sigma\sqrt{2\pi}} e^{-\frac{(t-\mu)^2}{2\sigma^2}} \tag{8.53}$$

where μ – mean, σ – standard deviation.

- **Exponential 2-parameter distribution**

$$f(t) = \lambda e^{-\lambda(t-\gamma)} \tag{8.54}$$

where λ – mean, γ – location parameter.

The results of the modified K-S test for the five distributions and the estimated parameters of the best-fitted PDF for TBF and TTR data are listed in Table 8.10. The best fit distribution was identified based on the lowest value of the K-S test.

The results shown in Table 8.10 indicate that the best-fit PDF for the TBF data is the Weibull 3-parameter distribution function. The shape parameter (α), the scale parameter (β) and the location parameter (γ) of Weibull 3-parameter distribution functions are found to be 2.4619, 88.113, and 4.4859, respectively. Thus, the estimated PDF for the TBF data can be derived from Eq. (8.49) as

$$f(t) = 2.4619(88.113)^{-2.4619}\left(t-4.4859\right)^{2.4619-1} e^{-\left(\frac{t-4.4859}{88.113}\right)^{2.4619}} \qquad t>\gamma;\ \alpha,\beta>0$$

The cumulative failure distribution function (CDF) is given by

$$F(t) = 1-e^{-\left(\frac{t-\gamma}{\beta}\right)^{\alpha}} = 1-e^{-\left(\frac{t-4.4859}{88.113}\right)^{2.4619}}$$

Therefore, the reliability function of the conveyor belt is given by

$$R(t) = 1-F(t) = e^{-\left(\frac{t-\gamma}{\beta}\right)^{\alpha}} = e^{-\left(\frac{t-4.4859}{88.113}\right)^{2.4619}}$$

Similarly, the results (Table 8.10) indicate that the best-fit PDF for the TTR data is a lognormal 3-parameter distribution function. The estimated values of standard deviation (σ),mean of the natural logarithm (μ), and the location parameter (γ) of the distribution function are respectively 1.23, 0.019, and 0.378. Thus, the estimated PDF for the TTR data can be derived from Eq. (8.51) as

TABLE 8.10
Kolmogorov-Smirnov goodness of fit test results for TBF and TTR data

PDF	K-S test (goodness of fit)	Parameters of best fit distribution	Remarks
TBF data			
Exponential 2 parameter	0.261	$\gamma = 15.0, \lambda = 67.68$	
Normal	0.072	$\mu = 82.68, \sigma = 33.86$	
Log-normal 2 parameter	0.105	$\mu = 4.31, \sigma = 0.48$	
Log-normal 3 parameter	0.056	$\mu = 5.47, \sigma = 0.14, \gamma = -158.28$	
Weibull 2 parameter	0.051	$\alpha = 2.62, \beta = 93.07$	
Weibull 3 parameter	0.050	$\alpha = 2.4619, \beta = 88.113, \gamma = 4.4859$	Best fit
TTR data			
Exponential 2 parameter	0.206	$\gamma = 0.42, \lambda = 2.12$	
Normal	0.272	$\mu = 2.54, \sigma = 3.50$	
Log-normal 2 parameter	0.134	$\mu = 0.471, \sigma = 0.861$	
Log-normal 3 parameter	0.074	$\mu = 1.23, \sigma = 0.019, \gamma = 0.378$	Best fit
Weibull 2 parameter	0.160	$\alpha = 0.997, \beta = 2.54$	
Weibull 3 parameter	0.1553	$\alpha = 0.84, \beta = 1.65, \gamma = 0.40$	

$$f(t)=\frac{1}{1.23\sqrt{2\pi}(t-0.378)}e^{-\frac{ln\left(\frac{t-0.378}{0.019}\right)^2}{2(1.23)^2}} \qquad t>\gamma;\ \mu,\sigma>0$$

The CDF is given by

$$F(t)=\int_0^T\frac{1}{1.23\sqrt{2\pi}(t-0.378)}e^{-\frac{ln\left(\frac{t-0.378}{0.019}\right)^2}{2(1.23)^2}}=\varphi\left[\frac{ln\left(\frac{t-0.378}{0.019}\right)}{1.23}\right]$$

Where $\varphi(z)$ denotes the standardized normal CDF.

8.11.5 Reliability analysis for estimation of maintenance schedule

The reliability of the conveyor belt was determined using an estimated reliability function for the different time periods. The reliability of the conveyor belt was determined from 10 hours to 145 hours for every 5-hour interval. The results are summarized in Table 8.11. The trend of the reliability with time is represented in Figure 8.23. The results indicate that the reliability of the belt decreases with time. For low reliability, chances of failure will be high, and hence the availability of the belt will be less. Thus, the performance of the belt can be improved by increasing the reliability through scheduled maintenance. The desired reliability of the belt can be achieved through estimated scheduled maintenance, as shown in Table 8.12. The scheduled maintenance hour is high for achieving higher reliability and vice-versa. That is, the maintenance frequency is high for higher desirable reliability.

TABLE 8.11
Reliability of the conveyor belt

Time (hour)	Reliability	Time (hour)	Reliability	Time (hour)	Reliability
10	0.999	60	0.726	110	0.21
15	0.995	65	0.673	115	0.174
20	0.986	70	0.617	120	0.143
25	0.973	75	0.561	125	0.115
30	0.954	80	0.505	130	0.092
35	0.929	85	0.449	135	0.072
40	0.899	90	0.395	140	0.056
45	0.863	95	0.344	145	0.043
50	0.821	100	0.295		
55	0.776	105	0.251		

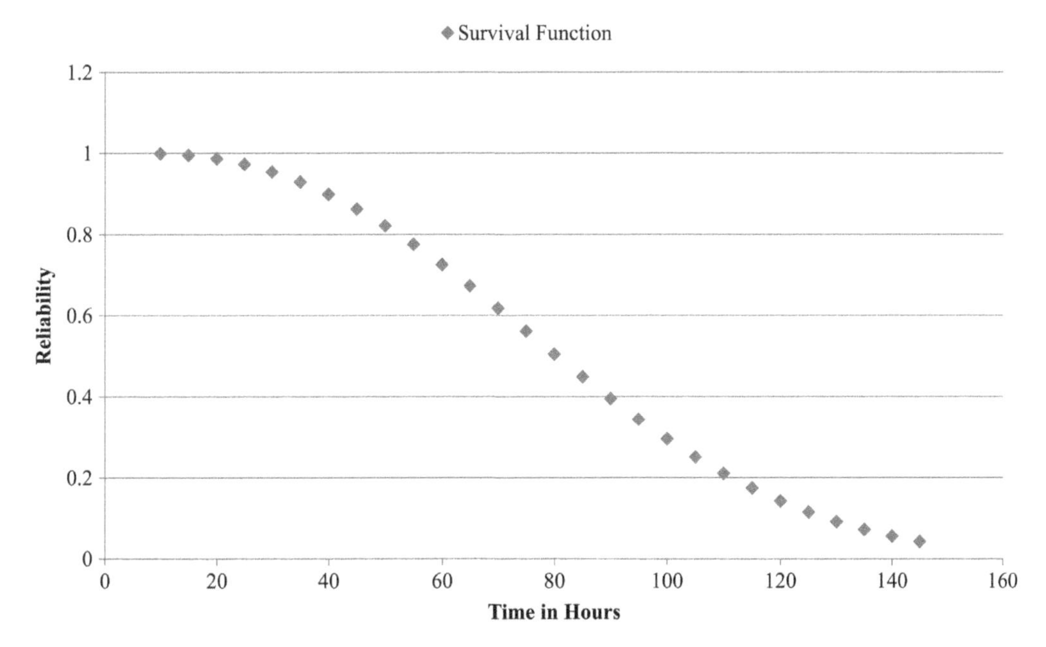

FIGURE 8.23 Reliability vs. time.

TABLE 8.12
Reliability-based time intervals for preventive maintenance

Desired level of reliability in %	75	80	85	90	95	98	99
Time in hours	57.55	52.39	46.50	39.7	30.8	22.5	18.0

TABLE 8.13
Availability of system

System	Frequency	MTBF	MTTR	Availability
C	53	82.68	2.548	0.97

The availability of the belt can be determined using the following formula:

$$\text{Availability} = \frac{\text{MTBF}}{\text{MTBF} + \text{MTTR}}$$

where, MTBF and MTTR represent the mean time between failure and mean time to repair, respectively.

The MTBF of the belt can be determined by dividing the CTBF by the total number (frequency) of failures. Similarly, the MTTR for the system was determined by dividing the CTTR by the total number (frequency) of failures. The respective values of MTBF and MTTR for the belt and the failure frequency are summarized in Table 8.13. It is clear from the results that the conveyor belt is available 97% of the time.

EXERCISE 8

Q1. It is observed from a pilot test that the failure rate of the new tyre of truck in mines linearly increases with time (hours), as shown below

$$\lambda(t) = 0.5\times10^{-8}t$$

Find the reliability of the tyre for 8000 hours of successful run.

[**Ans.** 0.852]

Q2. In an open-pit mine, one pump in installed for dewatering the pit. It was estimated that the mean time to failure of the pump is 200 hours and the failure time follows an exponential distribution. For continuous dewatering without any fail, another identical capacity pump is installed as a standby redundancy. While inactive, the mean time to failure of the standby pump is 1,000 hours, and the failure time of the standby pump is also exponentially distributed. Determine the reliability of the system for a mission of 300 hours and calculate the mean time to failure of the system.

[**Ans.** 0.87]

Q3. In an open-pit mine, materials are loaded by shovel and transported from the working area to the processing area by trucks. The mine has one shovel and six trucks for mine-to-mill transportation. The reliability of the shovel is 0.9, and all trucks have equal reliability of 0.8. What is the system reliability?

[**Ans.** 0.899]

Q4. A pumping station in a mine has four pumps, P_1, P_2, $P_{3,}$ and $P_{4,}$ which are connected functionally in a parallel configuration. The individual reliability of the pumps P_1, P_2, $P_{3,}$ and P_4 are 0.80, 0.90, 0.85, and 0.95, respectively. For an effective dewatering operation in the mine, at least three pumps should be in operation. Determine the reliability of the pumping system.

[**Ans.** 0.9245]

Q5. The lifetime of a shovel is exponentially distributed with a mean life of ten years. If the shovel has worked for ten years in an opencast mine, determine the probability that it will be working for an additional ten years.

[**Ans.** 0.3679]

Q6. The failure data analysis of a longwall mining system, consisting of three major subsystems, revealed that the failure rate of Shearer, face conveyor and stage loader is 0.0015/ hour, 0.002/ hour and 0.001/ hour, respectively. If all the subsystems must be running for a satisfactory system operation, determine the reliability of the longwall system for a period of 8 hours.

[**Ans.** 0.964]

Q7. The transportation system in a mine transfers the ore from the mine through two conveyor belts. The end of the conveyor system is connected with a bunker for further transferring into trucks, as shown in Figure 8.24. The transportation system works under the following scheme:

- Bunker and two conveyors running.
- Bunker and either of the two conveyors running.

If the bunker and the two conveyors have identical reliability of 0.9, determine the reliability of the system.

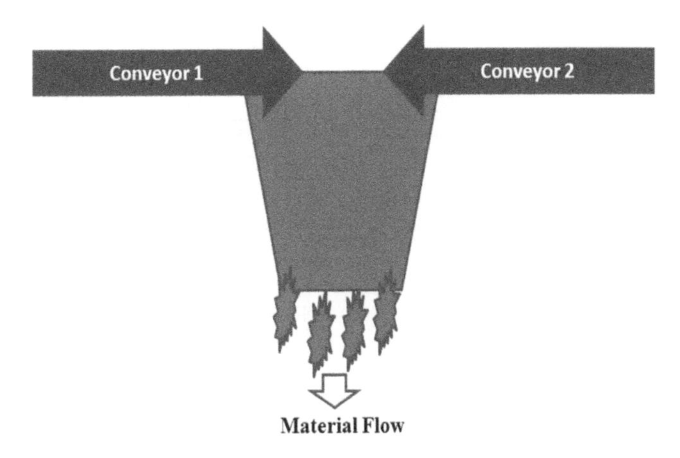

FIGURE 8.24 Material flows in a mine transportation system.

[**Ans.** 0.891]

Q8. Lifetime distribution of an equipment has a hazard function, $\lambda(t) = \beta_0 + \beta_1 t$. In the hazard function, β_0 *and* β_1 are constant. Find the reliability function and the probability density function.

[Ans.

$$R(t) = \exp\left[-\left(\beta_0 t + \frac{\beta_1 t^2}{2}\right)\right]$$

$$f(t) = (\beta_0 + \beta_1 t)\exp\left[-\left(\beta_0 t + \frac{\beta_1 t^2}{2}\right)\right]$$

Q9. Lifetime distribution of a mining machine has a hazard function, $\lambda(t) = \frac{1}{t+a}$. Find the reliability function and the probability density function.

[Ans.

$$R(t) = \exp\left[-\ln(t+a)\right]$$

$$f(t) = \frac{1}{(t+a)}\exp\left[-\ln(t+a)\right]$$

Q10. The time to wear out a drill bit follows a normal distribution with a mean of 2.8 hours and a standard deviation of 0.6 hours. Find

a. the probability that the drill bit will wear out in 1.5 hours
b. the reliability of the drill bit for successful run of 1.5 hours
c. replacement time of the drill bit in order to maintain the failure rate less than ten percentages?

[**Ans**. a. 0.0228; b. 0.9772; c. 1.932 hours]

Q11. A pump deployed in an underground mine for continuous dewatering operation. The failure time of pump follows an exponential distribution with a mean time to fail is 250 hours. An identical pump is deployed in standby redundancy with a mean time to fail

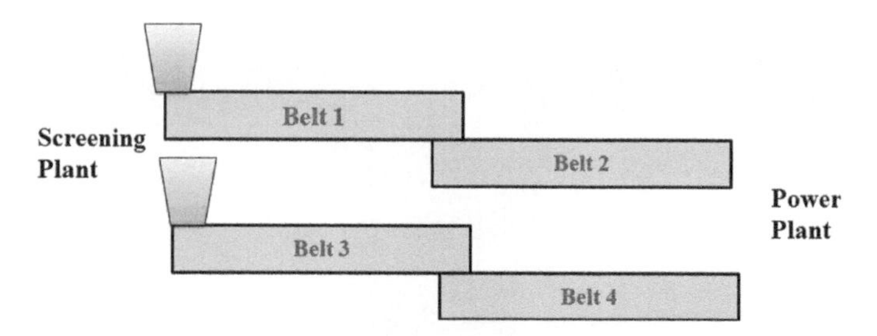

FIGURE 8.25 Arrangement of conveyor belt for transportation of coal from crushing plant to thermal power plant.

1,200 hours while inactive also follow exponential distribution. It is given that the standby pump automatically starts if the operating pump fails.

Determine the reliability of the system for a successful operation of 500 hours. Also, determine the mean time between failure of the system.

[**Ans.** 0.797; 1450 hours]

Q13. The transportation system in a mine consists of four identical conveyor belts. The arrangement of the belts is shown in Figure 8.25. Belt 1 can be coupled with Belt 4 and Belt 3 can be coupled with Belt 2. If more than two belts fail, the transportation system fails. If the failure times of each belt follow an exponential distribution with a failure rate of 0.000388 per hour, determine the reliability of the system for successful operation of 300 hours and mean time to fail the system.

[**Ans.** 0.9982]

Q14. The primary unit has a failure rate of 0.001 failure/hour. The unit in standby has a failure rate of 0.002 failure/hour. If the reliability of the switch can be considered to be 1 (perfect), what is the reliability for 200 hours?

[**Ans.** 0.967]

Q15. Four different machines (loader, haulage, winding, and truck) are deployed in a mine for loading and transportation of coal from face to stock piles. The reliabilities of loader, haulage, winding, and truck are 0.83, 0.89, 0.90, and 0.95, respectively. Determine the system's reliability. If the reliability needs to be increased to a value of 0.80, determine the optimum number of redundant unit of each machine?

[**Ans.** 2 Loaders, 2 Haulages, 1 Winding, 1-Truck]

9 Inventory management in mines

9.1 INTRODUCTION

Inventory refers to lists of materials kept in stock to prevent the delay in supplying to the customer or to avoid imperfections in the production process if production capacity stands idle for lack of raw materials. In the competitive global economy, an asset-intensive industry like mining should properly manage and optimize the inventory system for the organization's success.

For example, one of the challenges faced by the mining industry is proper maintenance of heavy and expensive machinery. These machines work in harsh environments, and the only way to achieve high performance from these machinery are by maintaining them in proper working condition. This can be achieved by periodic maintenance. Therefore, the project manager of the mine may adopt the reliability-based maintenance approach to these types of assets, which helps to ensure that critical parts of these machines are in running condition, especially those with long lead times or those that are scarce.

In the mining industry, inventory management solutions need both the raw materials required for mineral production and produced minerals. This is because the inventory in the mine is made up of commodity items or raw materials, and production materials run side by side. Because inventory costs have a significant share of the total annual mining cost, the mining industry needs to maintain an optimal inventory cost.

Moreover, mining industries are asset-heavy organizations, and hence they require a proper inventory system of spare parts to avoid the shutdown of the production line. The ultimate goal has the right parts, at the right price, in the right place, at just the right time. The challenge is maximizing value from this large investment, as rising costs can quickly consume production margins and take a significant bite out of profits. However, any inventory reductions must be achieved while mitigating the risk of production disruption from the potential unavailability of critical parts. Savings achieved through reduced inventory can quickly be lost if an unexpected production outage occurs because of non-availability of critical parts from inventory.

There are three basic reasons for keeping an inventory:

- **Time** – The time lags present in the supply chain, from supplier to user at every stage, require maintaining certain amounts of inventory to use in this 'lead time'.
- **Uncertainty** – Inventories are maintained as buffers to meet uncertainties in demand, supply, and movements of goods during lead time.
- **Economies of scale** – It is always desirable to buy or supply materials in bulk to reduce transportation, handling, and processing costs. So bulk buying or supplying, transportation, and handling makes the economics of scale.

Pairing inventory management with asset management automation

Despite an in-depth understanding of the inventory principles needed for identifying stock levels, it is always difficult for the inventory manager to maintain that due to the large volume of items. The inventory management in many mining organizations has often been conducted manually on an annual basis, which can be a manpower-intensive and prohibitively expensive proposition.

It is always desired to implement an automated replenishment of a stock item when the stock on hand has fallen below a designated minimum value for cost efficiency. For example, one of the largest alumina refineries, Queensland Alumina Limited (QAL), *saved more than $8 million* in three

DOI: 10.1201/9781003200703-9

years by reducing the risk of production loss due to stock-out conditions (Mincom, 2011). Available at: www.mining.com/optimizing-inventory-management-for-maximum-asset-performance/).

9.2 COSTS INVOLVED IN INVENTORY MODELS

The inventory system mainly consists of four types of cost, viz. purchasing cost; setup cost (or ordering cost); holding cost; and penalty cost (or shortage cost). Mathematically, it can be written as:

$$\text{Total Cost} = \text{purchasing cost} + \text{setup cost} + \text{holding cost} + \text{shortage cost}$$

The trends of the various cost functions with the inventory level are shown in Figure 9.1.

Purchasing cost represents the cost per unit of an item. Purchasing costs may or may not be constant depending on the type of item or supplier. For producing items, the purchasing cost represents the unit replenishment cost.

Setup cost represents the cost incurred for placing an order, including handling of materials. Ordering cost mainly involves preparing purchase orders, selecting suppliers, and handling costs of materials. In general, the ordering cost is fixed irrespective of the size of the order.

Holding cost represents the cost of maintaining inventory in stock before it is used (or sold). Holding cost covers the inventory financing costs, opportunity cost of the money invested in inventory, storage space costs, and inventory risk costs. Inventory financing costs include everything related to the investment made in inventory.

Shortage cost is the penalty incurred when the mines run out of stock. It involves the following:

- **Production interruption**: The mining industry producing minerals and sells these minerals to consumers. In the mines, the company has to pay for things like idle workers and machines and overhead, even when production is stopped due to a shortage of materials like explosive or spare parts of machinery.
- **Timely shipments**: In the mining industry, the stock-out condition leads to untimely shipment of materials to consumers. As a result, the mining authority may have to pay penalties and lose consumers.
- **Customer loyalty and reputation**: These are generally invisible costs and are hard to quantify. But, in the long run, these are certainly affecting the company's reputation when the consumers do not get desired material in time.

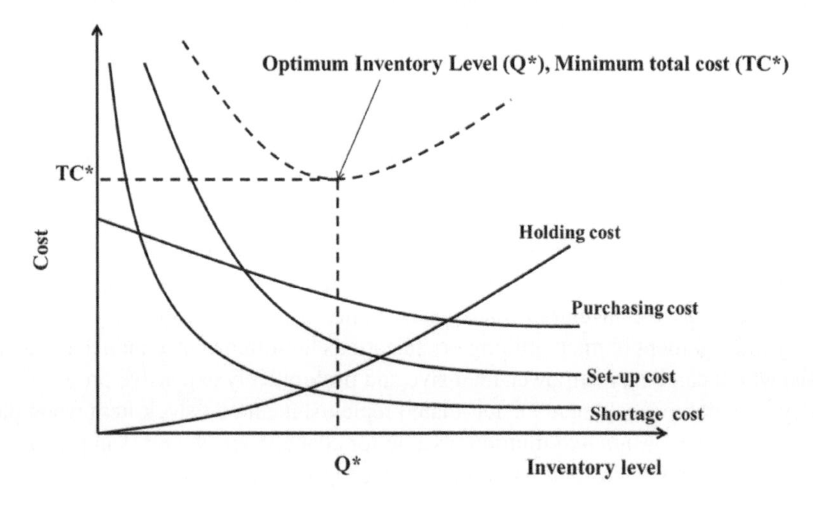

FIGURE 9.1 Trends of various cost functions with the inventory level.

9.3 INVENTORY MODELS

An inventory problem exists when it is necessary to stock physical goods or commodities for the purpose of satisfying demand over a specified time horizon (finite or infinite). Every mining business must stock goods to ensure the smooth and efficient running of its operation. Decisions regarding how much and when to order (or to produce) are typical for inventory problems. The required demand may be satisfied by stocking once for the entire time horizon or by stocking separately for every time unit of the horizon.

On one side, an overstock requires higher invested capital per unit time, reducing the chances of shortage and number of orders per unit time. On the other side, an understock decreases the invested capital per unit time, increases the risk of running out of stock and the number of orders per unit time. Thus, it is always desired to optimize the inventory system. A mathematical model helps the mining industry in determining the optimum inventory level for uninterrupted production or supply, the optimum number of orders, optimum storage of raw materials required for production, facilitates the timely supply of materials to consumers. The inventory models can be designed in two ways:

- **Fixed Reorder Quantity System (FOQS)**
- **Fixed Reorder Period System (FRPS)**

The parameters of a basic inventory model are represented in Figure 9.2. In the FOQS type inventory model, the order is placed immediately when the inventory level drops to a fixed specified **quantity** to replenish the inventory to an optimum level based on the demand. The point of a fixed specified **quantity, q**, at which the order is placed, is known as **Reorder Point, and the quantity of order, Q, is known as Reorder Level.**

In the FRPS type inventory model, an order is placed after every **fixed period of time** to replenish the inventory to an optimum level based on the demand. In this inventory management process, the inventory level is replenished in a continuous mode at a fixed interval of time irrespective of the remaining items in the inventory.

In Figure 9.2, the order is placed at times T_1, T_3, T_5. Thus, in the FRPS inventory system, $T_1 - 0 = T_3 - T_1 = T_5 - T_3$. The time between placing the orders and receipt of orders ($T_2 - T_1 = T_4 - T_3 = T_6 - T_5$) is called lead time, and the demand during the lead time is called lead time demand.

The selection of optimized inventory parameters can be made with the following models.

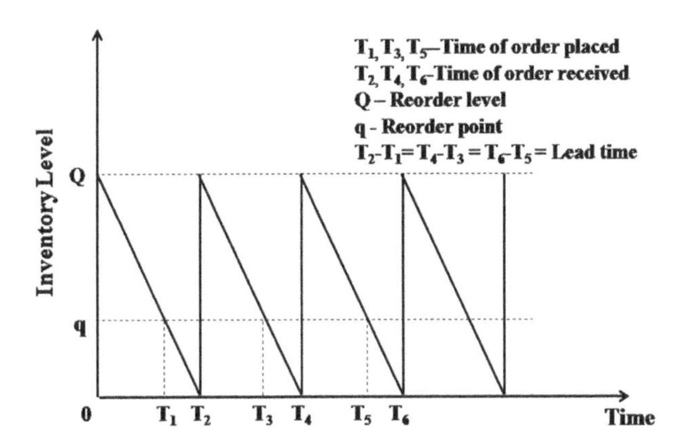

FIGURE 9.2 Parameters of a basic inventory model.

9.3.1 Deterministic model

The main characteristic of the deterministic inventory model is that it does not consider any random input. Therefore, the economic order quantity under the deterministic model can be determined in various ways depending on the situation. These are explained below.

9.3.1.1 Basic economic order quantity (EOQ) model

Assumptions of the basic EOQ model:

- The demand rate is constant with time.
- The inventory level is replenished immediately to its full capacity when the inventory level drops to zero.
- No shortage is allowed.

The behaviour of the inventory model is represented using a diagram, as shown in Figure 9.3. It is considered that the system has a capacity of Q units and the consumption or withdrawn rate is d unit per unit time. The lead time is zero or a constant. The optimal EOQ level can be determined with the above assumptions. In regard to the first assumption, the consumption of the inventory is uniform per unit time. The second assumption indicates that the lag time (time between placing an order and material receipt in the inventory) is zero or constant. Zero lag time means immediate replenishment of the inventory items after placing the order. The third assumption indicates that there is no shortage of material in the inventory system. Otherwise, penalty cost needs to be considered in the model, which is discussed in the next model.

Let's assume the costs involved in the inventory system are:

Setup cost per cycle – K
Purchasing cost per unit item – C
Holding cost per unit item per unit time – h

In Figure 9.3, the time between consecutive replenishments of inventory is referred to as a cycle. Thus, if the withdrawn rate is d units per unit time and Q is the order quantity in a batch, the cycle length is Q/d.

The total cost per cycle is obtained from the following components:

Ordering cost per cycle = K
Purchasing cost per cycle = CQ

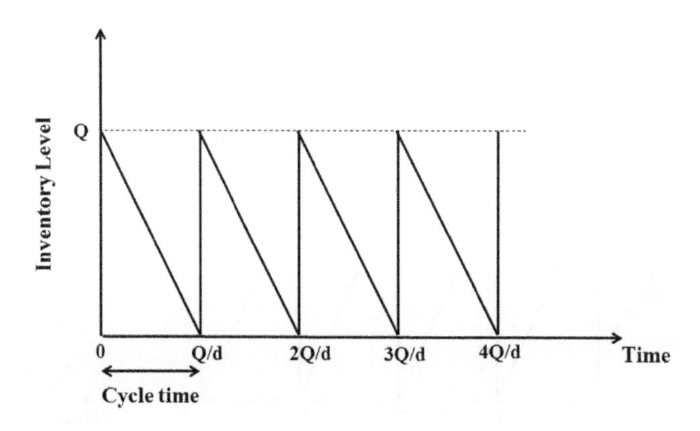

FIGURE 9.3 Basic Inventory System.

For a uniform demand, the level of inventory follows a linear equation. Thus, the average inventory level can be determined as

$$\text{Average inventory level during a cycle} = \frac{\text{Intial Inventory Level} + \text{Final Inventory Level}}{2} = \frac{Q+0}{2} = \frac{Q}{2}\text{ units}$$

The corresponding cost is given by

$$\text{Holding cost per unit time} = \text{Inventory Level} * \text{Holding cost per unit item per unit time}$$

$$= \frac{Q}{2} * h$$

$$\Rightarrow \text{Holding cost per cycle} = \text{Holding cost per unit item per unit time} * \text{Cycle time}$$

$$= \frac{Qh}{2} * \frac{Q}{d} = \frac{hQ^2}{d}$$

Therefore, the total cost expressed per cycle is given by

$$\text{Total cost per cycle} = \text{Purchasing cost per cycle} + \text{Ordering cost per cycle} + \text{Holding cost per cycle}$$

$$\Rightarrow \text{Total cost per cycle} = K + CQ + \frac{hQ^2}{2d}$$

Therefore, the total cost expressed per unit time is given by

$$\text{Total cost per unit time} = \text{TC}(\text{Q}) = \left(\frac{\text{Total cost per cycle}}{\text{Cycle time}}\right) = \frac{\left(K + CQ + \frac{hQ^2}{2d}\right)}{Q/d}$$

$$\Rightarrow TC(Q) = \frac{d}{Q}K + Cd + \frac{Q}{2}h$$

In the above equation, the total inventory cost is a function of inventory level, Q. All other parameters in the equation remain constant. The objective is to determine the EOQ level for which the total inventory cost is the lowest.

Mathematically, the objective is

$$\text{Minimum TC}(\text{Q}) = \frac{d}{Q}K + Cd + \frac{Q}{2}h$$

The problem is an unconstrained single variable optimization problem. To minimize $\text{TC}(\text{Q})$, we have

$$\frac{\partial TC(\text{Q})}{\partial \text{Q}} = 0$$

$$\Rightarrow \quad \frac{\partial}{\partial Q}\left(\frac{d}{Q}K + Cd + \frac{Q}{2}h\right) = 0$$

$$\Rightarrow \quad -\frac{d}{Q^2}K + 0 + \frac{h}{2} = 0$$

$$\Rightarrow \quad Q^* = \sqrt{\frac{2Kd}{h}} \tag{9.1}$$

Equation (9.1) is known as Wilson's Formula, which provides an optimum Q value. To solve the minimization problem

$$\text{we have,} \quad \frac{\partial^2 TC(Q)}{\partial Q^2} = \frac{2Kd}{Q^3} > 0$$

Thus, the function $TC(Q)$ is minimum at $Q = Q^*$.

The optimum total quantity, Q^* determined by Wilson's formula, is known as economic order quantity (EOQ).

Cycle time of placing an order is given by

$$t^* = \frac{Q^*}{d} = \sqrt{\frac{2K}{dh}} \tag{9.2}$$

If the cycle time, t^*, is more than the lead time, L_t, the effective lead time can be determined as

$$\text{Effective lead time} = \text{Lead time} - \text{cycle time} * \text{abs}\left(\frac{\text{Demand per cycle}}{\text{cycle time}}\right)$$

$$L_{et} = L_t - t^* \text{abs}\left(d / t^*\right) \tag{9.3}$$

The minimum total cost per unit time is given by

$$TC(Q^*) = \frac{d}{Q^*}K + Cd + \frac{Q^*}{2}h = \frac{d}{\sqrt{\frac{2Kd}{h}}}K + Cd + \frac{\sqrt{\frac{2Kd}{h}}}{2}h$$

$$\Rightarrow TC(Q^*) = \sqrt{\frac{dKh}{2}} + Cd + \sqrt{\frac{dKh}{2}} = \sqrt{2Kdh} + Cd \tag{9.4}$$

It is interesting to note that the holding cost and the setup cost are equal at the *EOQ* level. It is also observed from the above calculations that *EOQ* level (Q^*) and cycle time (t^*) change automatically with the change in the values of K, h, or d.

The optimal policy of this model does not depend on the unit purchasing or product cost. The *EOQ* level increases with increasing setup cost and consumption rate and decreases with increasing holding cost.

Example 9.1

In an underground mine, the demand for roof bolts for roof support is 700 units per week. Roof bolts are fixed at a constant rate. The setup cost for placing an order to replenish the bolts in the inventory system is US$400. The purchasing cost is US$20 per bolt, and the storage cost in the inventory is US$0.01 per bolt per day. Assuming no shortages of the bolts, determine the economic order quantity, inventory cost, and the frequency of order. If the lead time is four days, determine the reorder point.

Solution

Given data: Demand (d) = 700 per week = 700/7 per day = 100 per day
Setup cost (K) = US$400 per order
Purchasing cost of bolts (C) = US$20 per bolt
Holding cost (h) = US$0.01 per bolt per day
Lead time (L_t) = 4 days

Let the inventory level is Q.
Therefore, the total inventory cost per day is

$$TC = \frac{d}{Q}K + Cd + \frac{Q}{2}h$$

The economic order quantity (EOQ) at which minimum cost is obtained is

$$EOQ = Q^* = \sqrt{\frac{2Kd}{h}} = \sqrt{\frac{2*400*100}{0.01}} = \sqrt{\frac{80000}{0.01}} = 2828$$

$$\therefore \text{Total Invetory cost} = TC_{minimum} = \sqrt{2Kdh} + Cd$$

$$= \sqrt{2*400*100*0.01} + 20*100 = \text{US\$2028.28 per day}$$

$$\text{Cycle time}\left(t_c\right) = \frac{Q^*}{d} = \frac{2828}{100} = 28.28 \text{ days} \cong 28 \text{ days}$$

$$\text{Reorder point} = \text{Lead time} * \text{Demand per unit time} = L_t * d = 4*100 = 400 \text{ bolts}$$

That is, the order should be placed when the inventory level drops to 400 bolts.
All these parameters are represented graphically in Figure 9.4.

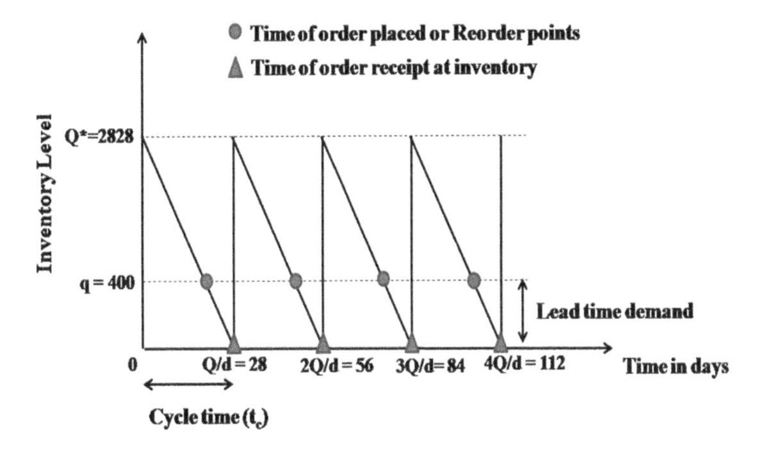

FIGURE 9.4 Inventory parameter of the defined system.

9.3.1.2 EOQ model with planned shortages

The nightmare of any mining inventory manager is the occurrence of an items shortage in the inventory system. This is also referred to as a stock-out condition. In the stock-out condition, demand cannot be met due to the unavailability of the desired items in the inventory. The stock-out condition leads to many problems in the mining industry, like an interruption of production, dealing with customers whose demands are not fulfilled. In the basic EOQ model, the shortage is not allowed, but in reality, it is difficult to figure out the exact inventory due to fluctuation in demand rate and replenishment time.

However, there are situations where a planned shortage of items in the inventory is more economical to the inventory manager. In this case, it is important to note that consumers should accept a reasonable delay in receiving their orders, or the mine production system should not affect the mineral supply.

Thus, the costs incurred due to shortages should not be significant. This type of management is generally suggested when the cost of holding inventory is high relative to these shortage costs.

Assumptions of the planned shortage model:

- The demand rate is constant with time.
- The inventory level is replenished immediately to its full capacity when the inventory level drops to zero.
- The planned shortage is allowed. For mines, the shortage of the required items does not hamper the supply of minerals to the consumer. Suppliers should ensure the affected consumers, due to shortage, that the material will become available after the wait time. The backorders are filled immediately when the order quantity arrives to replenish inventory.

The inventory pattern over time with the above assumptions is shown in Figure 9.5. The pattern is the same as that of the basic EOQ model, except the inventory levels extend down to negative values that reflect the number of units of the product that are back-ordered.

Let

p = shortage cost per unit item short per unit of time short,
S = inventory level just after replenishment of inventory with a batch of Q units,
$Q - S$ = shortage in inventory just before replenishment of inventory with a batch of Q units.

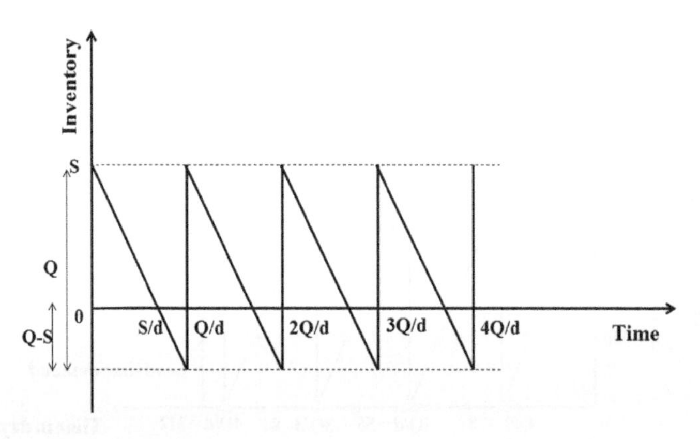

FIGURE 9.5 EOQ model with planned shortage.

The total cost per cycle is obtained from the following components:

Ordering cost per cycle = K
Purchasing or production cost per cycle = CQ
Holding cost per unit item per unit time = h

It is clear from Figure 9.5 that the inventory level is positive for a time S/d during each cycle.

Average inventory level during a cycle

$$= \frac{\text{Intial Inventory Level} + \text{Final Inventory Level}}{2} = \frac{S+0}{2} = \frac{S}{2} \text{ units}$$

The corresponding cost is given by

Holding or Inventory cost per unit time
= Inventory Level ∗ Holding cost per unit item per unit time

$$= \frac{S}{2} * h$$

⇨ Holding cost per cycle = Holding cost per unit item per unit time ∗ Cycle time

$$= \frac{hS}{2} * \frac{S}{d} = \frac{hS^2}{2d}$$

Similarly, shortages occur for a period of $\frac{Q-S}{d}$.

Average shortage level in the inventory during a cycle

$$= \frac{\text{Initial shortage Level} + \text{Final shortage Level}}{2} = \frac{(Q-S)+0}{2}$$

$$= \frac{(Q-S)}{2} \text{ units}$$

The corresponding shortage cost is given by

Shortage cost per unit time = Average shortage Level
∗ Shortage cost per unit item per unit time

$$= \frac{(Q-S)}{2} * p$$

Hence,

Shortage cost per cycle = Shortage cost per unit time ∗ shortage time

$$= \frac{p(Q-S)}{2} * \frac{(Q-S)}{d} = \frac{p(Q-S)^2}{2d}$$

Therefore, the total cost expressed per cycle is,

Total cost per cycle

= Setup cost per cycle + Purchasing (or production) cost per cycle + Holding cost per cycle + Shortage cost per cycle

$$\text{Total cost per cycle} = K + CQ + \frac{hS^2}{2d} + \frac{p(Q-S)^2}{2d}$$

The total cost expressed per unit time is,

$$TC(Q) = \left(\frac{\text{Total cost per cycle}}{\text{Cycle time}}\right) = \frac{\left(K + CQ + \frac{hS^2}{2d} + \frac{p(Q-S)^2}{2d}\right)}{Q/d}$$

$$\Rightarrow \quad TC(Q) = \frac{d}{Q}K + Cd + \frac{hS^2}{2Q} + \frac{p(Q-S)^2}{2Q}$$

In the above equation, the total inventory cost is a function of inventory level, Q, and amount of shortage, S. All other parameters in the equation remain constant. The objective is to determine the EOQ level for which the total inventory cost is the lowest.

Mathematically, the objective can be defined is

$$\text{Minimum } TC(Q) = \frac{d}{Q}K + Cd + \frac{hS^2}{2Q} + \frac{p(Q-S)^2}{2Q}$$

The problem is an unconstrained two variables optimization problem. To minimize $TC(Q)$, the partial derivatives $\frac{\partial TC(Q)}{\partial Q}$ and $\frac{\partial TC(Q)}{\partial S}$ equate to zero.

We have,

$$\frac{\partial TC(Q)}{\partial Q} = 0$$

$$\Rightarrow \frac{\partial}{\partial Q}\left(\frac{d}{Q}K + Cd + \frac{hS^2}{2Q} + \frac{p(Q-S)^2}{2Q}\right) = 0$$

$$\Rightarrow -\frac{d}{Q^2}K + 0 - \frac{hS^2}{2Q^2} + \frac{2Q * p * 2(Q-S) * 1 - p(Q-S)^2 * 2}{(2Q)^2} = 0$$

$$\Rightarrow -\frac{dK}{Q^2} - \frac{hS^2}{2Q^2} + \frac{p(Q^2 - S^2)}{2Q^2} = 0$$

Again,

$$\frac{\partial TC(Q)}{\partial S} = 0$$

$$\Rightarrow \frac{\partial}{\partial S}\left(\frac{d}{Q}K + cd + \frac{hS^2}{2Q} + \frac{p(Q-S)^2}{2Q}\right) = 0$$

$$\Rightarrow \frac{hS}{Q} + \frac{p(Q-S)(-1)}{Q} = 0$$

$$\Rightarrow \frac{hS}{Q} - p + \frac{pS}{Q} = 0$$

Solving the above two equations, we get

$$Q^* = \sqrt{\frac{2Kd}{h}\left(\frac{p+h}{p}\right)} \tag{9.5}$$

$$\text{And,} \quad S^* = \sqrt{\frac{2Kd}{h}\left(\frac{p}{p+h}\right)} \tag{9.6}$$

The optimal cycle time of placing an order is

$$t^* = \frac{Q^*}{d} = \sqrt{\frac{2K}{dh}\left(\frac{p+h}{p}\right)} \tag{9.7}$$

The maximum shortage is

$$Q^* - S^* = \sqrt{\frac{2Kd}{p}\left(\frac{h}{p+h}\right)} \tag{9.8}$$

The minimum total cost per cycle is

$$TC_{\text{minimum}} = \mathrm{TC}(\mathrm{Q}^*) = \frac{d}{Q^*}K + cd + \frac{hS^2}{2Q^*} + \frac{p(Q^* - S^*)^2}{2Q^*} =$$

$$= \frac{d}{\sqrt{\frac{2Kd}{h}\left(\frac{p+h}{p}\right)}}K + Cd + \frac{h\left(\sqrt{\frac{2Kd}{h}\left(\frac{p}{p+h}\right)}\right)^2}{2*\sqrt{\frac{2Kd}{h}\left(\frac{p+h}{p}\right)}} + \frac{p\left(\sqrt{\frac{2Kd}{h}\left(\frac{p+h}{p}\right)} - \sqrt{\frac{2Kd}{h}\left(\frac{p}{p+h}\right)}\right)^2}{2*\sqrt{\frac{2Kd}{h}\left(\frac{p+h}{p}\right)}}$$

After simplifying the above equation, the value of $\mathrm{TC}_{\text{minimum}}$ is

$$\mathrm{TC}_{\text{minimum}} = \sqrt{2Kdh\left(\frac{p}{p+h}\right)} + \mathrm{Cd} \tag{9.9}$$

Example 9.2

A coal mine in India purchases diesel for their heavy earth moving machinery (HEMM) in bulk and stores it in the inventory system. The demand for diesel is 730,000 litres per year. Diesel is consumed at a constant rate. The setup cost for placing an order to replenish the diesel in the inventory system is US$800. The purchasing cost is US$1.5 per litre, and the storage cost in the inventory is US$0.05 per litre per day. The lead time is 16 days. If shortages are allowed but cost US$0.04 per litre per day, determine how often to place an order and what size it should be. It is given that the working days in a year is 365.

Solution

Given data: Demand (d) = 730000 litres per year = 730000/365 litres per day
= 2000 litres per day
Setup cost (K) = US$800 per order
Purchasing cost of diesel (C) = US$1.5 per litre
Holding cost (h) = US$0.05 per litre per day
Shortage cost (p) = US$0.1 per litre per day
Lead time (L_t) = 16 days

Let the inventory level is Q.
Therefore, the total inventory cost per unit time or per day is

$$TC(Q) = \frac{d}{Q}K + Cd + \frac{hS^2}{2Q} + \frac{p(Q-S)^2}{2Q}$$

The economic order quantity (EOQ) at which minimum cost is obtained is

$$EOQ = Q^* = \sqrt{\frac{2Kd}{h}\left(\frac{p+h}{p}\right)} = \sqrt{\frac{2*800*2000}{0.05}\left(\frac{0.1+0.05}{0.1}\right)} = 19595.9 \cong 19596 \text{ litres}$$

$$\text{And } S^* = \sqrt{\frac{2Kd}{h}\left(\frac{p}{p+h}\right)} = \sqrt{\frac{2*800*2000}{0.05}\left(\frac{0.1}{0.1+0.05}\right)} = 6531.97 \cong 6532 \text{ litres}$$

$$\therefore \text{Total Inventory cost} = TC_{\text{minimum}} = \sqrt{2Kdh\left(\frac{p}{p+h}\right)} + Cd$$

$$= \sqrt{2*800*2000*0.05*\left(\frac{0.1}{0.1+0.05}\right)} + 1.5*2000$$

$$= 326.59 + 3000 = \text{US\$3326.59 per day}$$

$$\text{Cycle time}\left(t_c\right) = \frac{Q^*}{d} = \sqrt{\frac{2K}{dh}\left(\frac{p+h}{p}\right)} = \sqrt{\frac{2*800}{2000*0.05}\left(\frac{0.1+0.05}{0.1}\right)} = 4.89 \cong 5 \text{ days}$$

The maximum shortage is given by

$$Q^* - S^* = \sqrt{\frac{2Kd}{p}\left(\frac{h}{p+h}\right)} = \sqrt{\frac{2*800*2000}{0.1}\left(\frac{0.05}{0.1+0.05}\right)} \cong 3266$$

$$\text{Fraction of time, no shortage exists in the inventory} = \frac{S^*}{d} = \frac{6532}{2000}$$
$$\approx 3.26 \text{ days}$$

$$\text{Fraction of time, no shortage exists in the inventory} = \frac{Q^* - S^*}{d} = \frac{3266}{2000}$$
$$\approx 1.64 \text{ days}$$

$$\text{Effective lead time} = \text{Lead time} * \text{Demand per unit time} = L_t - t_c * \text{abs}\left(L_t / t_c\right)$$
$$= 16 - 5*\text{abs}\left(\frac{16}{5}\right) = 16 - 5*3 = 1 \text{ day}$$

All these parameters are represented graphically in Figure 9.6.

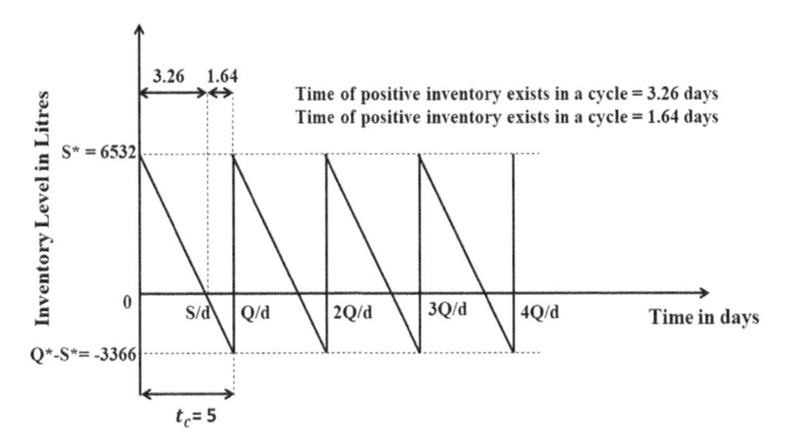

FIGURE 9.6 Inventory parameter of the defined system.

Example 9.3

In Example 9.1, if shortages are allowed, and the shortage cost is US$5 per bolt per day, recalculate the inventory system considering shortage.

Given data: Demand (d) = 700 per week = 700/7 per day = 100 per day
Setup cost (K) = US$400 per order
Purchasing cost of bolts (C) = US$20 per bolt
Holding cost (h) = US$0.01 per bolt per day
Lead time (L_t) = 4 days
Shortage cost (p) = US$5 per bolt per day

We know,

$$EOQ = Q^* = \sqrt{\frac{2Kd}{h}\left(\frac{p+h}{p}\right)} = \sqrt{\frac{2*400*100}{0.01}\left(\frac{5+0.01}{5}\right)} \cong 2831 \text{ bolts}$$

$$\text{And, } S^* = \sqrt{\frac{2Kd}{h}\left(\frac{p}{p+h}\right)} = \sqrt{\frac{2*400*100}{0.01}\left(\frac{5}{5+0.01}\right)} \cong 2825 \text{ bolts}$$

Thus, the maximum shortage allowed is given by

$$Q^* - S^* = 2831 - 2825 = 6$$

It is observed that the *EOQ* level in no shortage allowed condition was 2828 (refer to solution of Example 9.1). The results indicate the *EOQ* levels are not significantly varied with and without shortages. Also, the maximum shortage allowed is only six.

That is, if either p or h is significantly larger than the other, the *EOQ* level with and without shortage does not vary significantly.

On the other hand, if the shortage cost and inventory cost are close to each other, the maximum shortage allowed will have a significant value (see Example 9.2).

9.3.1.3 EOQ model with price discounts

In the preceding two models, the price discounts with order size in purchasing or selling the items are not considered. That is, in both the previous models, the price per unit item was assumed to be constant irrespective of the order size. But, in reality, there is a chance that bulk quantity orders may avail discount on the price for the items. The EOQ model with price discount works with the following assumptions.

- The demand rate is constant with time.
- The inventory level is replenished immediately to its full capacity when the inventory level drops to zero.
- No shortages are allowed.
- The unit cost of an item now depends on the order size in a batch.

The total cost per cycle is obtained from the following components:

Ordering cost per cycle = K
Purchasing or production cost per cycle = C_jQ
Holding cost per unit item per unit time = h

In this case, the total cost has three components (purchasing cost, setup cost, and holding cost). In contrast to the basic EOQ model, the purchasing cost in the price discount model is varying, as the cost per unit item is not constant.

Thus, the total cost expressed per unit time can be written as:

$$\begin{aligned} &\text{Total cost per cycle} \\ &\quad = \text{Purchasing cost per cycle} + \text{Setup cost per cycle} \\ &\quad\quad + \text{Holding cost per cycle} \end{aligned}$$

$$TC(Q) = C_jQ + K + \frac{hQ^2}{2d}$$

Therefore, the total cost per unit time is given by

$$TC(Q) = C_jd + \frac{d}{Q}K + \frac{Q}{2}h$$

Assuming there is (n–1) number of breakpoints $(q_1, q_2, \ldots, q_{n-1})$ for the price discount. That is, the price per unit item are $C_1, C_2, \ldots, C_n$. It is assumed that $C_1 > C_2 > \ldots > C_n$, that is higher the order size, less is the price per unit item. Mathematically, this can be represented as

$$C_{j=1 \text{ to } n} = \begin{cases} C_1, & \text{if } Q \le q_1 \\ C_2, & \text{if } Q \le q_2 \\ \vdots & \\ C_{n-1} & \text{if } Q \le q_{n-1} \\ C_n & \text{if } Q > q_{n-1} \end{cases}$$

To explain the model, assume there is only one breakpoint for the price discount. That is, the price per unit item is C_1 if the order size is less and equal to q and C_2 otherwise. It is assumed that $C_1 > C_2$, that is higher the order size, less is the price per unit item. Mathematically, this can be represented as

$$C_j = \begin{cases} C_1, & \text{if } Q \le q \\ C_2, & \text{if } Q > q \end{cases}$$

Hence,

$$\text{Purchasing cost per cycle} = \begin{cases} C_1 Q, & \text{if } Q \le q \\ C_2 Q, & \text{if } Q > q \end{cases}$$

$$\text{Purchasing cost per unit time} = \begin{cases} \dfrac{C_1 Q}{Q/d} = C_1 d, & \text{if } Q \le q \\ \dfrac{C_2 Q}{Q/d} = C_2 d, & \text{if } Q > q \end{cases}$$

Therefore, the total cost per unit time is given by

$$TC(Q) = \begin{cases} TC_1(Q) = \dfrac{d}{Q}K + C_1 d + \dfrac{Q}{2}h, & \text{if} Q \le q \\ TC_2(Q) = \dfrac{d}{Q}K + C_2 d + \dfrac{Q}{2}h, & \text{if} Q > q \end{cases}$$

The total cost function is represented graphically in Figure 9.7. Since the two functions ($TC_1(Q)$ and $TC_2(Q)$) differ only by a constant amount C_j, their minima will obtain at same Q. The Q value at which both the function offers minimum is given by

$$Q^* = \sqrt{\frac{2Kd}{h}}$$

The determination of the optimum order quantity Q^* depends on the price breakpoint q. The price breakpoint, q, maybe less than Q^* or more than Q^* or equal to Q^*. In Figure 9.7, it can be observed that the value of total cost (TC_1) at Q^* without discount is also equal to the total cost with discount at Q. Mathematically, this can be written as:

$$TC_2(Q) = TC_1(Q^*)$$

$$\Rightarrow \frac{d}{Q}K + C_2 d + \frac{Q}{2}h = \frac{d}{Q^*}K + C_1 d + \frac{Q^*}{2}h$$

$$\Rightarrow \frac{d}{Q}K + \frac{Q}{2}h = \frac{d}{Q^*}K + C_1 d - C_2 d + \frac{Q^*}{2}h$$

$$\Rightarrow Q^2 + \frac{2dK}{h} = Q\left\{\frac{2d}{hQ^*}K + \frac{2(C_1 d - C\, d)}{h} + Q^*\right\}$$

$$\Rightarrow Q^2 - mQ + \frac{2dK}{h} = 0$$

$$\text{Where } m = \left\{\frac{2d}{hQ^*}K + \frac{2(C_1 d - C_2 d)}{h} + Q^*\right\}$$

By solving the above quadratic equation, we can get two values of Q. The value of Q, which is less than Q^*, should be discarded, and the other one, which is greater than Q^*, should be considered for defining the region, as represented in Figure 9.7. The entire range of Q can be divided into three regions (Region 1: 0 to Q^*, Region II: Q^* to Q, and Region III: Q to ∞).

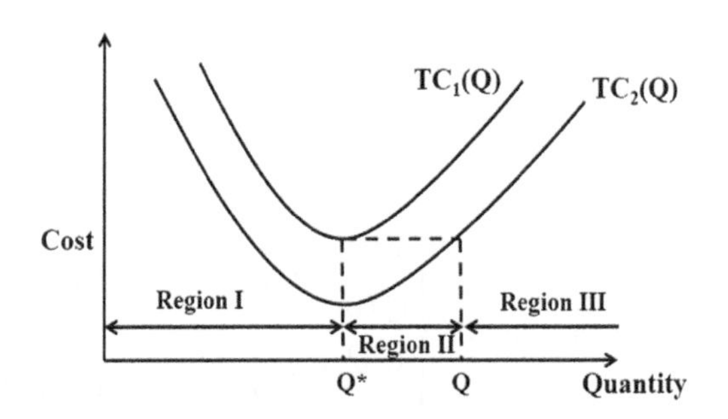

FIGURE 9.7 Total cost function for an inventory model with price discounts.

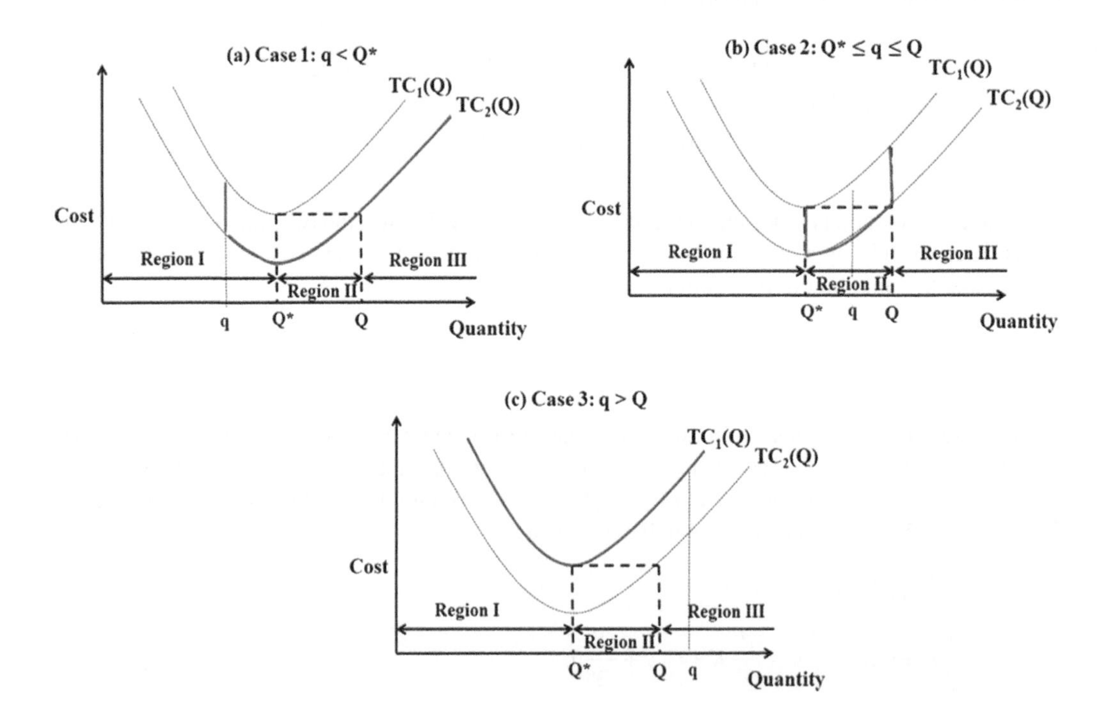

FIGURE 9.8 Cost function for price discounts inventory model in different conditions.

If the price breakpoint, q, lies in Region I, the economic order quantity should be Q^*.
If the price breakpoint, q, lies in Region II, the economic order quantity should be q.
If the price breakpoint, q, lies in Region III, the economic order quantity should be Q^*.

That is,

$$\text{Economic order quanity}\left(\text{EOQ}\right)=\begin{cases} Q^*, & \text{if } q<Q^* \\ q, & \text{if } Q^*\leq q\leq Q \\ Q & \text{if } q>Q \end{cases} \tag{9.10}$$

Therefore, the function follows to determine the total cost in the price discount inventory model is represented by the red line in Figure 9.8 for different values of price breakpoint.

The determination of the optimum order quantity Q^* depends on the price breakpoint q. The price discount breakpoints may be more than two, depending on the order size. The breakpoint of price discount is either less than Q* or greater than Q^* or equal to Q^*. If $q < Q^*$, the cost function TC_1 can be eliminated, and the total inventory cost will be determined using the TC_2 cost function, as shown in Figure 9.8(a). Therefore, the optimal order quantity is

$$EOQ = Q^*$$

On the other hand, if $q > Q^*$, the cost function, $TC_{1,}$ is followed to determine the total inventory cost, as shown in Figure 9.8(c). Moreover, if $Q^* \leq q \leq Q$, the cost function TC_2 should be followed, as shown in Figure 9.8(b).

The price discount inventory model can also be solved using backward calculation by determining the total cost using a different cost function. The number of breakpoints of price discount in an inventory model may be more than two, depending on the order size. The optimal order quantity should be decided based on the minimum total inventory cost. The method is explained in Example 9.4.

Example 9.4

In an underground mine, the demand for roof bolts for roof support is 700 units per week. Roof bolts are installed at a constant rate. The setup cost for placing an order to replenish the bolts in the inventory system is US$400. The purchasing cost of the bolts depends on the lot size ordered. The cost per unit bolt is as follows:

$$C_{j=1\ to 3} = \begin{cases} US\$20 & \text{for} Q < 2000 \\ US\$18 & 2000 \leq Q < 3000 \\ US\$16 & Q \geq 3000 \end{cases}$$

The storage cost in the inventory is US$0.01 per bolt per day. Assuming no shortages of the bolts, determine the economic order size, inventory cost, and the frequency of order. If the lead time is four days, determine the reorder point.

Solution

Given data: Demand (d) = 700 per week = 700/7 per day = 100 per day
Setup cost (K) = US$400 per order
Holding cost (h) = US$0.01 per bolt per day
Lead time (L_t) = 4 days

The economic order quantity (EOQ) at which minimum cost is obtained is

$$EOQ = Q^* = \sqrt{\frac{2Kd}{h}} = \sqrt{\frac{2*400*100}{0.01}} = \sqrt{\frac{80000}{0.01}} \cong 2828 \text{ units}$$

The cost functions for different order sizes are shown in Figure 9.9.

Thus, the economic order quantity (EOQ) value obtained for the given condition is 2828. As the EOQ level is more than the quantity of the first price discount breakpoint, the purchasing price per unit item is US$18. That is, the cost function TC_2 can be used to determine the inventory cost. For the calculated EOQ, $TC_2 < TC_1$ and hence TC_1 can be eliminated from further consideration. However, TC_3 cannot be immediately discarded as this may offer lesser total inventory cost with further availing the discount of US$16 per unit. To avail the further discount, the minimum order size is 3000. Thus, the inventory cost per day with calculated EOQ level (Q* = 2828) should be compared with the inventory cost per day with minimum order size for availing of the further discount.

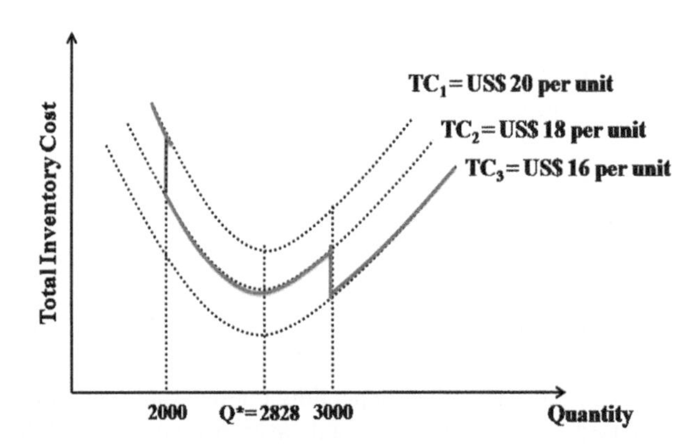

FIGURE 9.9 Total cost function for an inventory model with price discounts.

∴ Inventory cost per unit time for Q = 2828 is given by

$$TC_2(Q = 2828) = \frac{d}{Q}K + C_2 d + \frac{Q}{2}h = \frac{100}{2828} * 400 + 18 * 100 + \frac{2828}{2} * 0.01$$

$$= \text{US\$1828.28 per day}$$

Similarly, the Inventory cost per unit time for Q = 3000 is given by

$$TC_3(Q = 3000) = \frac{d}{Q}K + C_3 d + \frac{Q}{2}h = \frac{100}{2828} * 400 + 16 * 100 + \frac{3000}{2} * 0.01$$

$$= \text{US\$1629.14 per day}$$

The inventory cost per day for ordering 3000 units was estimated using the cost function TC_3 and found to be US$1629.14. On the other hand, the inventory cost per day for ordering 2828 units was estimated using the cost function TC_2 and found to be US$1828.28. Therefore, it is better to order quantities of 3000 to avail the discount, and this is the most economical.

9.3.1.4 Multi-item EOQ model with no storage limitation

In the mining industry, similar to other industries, mines store multiple items in their inventory. Therefore, the single-item inventory model may not always be very useful to understand the cost and benefit of managing inventory. The multi-item inventory model can eliminate the limitation of the single-item inventory model. The multi-item element inventory model is a natural extension of the single-item inventory model. In a multi-item EOQ model with no storage limitation, all the conditions are the same as that of the basic EOQ model, except the number of items. There are *n* items that need to be stored in the inventory without any competition for area availability.

Assumptions of the model are as follows:

- The demand rate is constant with time for each item.
- The inventory level of each individual item is replenished immediately to its full capacity when the inventory level drops to zero.
- No shortage is allowed for any item.

Let the cost involved in the inventory system for i^{th} items (i= 1, 2…,n), size of the order, and consumption rate are as

Setup cost per cycle for i^{th} item – K_i
Purchasing or production cost per unit of ith item – C_i
Holding cost per unit of ith item per unit time – h_i
Size of the order of ith item – Q_i
The consumption rate of ith item per unit time – d_i

Thus, the total inventory cost involved per unit time for n number of items is given by

$$TC\left(Q_1,\ Q_2,\ldots,Q_n\right)=\sum_{i=1}^{n}\frac{d_i}{Q_i}K_i+C_id_i+\frac{Q_i}{2}h_i$$

The objective is to minimize the total cost function, $TC\left(Q_1,Q_2,\ldots,Q_n\right)$.

To minimize $TC\left(Q_1,Q_2,\ldots,Q_n\right)$, the partial derivative of TC with respect to Q_i ($i= 1, 2,\ldots,n$) are equated to zero. Thus, we have,

$$\frac{\partial TC\left(Q_1,\ Q_2,\ldots Q_n\right)}{\partial Q_i}=0$$

$$\Rightarrow\frac{\partial}{\partial Q_i}\left(\sum_{i=1}^{n}\frac{d_i}{Q_i}K_i+C_id_i+\frac{Q_i}{2}h_i\right)=0$$

After performing the above partial differentiation with respect to Q_i, and simplifying the equations, we get

$$Q_i^*=\sqrt{\frac{2K_id_i}{h_i}}\quad (i=1,2,\ldots,n) \tag{9.11}$$

This formulation of Q_i is very similar to the single-item EOQ model, with the only difference it provides quantity for multiple items i ($i = 1, 2,\ldots,n$).

Example 9.5

A mine workshop has an inventory size of unlimited storage for the HEMM accessories. The other inventory parameters are listed in Table 9.1.

TABLE 9.1
Inventory parameters

	Item		
Costs	**Tires**	**Lubricant oil**	**Miscellaneous parts**
Demand of items (unit per year)	730	200000	10000
Setup cost (US$ per order)	1000	800	1200
Purchasing cost (US$ per unit)	4000	4	60
Holding cost (US$ per unit per unit time)	0.5	0.1	0.05

Solution

Given data:
Demand of tire (d_1) = 730/365 = 2 units per day
Demand of lubricant oil (d_2) = 200000/365 = 548 gallons per day
Demand of miscellaneous items (d_3) = 10000/365 = 27 units per day

Setup cost of tire (K_1) = US\$1000 per order
Setup cost of lubricant oil (K_2) = US\$800 per order
Setup cost of miscellaneous items (K_3) = US\$1200 per order

Purchasing cost of tire (C_1) = US\$4000 per unit
Purchasing cost of lubricant oil (C_2) = US\$4 per gallon
Purchasing cost of miscellaneous items (C_3) = US\$60 per unit

Holding cost of tire (h_1) = US\$0.5 per unit per day
Holding cost of lubricant oil (h_2) = US\$0.1 per gallons per day
Holding cost of miscellaneous items (h_3) = US\$0.05 per unit per day

The economic order quantity (EOQ) for i^{th} item can be determined as

$$EOQ \text{ of } i^{th} \text{ item} = Q_i^* = \sqrt{\frac{2K_i d_i}{h_i}}$$

Therefore,

$$EOQ \text{ level of tire} \left(Q_1^*\right) = \sqrt{\frac{2K_1 d_1}{h_1}} = \sqrt{\frac{2*1000*2}{0.5}} = 89.44 \cong 89 \text{ units}$$

$$EOQ \text{ level of lubricant oil} \left(Q_2^*\right) = \sqrt{\frac{2K_2 d_2}{h_2}} = \sqrt{\frac{2*800*548}{0.1}} = 2960.93 \cong 2960 \text{ litres}$$

$$EOQ \text{ level of miscellaneous items} \left(Q_3^*\right) = \sqrt{\frac{2K_3 d_3}{h_3}} = \sqrt{\frac{2*1200*27}{0.05}} \cong 1315 \text{ units}$$

The total cost of inventory per unit time is given by

$$TC\left(Q_1,\ Q_2, Q_3\right) = \sum_{i=1}^{3} \frac{d_i}{Q_i^*} K_i + C_i d_i + \frac{Q_i^*}{2} h_i$$

$$\Rightarrow \quad TC\left(Q_1, Q_2, Q_3\right) = \left(\frac{2}{89} * 1000 + 4000 * 2 + \frac{89}{2} * 0.5\right) + \left(\frac{548}{2960} * 800 + 4 * 548 + \frac{2960}{2} * 0.1\right) + \left(\frac{27}{1315} * 1200 + 60 * 27 + \frac{1315}{2} * 0.05\right)$$

$$\Rightarrow \quad TC\left(Q_1,\ Q_2, Q_3\right) = 8044.72 + 2488.10 + 1617.51 = \text{US\$}12210.33 \text{ per day}$$

Since there is no constraint in the availability of area in the store, the optimal lot size of tire, lubricant oil, and miscellaneous parts are 89 units, 2960 gallons, and 1315 units, respectively. The cycle time of the respective items can be determined as

$$t_1^* = \frac{Q_1^*}{d_1} = \frac{89}{2} = 44.5 \cong 45 \text{ days}$$

$$t_2^* = \frac{Q_2^*}{d_2} = \frac{2960}{548} = 5.4 \cong 6 \text{ days}$$

$$t_3^* = \frac{Q_3^*}{d_3} = \frac{1315}{27} = 48.7 \cong 49 \text{ days}$$

9.3.1.5 Multi-item EOQ model with storage limitation

All the assumptions remain same as that of multi-item EOQ with no storage limitation except the following:

- Availability of storage area for i^{th} item – *ai*
- Maximum available storage – *A*

This leads to the following constraint

$$a_1Q_1 + a_2Q_2 + \cdots a_nQ_n \leq A$$

$$\Rightarrow \sum_{i=1}^{n} a_iQ_i \leq A$$

Therefore, the model can be modified as

$$TC(Q_1,\, Q_2,\ldots,Q_n) = \sum_{i=1}^{n} \frac{d_i}{Q_i}K_i + c_id_i + \frac{Q_i}{2}h_i$$

$$\text{s.t.} \sum_{i=1}^{n} a_iQ_i \leq A$$

The above optimization problem is a non-linear problem, and there are multiple ways to solve a non-linear optimization problem, including Lagrange's multiplier method, dynamic programming, evolutionary algorithms, etc.

Applying Lagrange's multiplier, the above model can be written as

$$L(\lambda, Q_1,\, Q_2,\ldots,Q_n) = TC(Q_1,\, Q_2,\ldots,Q_n) + \lambda\left(\sum_{i=1}^{n} a_iQ_i - A\right)$$
$$= \sum_{i=1}^{n} \frac{d_i}{Q_i}K_i + c_id_i + \frac{Q_i}{2}h_i + \lambda\left(\sum_{i=1}^{n} a_iQ_i - A\right)$$

To find out the minimum *L*, we have

$$\frac{\partial L(\lambda, Q_1,\, Q_2,\ldots,Q_n)}{\partial Q_i} = 0, \qquad i = 1,2,\ldots,n$$

$$\text{And,} \quad \frac{\partial L(\lambda, Q_1,\, Q_2,\ldots,Q_n)}{\partial \lambda} = 0$$

Solving the above equations, we get,

$$Q_i^* = \sqrt{\frac{2K_i d_i}{h_i + 2\lambda a_i}} \tag{9.12}$$

This is a minimization problem, so $\lambda \leq 0$. The value of λ can be determined by solving the above partial derive equations. For a detailed explanation of Lagrange's multiplier method, refer to Chapter 11.

Example 9.6

In Example 9.5, if the mine workshop has an inventory size of 3600 sq. ft. for storage of the items. The space requirement for each item, along with other inventory parameters, are listed in Table 9.2.

Determine the EOQ level for each item for optimal inventory cost.

Solution

Given data:
Demand of tire (d_1) = 730/365 = 2 units per day
Demand of lubricant oil (d_2) = 200000/365 = 548 gallons per day
Demand of miscellaneous items (d_3) = 10000/365 = 27 units per day

Setup cost of tyre (K_1) = US$1000 per order
Setup cost of lubricant oil (K_2) = US$800 per order
Setup cost of miscellaneous items (K_3) = US$1200 per order

Purchasing cost of tire (C_1) = US$4000 per unit
Purchasing cost of lubricant oil (C_2) = US$4 per gallon
Purchasing cost of miscellaneous items (C_3) = US$60 per unit

Holding cost of tire (h_1) = US$0.5 per unit per day
Holding cost of lubricant oil (h_2) = US$0.1 per gallons per day
Holding cost of miscellaneous items (h_3) = US$0.05 per unit per day

Space required for tire (a_1) = 12 sq. ft. per unit
Space required for tire (a_2) = 0.25 sq. ft. per gallon
Space required for tire (a_3) = 0.4 sq. ft. per unit

Total available area (A) = 3600 sq. ft.

TABLE 9.2
Inventory parameters

	Item		
Costs	Tires	Lubricant oil	Miscellaneous parts
Demand of items (unit per year)	730	200000	10000
Setup cost (US$ per order)	1000	800	1200
Purchasing cost (US$ per unit)	4000	4	60
Holding cost (US$ per unit per unit time)	0.5	0.1	0.05
Space required (sq. ft. per unit)	12	0.25	0.4

TABLE 9.3
Optimal solution found using excel solver

Variables	Q1		87.82
	Q2		5592.26
	Q3		1050.41
Objective Function Minimum	12271.84		
Constraints 1	2872.115	<=	3600

We have,

$$TC(Q_1, Q_2, Q_3) = \sum_{i=1}^{3} \frac{d_i}{Q_i} K_i + C_i d_i + \frac{Q_i}{2} h_i$$

$$\text{s.t.,} \sum_{i=1}^{n} a_i Q_i \le A$$

$$\Rightarrow \; TC(Q_1, Q_2, Q_3) = \left(\frac{2}{Q_1} * 1000 + 4000 * 2 + \frac{Q_1}{2} * 0.5\right) + \left(\frac{548}{Q_2} * 800 + 4 * 548 + \frac{Q_2}{2} * 0.1\right) + \left(\frac{27}{Q_3} * 1200 + 60 * 27 + \frac{Q_3}{2} * 0.05\right)$$

$$\text{s.t.,} 12Q_1 + 0.25Q_2 + 0.4Q_3 \le 3600$$

Since the problem is a complex problem, it is difficult to solve manually, and thus the problem is solved in Excel Solver. The optimal solution is shown in Table 9.3.

Thus, the following EOQs of the items offer the minimum inventory cost of the system.

EOQ level of tire $(Q_1^*) \cong 87$ units

EOQ level of lubricant oil $(Q_2^*) \cong 5592$ gallons

EOQ level of miscellaneous items $(Q_3^*) \cong 1050$ units

The inventory cost is US$12271.84.

9.3.2 Fixed time-period model

In fixed time-period models, orders are placed at fixed periods without reviewing the status of the current inventory level, unlike earlier models. An order quantity replenishes the inventory level after a fixed interval of time irrespective of the demand and usage patterns. As the demand pattern is uncertain, a safety stock (minimum stock levels maintained in the inventory system without consideration of the order quantity) is usually maintained in this type of inventory system. The EOQ

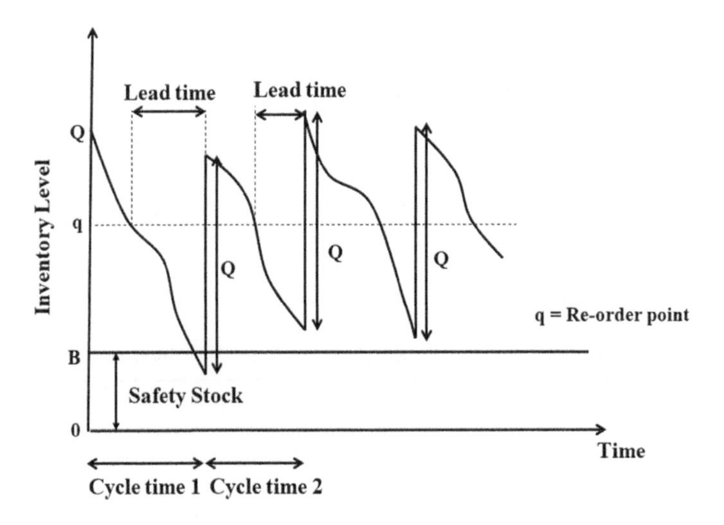

FIGURE 9.10 Inventory parameters of a probabilistic EOQ model.

is always over and above the safety stock level. The characteristics of the model are represented in Figure 9.10. This model assumes that demand is known with uncertainty but follows a known probability distribution function.

The optimal inventory policy is determined in terms of the probability distribution function. It is important to decide the type of distribution function which follows the demand (or consumption) rate. This section considers a normal probability distribution function (pdf) to determine the optimal inventory policy.

Followings are the assumptions of the model:

- The demand level during the lead time follows a probability distribution function (normal distribution function) with a known mean and variance.
- The inventory level increases by the amount of order quantity, Q with the receipt of an order at fixed interval of time.
- It is assumed that shortages are fulfilled from the safety stock maintained in the inventory system. Any consumption from safety stock is replenished in the next order. Thus, no shortage or stock-out situation occurs in this type of inventory system due to safety stock. The safety stock can be determined from the probabilities of shortages or stock-out conditions and the expected demand during the lead time. Thus, the demand is assumed a random variable, X, and follows a normal distribution. The random variable, X, has pdf, $f(X)$.

Assuming that the demand per unit time is normal with mean μ_d and standard deviation σ_d, the mean and standard deviation during the lead time, L_t, can be computed as:

$$\text{Mean demand during the lead time } (\mu_L) = \mu_d L_t$$

$$\text{The standard deviation of demand during the lead time } (\sigma_L) = \sqrt{\sigma_d^2 L_t}$$

The probability of determining the safety stock to avoid the shortage or stock-out condition is given by

$$P\left(X_L \geq B + \mu_L\right) \leq \alpha$$

where X_L is the demand during lead time.
B is the safety stock.
μ_L is the mean demand during lead time.
α is the maximum allowed probability of running out of a stock.

$$\Rightarrow P\{(X_L - \mu_L) \geq B\} \leq \alpha$$

$$\Rightarrow P\left\{\frac{(X_L - \mu_L)}{\sigma_L} \geq \frac{B}{\sigma_L}\right\} \leq \alpha$$

$$\Rightarrow P\left\{z \geq \frac{B}{\sigma_L}\right\} \leq \alpha$$

$$\Rightarrow P\{z \geq K_\alpha\} = A_\alpha \quad (9.13)$$

In Eq. (9.13), A_α represents the area of the shaded region in Figure 9.11. For a normal distribution function, A_α value for different K_α can be determined from the statistical table. Few data from the statistical table are represented in Table 9.4 for reference.

Hence, the safety stock must be determined by multiplying the standard deviation of the lead time demand (σ_L) and probability of stock-out condition during the lead time at a certain confidence level (K_α). The mathematical form to derive the safety stock from avoiding the shortage or stock-out condition is shown in Eq.(9.14)

$$\text{Safety stock}(B) \geq \sigma_L K_\alpha \quad (9.14)$$

The value of service factors for a different level of service are summarized in Table 9.4.

Example 9.7

To deal with the inventory policy of roof bolt in a mine, EOQ = 1000 units. If the daily demand follows a normal distribution with mean μ_d = 100 units and standard deviation, σ_d =10 units, then determine the safety stock so that the probability of running out of stock is below 0.05. It is given that the effective lead time is two days.

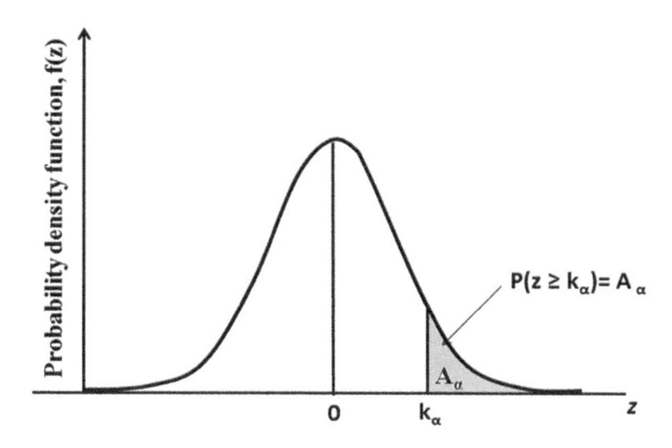

FIGURE 9.11 Probability of stock-out condition.

TABLE 9.4
Confidence level and their respective probability extracted from the normal distribution statistical table

Confidence Level (K_α)	$P\{z \geq K_\alpha\} = A_\alpha$	Confidence Level (K_α)	$P\{z \geq K_\alpha\} = A_\alpha$
0.50	0.00	0.08	1.41
0.45	0.13	0.07	1.48
0.40	0.25	0.06	1.55
0.35	0.39	0.05	1.64
0.30	0.52	0.04	1.75
0.25	0.67	0.03	1.88
0.20	0.84	0.02	2.05
0.15	1.04	0.08	2.33
0.10	1.28	0.05	2.58
0.09	1.34	0.01	3.72

Solution

Given data:

Daily mean demand $(\mu_d) = 100$ units
Standard deviation $(\sigma_d) = 10$ units
Lead time $(L_t) = 2$ days
Desired confidence level $(K_\alpha) = 0.05$

Therefore,

$$\mu_L = \mu_d L_t = 100 * 2 = 200 \text{units}$$

Standard deviation during the lead time is given by

$$\sigma_L = \sqrt{\sigma_d^2 L_t} = \sqrt{10^2 * 2} = 14.41$$

From Table 9.4, we have

$$K_{0.05} = 1.645$$

Thus, the safety stock can be computed as

$$B \geq \sigma_L K_\alpha$$

$$\Rightarrow B \geq 14.41 * 1.64$$

$$\Rightarrow B \geq 23.70 \approx 24$$

Therefore, the optimal inventory policy with safety stock, 24 units, should be maintained in the inventory system to avoid the stock-out condition with a 95% probability. The maximum inventory level in the system is equal to B + EOQ (= 24 + 1000 = 1024).,

9.3.3 Probabilistic EOQ model

The probabilistic EOQ model considers probabilistic demand to determine the EOQ level. In the probabilistic inventory model, the demand is considered to be probabilistic in nature in contrast to

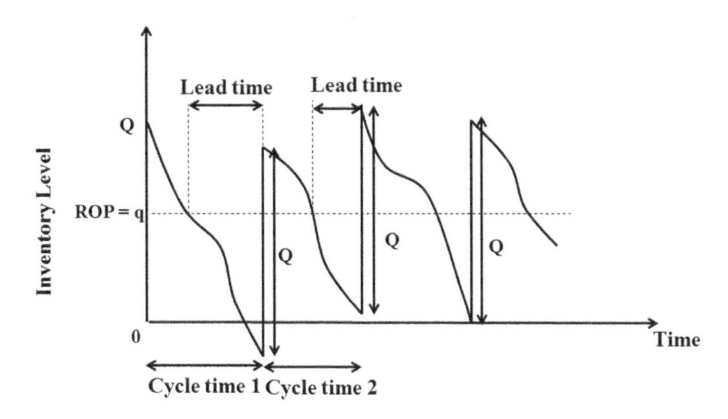

FIGURE 9.12 Inventory parameters of a Probabilistic EOQ model.

the deterministic model, where the demand is assumed to be constant. Following are the assumptions of the model:

- Unfilled demand during lead time is backlogged
- Maximum one outstanding order is allowed
- The distribution of the demand during lead time remains stationary with time

The model allows shortage of demand. The order of Q is placed whenever the inventory level drops to reorder level, q. In the deterministic inventory models, the reorder point, q, is the lead time demand, which is constant, but in the probabilistic inventory model, the lead time demand is variable, as the demand is not known with certainty. The objective is to determine the Q and q for which the total expected cost is the lowest. The nature of a typical probabilistic EOQ model is shown in Figure 9.12.

The total cost per cycle is obtained from the following components:

Ordering cost per cycle = K
Purchasing or production cost per unit = C
Holding cost per unit item per unit time = h
Shortage cost per inventory unit per unit time = p
Let pdf of demand, X during lead time L *is* $f(X)$
Expected demand per unit time is d

Let the expected shortage level in a cycle is S. Therefore, the inventory level is positive for a time S/d during each cycle.

$$\text{Average inventory level during a cycle } (I_{av}) = \frac{\text{Intial Inventory Level} + \text{Final Inventory Level}}{2}$$

$$\Rightarrow I_{av} = \frac{\{Q + E(q - X)\} + E(q - X)}{2} = \frac{Q}{2} + q - E(X) \tag{9.15}$$

[**Note**: $E(c\text{-}X) = c - E(X)$, c is constant]

The different inventory costs can be determined as follows:

Holding or Inventory cost per unit time
= Average Inventory Level * Holding cost per unit item per unit time

$$= \left\{ \frac{Q}{2} + q - E(X) \right\} * h \tag{9.16}$$

$$\text{Expected ordering cost per unit time} = \frac{\text{Ordering cost per cycle}}{\text{Expected cycle time}} = \frac{K}{Q/d} = \frac{d}{Q}\text{K} \tag{9.17}$$

The shortage will occur when demand during lead time, X, exceeds the reorder point, q. Thus the expected shortage quantity per cycle is given by

$$S = \int_{q}^{\infty} (X - q) f(X) dX \tag{9.18}$$

Since shortage cost, p is proportional to the shortage quantity per cycle, S. Therefore, the expected shortage cost per cycle is Sp.

The expected shortage cost per unit time is given by

$$\text{Expected shortage cost} = \frac{\text{shortage cost per cycle}}{\text{expected cycle time}}$$

$$\Rightarrow \text{Expected shortage cost} = \frac{Sp}{Q/d} = \frac{dSp}{Q} = \frac{dp}{Q}\int_{q}^{\infty} (X - q) f(X) dX \tag{9.19}$$

The total cost expressed per unit time is,

$$TC(Q,q) = \frac{d}{Q}K + h\left\{ \frac{Q}{2} + q - E(X) \right\} + \frac{dp}{Q}\int_{q}^{\infty} (X - q) f(X) dX \tag{9.20}$$

To find out the optimal value of $TC(Q,q)$

$$\frac{\partial TC(Q,q)}{\partial Q} = 0$$

$$\Rightarrow -\frac{d}{Q^2}K + h\frac{Q}{2} - \frac{dp}{Q^2}\int_{q}^{\infty} (X - q) f(X) dX = 0$$

$$\Rightarrow -\frac{d}{Q^2}K + h\frac{Q}{2} - \frac{dp}{Q^2}S = 0 \tag{9.21}$$

$$\text{And,} \frac{\partial TC(Q,\ q)}{\partial q} = 0$$

$$\Rightarrow h - \frac{dp}{Q}\int_{q}^{\infty} f(X) dX = 0 \tag{9.22}$$

From Eq. (9.21), we have

$$Q^* = \sqrt{\frac{2d(K + pS)}{h}} \tag{9.23}$$

Putting the value of Q^* in Eq. (9.22), we have

$$\int_{q^*}^{\infty} f(X)\,dX = \frac{hQ^*}{dp} \tag{9.24}$$

Eq. (9.23) and Eq. (9.24) are in closed form, and thus the optimal values of Q and q cannot be determined. An iterative algorithm (Hadley and Whitin 1963) can be used to solve the equations for obtaining the optimal values.

If the reorder point is zero (immediate replenishment of items after placing the order), the lead time is zero, and hence the value of q is equal to zero. Therefore, Eq. (9.18) becomes

$$S = \int_{0}^{\infty} (X-0) f(X)\,dX = \int_{0}^{\infty} X f(X)\,dX = E(X)$$

$$\text{And, } \int_{0}^{\infty} f(X)\,dX = 1$$

The simultaneous Eq. (9.23) and Eq. (9.24) becomes

$$Q^* = \sqrt{\frac{2d\{K + pE(X)\}}{h}} \tag{9.25}$$

$$\bar{Q} = \frac{dp}{h} \tag{9.26}$$

The problem has a feasible solution only when $Q^* \leq \bar{Q}$.

The steps of the algorithm suggested by Hadley and Whitin (1963) to solve the simultaneous Eqs. (9.9) and (9.10) are as follows:

Step 1: Determine the initial solution, Q_1 using the smallest possible Q ∗ value using Eq. (9.13).

The smallest value Q^* is $\sqrt{\frac{2dK}{h}}$, which is achieved for S = 0.

$$\therefore\ Q^* = Q_1 = \sqrt{\frac{2dK}{h}} \tag{9.27}$$

Step 2: Determine q_i using Q_i from Eq. (9.24).

Step 3: Step 2 repeats till $q_i = q_{i-1}$

$$Q^* = Q_i$$

$$\text{And, } q^* = q_i$$

Example 9.8

In an underground mine, production is done using drilling and blasting. The demand for explosives in the mine is 3000 kg per month. It costs US$400 to place an order for a new batch. The holding cost of the explosive in the magazine room is US$0.1 per kg per day, and the shortage cost is US$4 per kg. According to past statistics, consumption of the explosives during lead time is uniform over the range of 0 to 200 kg. Determine the optimum ordering policy of the explosive. The working day in a month is 30.

Solution

Given data:

Demand (d) = 3000 kg/month = 300 kg/day
Ordering cost (K) = US$400 per order
Holding cost (h) = US$0.1 per kg per day
Shortage cost (p) = US$4 per kg per day

The demand or consumption rate is a random variable, X, which follows uniform distribution in the range of 0 to 200 kg. The pdf for the random variable, X, can be determined as

$$f(X) = \frac{1}{200} \qquad 0 \le X < 200$$

Therefore, the expected or mean demand in a day is given by

$$\therefore E(X) = \int_0^{200} Xf(X)dX = \int_0^{200} X*\frac{1}{200}dX = \frac{1}{200}\left[\frac{X^2}{2}\right]_0^{200} = 100 \text{ kg}$$

First, we need to check whether the problem has a feasible solution. Using Eq. (9.25) and Eq. (9.26), we have

$$Q^* = \sqrt{\frac{2d\{K + pE(X)\}}{h}} = \sqrt{\frac{2*300(400+4*100)}{0.1}} \cong 2190 \text{ kg}$$

$$\bar{Q} = \frac{pd}{h} = \frac{4*300}{0.1} = 12000$$

As $Q^* \le \bar{Q}$, a feasible solution can be obtained.

The expected shortage can be derived as

$$S = \int_q^{200} (X-q)f(X)dX = \int_q^{200} (X-q)\frac{1}{200}dX = 100 - q + \frac{q^2}{400}$$

From Eq. (9.23) and Eq. (9.24), we have

$$Q_i = \sqrt{\frac{2d(K+pS)}{h}} = \sqrt{\frac{2*300(400+4S)}{0.1}} = \sqrt{2400000 + 24000S} \tag{9.28}$$

And,

$$\int_q^{200} f(X)dX = \frac{hQ^*}{dp}$$

$$\Rightarrow \int_q^{200} \frac{1}{200}dX = \frac{0.1*Q_i}{300*4}$$

$$\Rightarrow 1-\frac{q_i}{200}=\frac{Q_i}{12000}$$

$$\Rightarrow q_i = 200-\frac{Q_i}{60} \tag{9.29}$$

Eq. (9.28) and Eq. (9.29) can be used to determine the optimal solution.

Iteration 1

$$Q_1=\sqrt{\frac{2dK}{h}}=\sqrt{\frac{2*300*400}{0.1}}\cong 1549$$

Therefore,

$$q_1 = 200-\frac{Q_1}{60}=200-\frac{1549}{60}=174.18$$

Iteration 2

$$S=100-q_1+\frac{q_1^2}{400}=100-174.18+\frac{174.18^2}{400}=1.16$$

$$Q_2=\sqrt{2400000+24000S}=\sqrt{2400000+24000*1.16}=1558.15$$

$$q_2 = 200-\frac{Q_2}{60}=200-\frac{1558.15}{60}=174.03$$

Iteration 3

$$S=100-q_2+\frac{q_2^2}{400}=100-174.03+\frac{174.03^2}{400}=1.18$$

$$Q_3=\sqrt{2400000+24000S}=\sqrt{2400000+24000*1.18}=1558.35$$

$$q_3 = 200-\frac{Q_2}{60}=200-\frac{1558.35}{60}=174.02$$

The values of Q and q remain very close in the second and third iteration. That is, there is very little scope to improve the results further. Therefore, the iteration can be terminated, and the solution from the third iteration can be used as an optimal solution. Thus the optimal value of Q and q are $Q^*=1558.35$ and $q^*=174.02$. For selecting the iteration number and optimal solution, the user-defined threshold value for ΔQ and Δq, where $\Delta Q = Q_i - Q_{i-1}$, $\Delta q = q_i - q_{i-1}$, and i is iteration number, can be used. If the calculated ΔQ and Δq values in any iteration i are smaller than user define values, the algorithm will stop, and the estimated values of Q_i and q_i are considered optimal solutions.

EXERCISE 9

Q1. In an inventory system in the mine workshop, **t**he setup cost of dump truck tires is $600 per order with an annual carrying cost of US$40 per unit. Determine the EOQ level for an annual demand of 2000 units.

[**Ans.** 245 units]

Q2. Coal mining consumes explosives for blasting operations. The annual demand is 10,00,000 kg. The holding cost is US$10 per kg per year, and the setup cost is US$400 per order. Determine the EOQ level and the total annual inventory cost, excluding the purchasing cost.

[**Ans**. 8945 kg; US$89442 per year]

Q3. A mining company uses 25,000 pcs of roof bolts per year, each costing US$30. The ordering cost is US$500 per order, and the inventory cost is US$60 per unit per year. Based on the given data determine the followings:

a. EOQ
b. number of orders should be placed per year
c. the time interval between two orders
d. the total annual costs associated with inventory.

[**Ans.** a. 645 pcs; b. ≈ 38 days c. ≈ 9 days; d. US$ 788729.8 per year]

Q4. A mining industry requires HEMM machine parts with an average cost of $300 each. The lead-time is six weeks, and the mean demand per week is 90 units with a standard deviation of 5 units per week. The company stipulates the policy of maintaining a service level of 95% ($K_{0.05}$ = 1.64). The ordering cost is $800 per order, and the inventory carrying or holding cost is 1% of the purchasing cost per week. Based on the above information, design a fixed order quantity inventory system and explain how it operates.

[**Ans.** safety stock ≈ 20 units; EOQ = 219 units; Cycle time of order ≈17 days]

10 Queuing theory and its application in mines

10.1 INTRODUCTION

Queue is a common phenomenon in mining. When multiple dump trucks come to the loading point, due to the limited number of loaders, dump truck needs to wait when other trucks are either being loaded or are waiting for the loader. The long waiting time of trucks or any other mining machines may incur a significant financial loss. Queuing theory deals with the optimization of the service and waiting time through mathematical models. Queues contain customers (or 'items') such as dump trucks.

Queuing theory is applicable to optimize the combination of various machines. For example, in a particular mining operation, the number of dump trucks needs to be allocated for a single shovel such that the maximum utilization of both the dump trucks and shovel can be achieved, and thus, the overall operating cost can be reduced. There are two kinds of costs involved in optimizing the queuing problem: waiting cost and service cost. The typical characteristics of waiting and service cost are shown in Figure 10.1.

To evaluate a queuing system, the characteristics of the arrival patterns, service times, and service disciplines need to be identified. All these processes (arrival patterns and service times) are generally stochastic. The components of a queuing system are:

- **Arrival pattern**: This denotes the nature of the arrival of the machines in a service station. For example, how a truck is coming to the working bench for being loaded or how a truck is arriving at the waste dump site to unload the waste materials. The arrival process can be characterized by the distribution of the inter-arrival times of the mining machines. Usually, it is assumed that the inter-arrival times are independent and have a common distribution. However, in many practical situations, machines arrive according to a Poisson distribution (i.e., exponential inter-arrival times).
- **Service**: This indicates the nature or capacity of service offers by the server. For example, how an electric shovel is loading a truck on the working bench. Service time is a random variable and follows a typical probability distribution like exponential distribution. Thus, the service times are independent and identically distributed and independent of the inter-arrival times of machines.
- **Server**: This indicates the nature of the service facility in the system. A system can have a single server or a group of servers that offer the service to customers. For example, there might be separate loading shovels for ore and waste materials in an open-pit bench face.
- **Queuing capacity:** It indicates the capacity of the queue where customers (dump trucks) can wait to get the service. The waiting line may have the infinite capacity or finite capacity.
- **Size of the calling population:** The population of customers (dump trucks) can be assumed to be finite or infinite in a queuing model. In a finite calling population queuing model, the arrival rate depends on the number of customers being served and waiting. On the other hand, in the infinite calling population queuing model, the arrival rate does not depend on the number of customers being served and waiting as there is a large number of potential customers exists in the system.

DOI: 10.1201/9781003200703-10

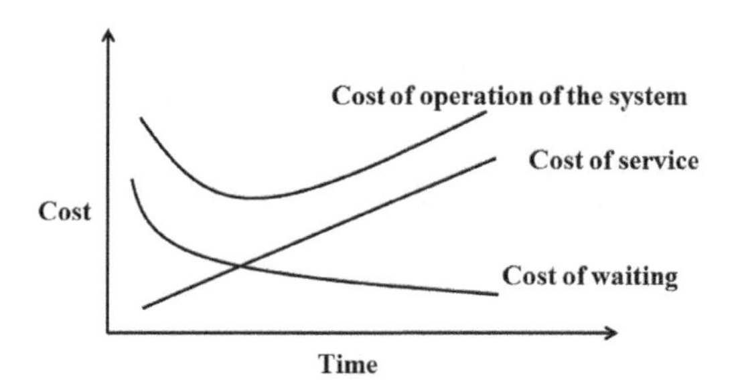

FIGURE 10.1 Characteristics of a Queuing System.

- **The service discipline:** In any system, the customers can be served one by one or in groups. The order of entering into the service facility and leaving the facility can have multiple options, as follows:
  - First In First Out (FIFO): Customers arrive first and leave earlier.
  - Last In First Out (LIFO): Customers comes later but leaves earlier.
  - Service In Random Order (SIRO): Customer is selected randomly for the service.
  - Priority-based service (PQ): A particular type of customer gets priority-based service.

10.2 KENDALL NOTATION

Kendall (1953) introduced an abbreviated notation to represent the different types of queuing models. Kendall's notation for a queuing system is given by

$$(a/b/c):(d/e/f)$$

Where

a: Arrival patterns,
b: Service patterns,
c: number of servers,
d: capacity of the queue,
e: size of the calling population,
f: queue or service discipline.,

The queuing parameters mentioned in the left side of the colon (:) are always defined to indicate the characteristics of the queuing model, but the parameters mentioned after the colon are not mandatory to define for default characteristics. The codes used to describe the different queuing characteristics are as follows:

Arrival patterns (*a*)

Markovian or memoryless (M) – Poisson arrival process, which has exponential inter-arrival times.
Degenerate distribution (D) – A deterministic or fixed inter-arrival time.
Erlang distribution (E_k) – An Erlang distribution with k as the shape parameter.
General distribution (G) – It usually refers to independent arrivals.

Service patterns (*b*)

The codes used for representing the distribution of the service times are similar to that of arrival patterns like *M*, *D*, E_k, or *G*. But, in a typical queuing model, the distribution of arrival pattern and service pattern may be different.

Number of the server (c)

The number of servers in a queuing system is represented by a natural number, *n*. ($n = 1, 2, \ldots$, n).

Capacity of the queue (*d*)

The maximum number of customers or dump trucks allowed in a queue may be finite or infinite. The finite capacity is represented by a finite number N, and the infinite capacity is represented by ∞.

Size of the calling population (*e*)

The finite calling population of customers (dump trucks) is by a finite number N, and the infinite capacity is represented by ∞.

Queue or service discipline (*f*)

The service discipline is denoted by *FIFO*, *LIFO*, *SIRO*, and *PQ*.

In a basic model, only the first three notations are used to explain the type of model, and the rest three are assumed to be default types. The default type of fourth, fifth, and sixth letters are infinite queue capacity (N), and infinite size of calling population (∞), and *FIFO* (Gautam, 2007). Any system that has infinite queue capacity means customers arrive one by one, and they are always allowed to enter into the system. The system has an infinite population size means there is no end for the customers. *FIFO* service indicates there are no priority rules, and customers are served in order of their arrival. Some examples are *M*/*M*/1, *M*/*M*/*c*, *M*/*G*/1, *G*/*M*/1, and *M*/*D*/1. If any of these assumptions do not hold, the notation is extended with an extra letter to indicate the nature of queuing models. For example, a system with exponential inter-arrival times, exponential service times, one server ($c = 1$), queuing capacity *K* (including the one in service) is abbreviated as (*M*/*M*/1): (*K*/∞/ *FIFO*).

10.3 PROBABILITY DISTRIBUTIONS COMMONLY USED IN QUEUING MODELS

In most queuing systems, arrivals occur randomly. Randomness means that the occurrence of an event (arrival of a dump truck or loading by a shovel) is not influenced by the amount of time that has elapsed since the last event. Random inter-arrival and service times are described quantitatively in queuing models by different distributions. Geometric, Poisson, Exponential, and Erlang distributions are commonly used distributions in queuing models. Here, these distributions are briefly described. To know more about the distribution, please refer to Probability and Statistics (Chapter 2).

10.3.1 Geometric distribution

The geometric distribution is either one of two discrete probability distributions. A geometric random variable, *T* with parameter *p* has a probability distribution

$$P(T = n) = (1 - p)p^n, \qquad n = 0, 1, 2, \ldots$$

For geometric distribution, the mean is equal to

$$\text{Mean} = E(T) = \frac{p}{1-p}$$

10.3.2 Poisson distribution

A Poisson random variable, *T*, with parameter μ has a probability distribution

$$P(T = n) = \frac{\lambda^n}{n!} e^{-\mu}, \qquad n = 0, 1, 2, \ldots$$

For Poisson distribution, the mean can be presented as:

$$\text{Mean} = E(T) = \sigma^2(T) = \lambda$$

10.3.3 Exponential distribution

The density function of an exponential distribution with parameter μ is given by

$$f(T) = \lambda e^{-\lambda T}, \qquad T \geq 0$$

For the exponential distribution, we have

$$P(T \leq t) = \int_0^t f(T)\,dT = \int_0^t \lambda e^{-\lambda T}\,dT = 1 - e^{-\lambda t}$$

For exponential distribution, the mean can be presented as:

$$\text{Mean} = E(T) = \frac{1}{\lambda}$$

An important property of an exponential distribution is the memoryless property (Refer Section 2.6 of Chapter 2). The memory property of the exponential distribution implies that

$$P\{T > t + \Delta t \mid T > \Delta t\} = P\{T > t\}$$

10.3.4 Erlang distribution

The Erlang distribution is the sum of *k* independent random variables T_1, T_2, …, T_k, which have a common exponential distribution with a mean of $1/\lambda$. It is a two-parameter [shape (*k*) and rate (λ)] continuous probability distribution function. The Erlang distribution with shape parameter, $k = 1$ simplifies to the exponential distribution. The probability density function of an Erlang distribution is given by

$$f(T;\ k, \lambda) = \lambda^k \frac{(\lambda T)^{k-1}}{(k-1)!} e^{-\lambda T}, \qquad T, \lambda \geq 0$$

For $k = 1$, the function becomes

$$f(T;\ 1, \lambda) = \lambda e^{-\lambda T}, \qquad T \geq 0$$

For the Erlang distribution, we have

$$P(T \le t) = \int_0^t f(T)dT = \int_0^t \lambda^k \frac{(\lambda T)^{k-1}}{(k-1)!} e^{-\lambda T} dT = 1 - \sum_{i=1}^{k-1} \frac{(\lambda t)^i}{(i)!} e^{-\lambda t}$$

The mean is given by

$$\text{Mean} = E(T) = \frac{k}{\lambda}$$

10.4 RELATION BETWEEN THE EXPONENTIAL AND POISSON DISTRIBUTIONS

If the inter-arrival time, T is exponential and the arrival rate is λ (i.e., λ customers are arriving per unit time) (shown in Figure 10.2), then the probability of no arrivals during a period of time t can be determined as

$$p_0(t) = P[\mathrm{T} > t] = 1 - P[T \le t] = 1 - (1 - e^{-\lambda t}) = e^{-\lambda t}$$

Where, $p_0(t)$ = probability of no arrival of the customer during the time t.

For a sufficiently small time interval $\Delta t > 0$, we have

$$p_0(\Delta t) = e^{-\lambda \Delta t} = 1 - \lambda \Delta t + \frac{(\lambda \Delta t)^2}{2!} - \frac{(\lambda \Delta t)^3}{3!} + \dots \approx 1 - \lambda \Delta t$$

[The higher-order terms are significantly small and close to zero, and thus deleted]

The exponential distribution is based on the assumption that during $\Delta t > 0$, at most, one arrival can occur. Thus, as $\Delta t \to 0$, the probability of one arrival in Δt time is given by

$$p_1(\Delta t) = 1 - p_0(\Delta t) = \lambda \Delta t$$

This indicates that the probability of arrival during Δt is directly proportional to Δt, with the arrival rate, λ, being the constant of proportionality.

Let $p_N(t)$ = probability of N arrival during a period t.

For a sufficiently small $\Delta t > 0$, it is assumed that at most, one arrival can occur during a very small period Δt follows an exponential distribution. Therefore, the probability of n-customer in the system is given by

$$p_N(t + \Delta t) = p_N(t) * \text{no arrival during } \Delta\text{t time} + p_{N-1}(t) * \text{one arrival during } \Delta\text{t time}, \; N > 0 \tag{10.1}$$

The above equation indicates that there are two possibilities of N arrival occurrence during $t + \Delta t$. These are either N arrivals that occurred during t and no arrivals during Δt, and $N - 1$ arrivals occurred during t and one that occurred during Δt.

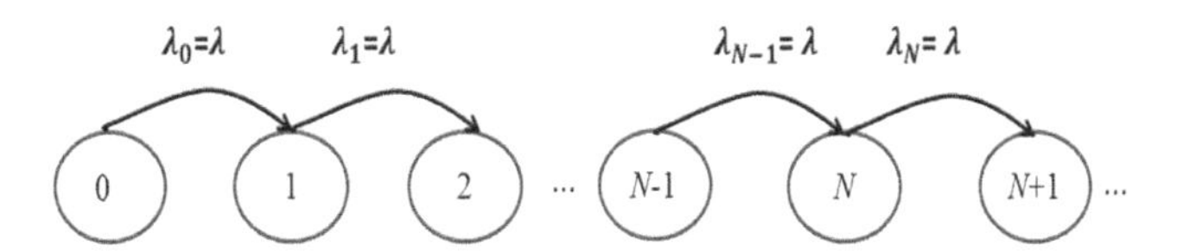

FIGURE 10.2 Flow diagram of customers in an exponential rate.

$$p_N(t+\Delta t) = p_N(t) * (1-\lambda\Delta t) + p_{N-1}(t) * \lambda\Delta t \quad (10.2)$$

For $N = 0$,

$$p_0(t+\Delta t) = p_0(t) * (1-\lambda\Delta t)$$

$$\Rightarrow p_0(t+\Delta t) = p_0(t) * (1-\lambda\Delta t) \quad (10.3)$$

[In zero state, zero arrivals during $(t+\Delta t)$ can occur only if no arrivals occur during $(t+\Delta t)$]

Eq. (10.2) and Eq. (10.3) can be rearranged as:

$$\frac{p_N(t+\Delta t) - p_N(t)}{\Delta t} = -\lambda p_N(t) + \lambda p_{N-1}(t), \quad N > 0 \quad (10.4)$$

$$\frac{p_0(t+\Delta t) - p_0(t)}{\Delta t} = -\lambda p_0(t), \quad N = 0 \quad (10.5)$$

Taking the limits as $\Delta t \to 0$, Eq. (10.4) and Eq. (10.5) becomes

$$\lim_{\Delta t \to 0} \frac{p_N(t+\Delta t) - p_N(t)}{\Delta t} = p'_N(t) = -\lambda p_N(t) + \lambda p_{N-1}(t), \quad N > 0$$

$$\lim_{\Delta t \to 0} \frac{p_0(t+\Delta t) - p_0(t)}{\Delta t} = p'_0(t) = -\lambda p_0(t), \quad N = 0$$

The solution of the preceding differential equations yields

$$p_N(t) = \frac{(\lambda t)^N}{N!} e^{-\lambda t}, \quad N = 0, 1, 2\ldots \quad (10.6)$$

Eq. (10.6) represents Poisson distribution with the mean number of arrival during t is given by $E(N) = \lambda t$.

Thus, the results indicate that if the time between arrivals is exponential with mean $1/\lambda$, then the number of arrivals during a specific period t is Poisson with mean λt. The vice-versa is also true.

The summarized relationship between the exponential and Poisson, given the arrival rate (λ) is shown in Table 10.1.

TABLE 10.1
Relationship between Exponential and Poisson Arrival Rate

	Exponential	**Poisson**
Random variable	Inter-arrival time, T	Number of arrivals, N, during a specified period t
Range	$T \geq 0$	$N = 0, 1, 2, \ldots$
Density function	$f(T=t) = \lambda e^{-\lambda t}, \quad t \geq 0$	$p_N(T=t) = \frac{(\lambda t)^n}{n!} e^{-\lambda t}, \quad n = 0,1,2,\ldots$
Mean value	$1/\lambda$ time units	λt arrivals during t
Cumulative probability	$P(T \leq t) = 1 - e^{-\lambda t}$	$p_{N \leq n}(T=t) = p_0(t) + p_1(t) \ldots p_n(t)$
P{no. arrival during period t}	$P(T > t) = e^{-\lambda t}$	$p_0(T=t) = e^{-\lambda t}$

10.5 LITTLE'S LAW

In queuing theory, Little's law established the relationship between the mean number of customers waiting in a stationary queuing system (L_s), the average waiting time of a customer within a system (W_s) and the average number of customer arriving at the system per unit of time (λ) (Little, 1961). Mathematically, it can be represented as:

$$L_s = \lambda W_s \tag{10.7}$$

The relationship represented in Eq. (10.7) is not influenced by the arrival patterns, the service pattern, the service discipline, and so on (Simchi-Levi and Trick, 2013). The law assumes that the capacity of the system is sufficient to deal with the customers. It means that the number of customers in the system does not grow to infinity.

Example 10.1
In a mine, the dump trucks are arriving at a rate of 6 per hour in the washing chamber for cleaning. Thus, the average time a dump truck spends is around 6 minutes for washing. Determine the average number of dump trucks queuing for washing by applying Little's law.

Solution
Given data:

Mean arrival rate (λ) = 6 dump trucks per hour
The average waiting time in the system (W_s) = 20 minutes = 1/3 hour

$$\therefore \text{The average number of dump trucks in the system}\left(L_s\right) = \lambda W_s = 6 * \frac{1}{3} = 2$$

Therefore, the average number of trucks in the queue is 2.

10.6 QUEUING MODEL

10.6.1 M/M/1 Model

M/M/1 is one of the basic queuing models in which the distributions of the inter-arrival times and distribution of service times are considered to be exponential distribution. In this model, the number of servers is one. In open-pit mines, this type of queuing characteristic is commonly observed in the loading face where only one loader is deployed to load the excavated material into the dump truck (Figure 10.3). The dump trucks are arriving from an infinite population following a single channel queue for being loaded by the loader. The loader offers the service on a first-come, first-serve basis, that the dump truck arrives first, gets loaded first, and exits the system. Thus, the characteristics of the M/M/1 queuing model is described below:

Characteristics of the M/M/1 model

- Inter-arrival times of dump truck follow exponential distribution function with mean arrival rate is λ per unit time.
- Service times of the loader follow exponential distribution function with mean service rate is μ per unit time.

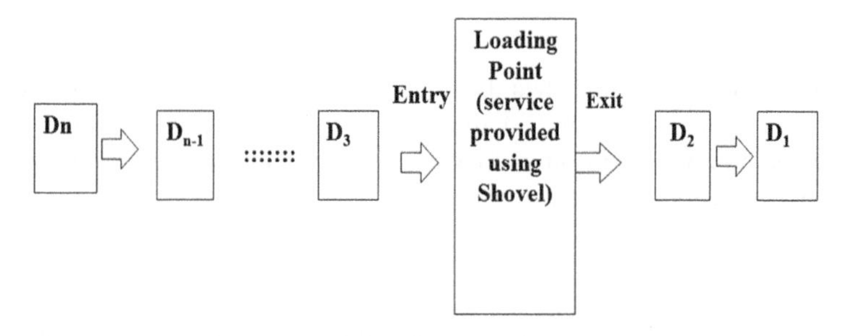

FIGURE 10.3 Queue system in M/M/1 queuing model.

- The number of loaders deployed in the face for loading operation is 1.
- Queuing capacity is infinite; that is, all the dump trucks that arrive at the loading point can wait in the queue for being loaded.
- Dump trucks are assumed to be coming from an infinite population.
- Dump trucks are served on a FIFO basis.

The basic assumption of the model is that the mean service (loading) rate is more than the mean arrival rate of dump trucks ($\mu > \lambda$) to avoid the infinite queue length. The condition can be represented mathematically as follows:

$$\rho = \frac{\lambda}{\mu} < 1; \tag{10.8}$$

where, ρ is the fraction of time the loader is busy.

10.6.1.1 Time-dependent behaviour of the flows of dump trucks

The exponential distribution has the memory-less property, and thus the future arrival time does not depend on the past arrival of the dump trucks. An exponential distribution offers a simple description of the state of the system at time *t*, like the number of trucks in the loading system in an open-pit mining bench (number of trucks waiting in the queue and the one being served by shovel). The flow behaviour of dump trucks in an M/M/1 queuing system is shown in Figure 10.4.

Let $p_N(t)$= probability of *N* dump trucks arrives during a period of time *t*.

For a sufficiently small Δt>0, it is assumed that at most one arrival and one exit can occur during a very small period Δt, following an exponential distribution. Therefore, the probability of *N*-dump trucks in the loading point is given by

$$\begin{aligned} p_N(t+\Delta t) = p_N(t) &* \text{no arrival during } \Delta t \text{ time} * \text{no leaving during } \Delta t \text{ time} + \\ & p_{N-1}(t) * \text{one arrival during } \Delta t \text{ time} * \text{no leaving during } \Delta t \text{ time} + p_{N+1}(t) * \\ & \text{no arrival during } \Delta t \text{ time} * \text{one leaving during } \Delta t \text{ time}, \qquad N > 0 \end{aligned} \tag{10.9}$$

The above equation indicates that there are three possibilities of *N* arrival occurrence of dump trucks during *t*+Δt. These are *N* arrivals occurred during *t*, and zero arrival and zero exits of dump truck occurred during Δt, *N*–1 arrivals occurred during *t*, and one arrival and zero exits of dump truck occurred during Δt, and *N*+1 arrivals occurred during *t*, and zero arrival and one exit of dump truck occurred during Δt.

Therefore, Eq. (10.9) can be rewritten as:

$$\begin{aligned} p_N(t+\Delta t) = p_N(t) * (1-\lambda\Delta t)(1-\mu\Delta t) + p_{N-1}(t) * \lambda\Delta t * (1-\mu\Delta t) \\ + p_{N+1}(t) * (1-\lambda\Delta t) * \mu\Delta t \end{aligned}$$

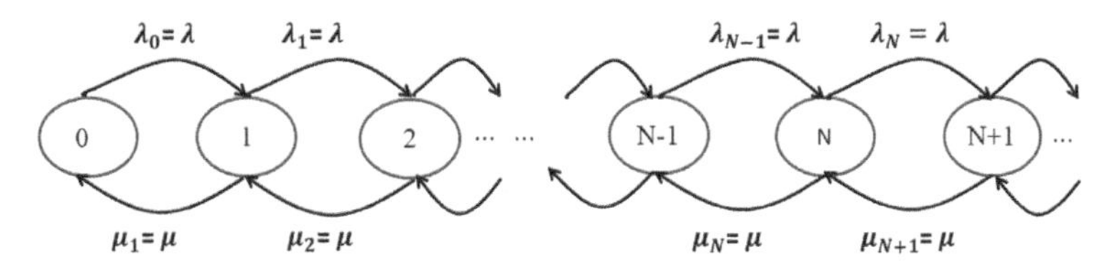

FIGURE 10.4 Flows of dump trucks in M/M/1 queuing system.

For $N = 0$,

$$\begin{aligned} p_0(t+\Delta t) = p_0(t) &* \text{no arrival during } \Delta t \text{ time} * \text{no leaving during } \Delta t \text{ time} \\ &+ p_1(t) * \text{no arrival during } \Delta t \text{ time} \\ &* \text{one leaving during } \Delta t \text{ time}, \quad N = 0 \end{aligned}$$

$$\Rightarrow p_0(t+\Delta t) = p_0(t)*(1-\lambda\Delta t)*1 + p_1(t)*(1-\lambda\Delta t)*\mu\Delta t, \quad N = 0$$

The above equations can be rearranged as:

$$\frac{p_N(t+\Delta t) - p_N(t)}{\Delta t} = -(\lambda+\mu)p_N(t) + \lambda p_{N-1}(t) + \mu p_{N+1}(t), \quad N = 1,2,\ldots \tag{10.10}$$

$$\frac{p_0(t-\Delta t) - p_0(t)}{\Delta t} = -\lambda p_0(t) + \mu p_1(t), \quad N = 0 \tag{10.11}$$

Taking the limits as $\Delta t \rightarrow 0$, we get

$$\lim_{\Delta t \to 0} \frac{p_N(t+\Delta t) - p_N(t)}{\Delta t} = p'_N(t)$$

$$\lim_{\Delta t \to 0} \frac{p_0(t-\Delta t) - p_0(t)}{\Delta t} = p'_0(t) = -\lambda p_0(t) + \mu p_1(t), \quad N = 0$$

Hence, Eq. (10.10) and Eq. (10.11) becomes

$$p'_N(t) = -(\lambda+\mu)p_N(t) + \lambda p_{N-1}(t) + \mu p_{N+1}(t), \quad N = 1,2,\ldots \tag{10.12}$$

$$p'_0(t) = -\lambda p_0(t) + \mu p_1(t), \quad N = 0 \tag{10.13}$$

The differential Eq. (10.12) and Eq. (10.13) cannot be solved using a simple method. For a detailed solution to derive the state probabilities $p_n(t)$, refer to the literature of Kleinrock (1975). It is observed that the simplest queuing models lead to a complex equation for the time-dependent behaviour to determine the state probabilities. Thus, the limiting (or equilibrium) behaviour approach is used for solving the differential equations.

The characteristics of the equilibrium behaviour are that the rate of change of the number of dump trucks in the loading point of mine is tending to zero, and the probability of the number of customers in a typical system is constant with time. These can be mathematically represented as $t \rightarrow \infty$, $p'_N(t) \rightarrow 0$ and $p_N(t) \rightarrow p_N$. Hence, Eq. (10.12) and Eq. (10.13) become:

$$0 = -(\lambda+\mu)p_N + \lambda p_{N-1} + \mu p_{N+1}, \quad N = 1, 2,\ldots \tag{10.14}$$

$$0 = -\lambda p_0 + \mu p_1, \quad N = 0 \tag{10.15}$$

For solving the M/M/1 model equations, a recursion method is on Eq. (10.14) and Eq. (10.15). From Eq. (10.15), we have,

$$p_1 = \frac{\lambda}{\mu} p_0$$

Substitution of the value of p_1 in Eq. (10.14) for $N = 1$ gives

$$p_2 = \left(\frac{\lambda}{\mu}\right)^2 p_0$$

Similarly, by substituting the value of p_2 in Eq. (10.14) for $N = 2$, the value of p_3 can be derived, and so on. Hence, we can recursively express all probabilities in terms of p_0, yielding

$$p_n = \left(\frac{\lambda}{\mu}\right)^N p_0, \qquad N = 1,2,\ldots \tag{10.16}$$

We know the sum of the total probability for all the possible events is equal to 1. Therefore, we have,

$$\sum_{N=0}^{\infty} p_N = 1$$

$$\Rightarrow p_0 + \sum_{n=1}^{\infty} p_N = 1$$

$$\Rightarrow p_0 + \sum_{n=1}^{\infty} \left(\frac{\lambda}{\mu}\right)^N p_0 = 1 \qquad \text{[Putting the value of } p_N \text{ from Eq. (10.16)]}$$

$$\Rightarrow p_0 \left[1 + \sum_{n=1}^{\infty} \left(\frac{\lambda}{\mu}\right)^N\right] = 1$$

$$\Rightarrow p_0 = \frac{1}{\left[1 + \sum_{N=1}^{\infty} \left(\frac{\lambda}{\mu}\right)^N\right]}$$

Putting the value of $\frac{\lambda}{\mu}$ from Eq. (10.8), we have

$$p_0 = \frac{1}{\left[(\rho)^0 + \sum_{n=1}^{\infty} (\rho)^N\right]} = \frac{1}{\sum_{n=0}^{\infty} \rho^N}$$

$$\Rightarrow p_0 = \frac{1}{(1-\rho)^{-1}} = 1 - \rho = 1 - \frac{\lambda}{\mu} \tag{10.17}$$

[The binomial expansion of $(1-\rho)^{-1} = 1 + \rho + \rho^2 + \cdots \infty = \sum_{n=0}^{\infty} \rho^N$]

Putting the value of p_0 in Eq. (10.16), we have

$$\Rightarrow p_N = \left(\frac{\lambda}{\mu}\right)^N p_0 = \left(\frac{\lambda}{\mu}\right)^N \left(1-\frac{\lambda}{\mu}\right) = \rho^N (1-\rho), \qquad N = 1,2,\ldots \tag{10.18}$$

Therefore, the performance measures of the queuing system can be summarized as follows:

- **Probability of at least one dump truck in the system**

The probability of at least one dump truck at the loading point is given by

$$\Rightarrow P(N>0) = 1 - p_0 = 1-(1-\rho) = \rho = \frac{\lambda}{\mu} \tag{10.19}$$

- **Probability of zero dump truck in the system**

The probability of zero dump truck at the loading point is given by

$$P(N=0) = p_0 = 1-\rho = 1-\frac{\lambda}{\mu} \tag{10.20}$$

- **Average number of customers in the system**

The probability of N dump trucks at the loading point is given by

$$P(N=N) = p_N = \rho^N (1-\rho) = \left(\frac{\lambda}{\mu}\right)^N \left(1-\frac{\lambda}{\mu}\right) \tag{10.21}$$

The average number of dump trucks at the loading point, L_s (including one being loaded) in the equilibrium state is given by the sum of all the expected number of dump trucks. This is given by

$$L_s = E(N) = \sum_{N=0}^{\infty} N p_N = \sum_{N=0}^{\infty} N \rho^N (1-\rho)$$

$$\Rightarrow L_s = (1-\rho)\rho \sum_{N=0}^{\infty} \frac{d}{d\rho}\rho^N = (1-\rho)\rho \frac{d}{d\rho}\sum_{N=0}^{\infty}\rho^N = (1-\rho)\rho\frac{d}{d\rho}(1-\rho)^{-1}$$

$$\Rightarrow L_s = (1-\rho)\rho * \frac{1}{(1-\rho)^2} = \frac{\rho}{1-\rho} = \frac{\lambda}{\mu-\lambda} \tag{10.22}$$

- **Average number of customers in the queue**

Similarly, the average number of dump trucks waiting in the queue, L_q (excluding one being loaded) can be obtained by the sum of the expected number of dump trucks waiting in the queue for loading. This can be determined as

$$L_q = E(N-1) = \sum_{N=1}^{\infty} (N-1) p_N = \sum_{N=1}^{\infty} N p_N - \sum_{N=1}^{\infty} p_N = \sum_{N=0}^{\infty} N p_N - \left(\sum_{N=0}^{\infty} p_N - p_0\right)$$

$$\Rightarrow L_q = L_s - (1 - p_0) = \frac{\rho}{1-\rho} - [1 - (1-\rho)] = \frac{\rho}{1-\rho} - \rho = \frac{\rho^2}{1-\rho} = \frac{\lambda^2}{\mu(\mu-\lambda)} \tag{10.23}$$

- **Average waiting time in the system**

The average time spent by the dump truck at the loading point, W_s (including the loading time) is given by

$$W_s = \frac{\text{Average number of dump truck at the loading point}}{\text{Mean arrival rate}} = \frac{L_s}{\lambda} = \frac{1}{\mu-\lambda} \tag{10.24}$$

- **Average waiting time in the queue**

The average waiting time in the queue, W_q (excluding the loading time) can be obtained by subtracting the mean loading time from mean time spent at the loading point. Since the service or loading rate is μ per unit time, the mean loading time of one dump truck is $1/\mu$. Therefore, the average waiting time of the dump truck in the queue is given by

$$W_q = W_s - \frac{1}{\mu} = \frac{1}{\mu-\lambda} - \frac{1}{\mu} = \frac{\lambda}{\mu(\mu-\lambda)} \tag{10.25}$$

- **Probability that an arriving dump truck finds *N* dump trucks in the system**

The probability that an arriving dump truck finds *N* dump trucks in the queuing system can be determined using the PASTA (Poisson Arrivals See Time Averages) rule (Wolff, 1982). PASTA states that the fraction of dump trucks finding on arrival N customers in the system is equal to the fraction of time *N* customers are in the system. Therefore, the arriving dump truck finds *N* dump trucks in the queuing system is given by

$$P(L_s = N) = p_N = \rho^N (1-\rho) \tag{10.26}$$

Example 10.2

A service engineer repairs dump trucks in a mining workshop. The repair time is exponentially distributed with a mean of 18 minutes. According to a Poisson distribution, Dump trucks arrive at the workshop on average 10 dump trucks per 8-hour shift.

a. What is the fraction of time in the 8-hour shift that the service engineer has no work to do?
b. How many dump trucks are, on average, at the workshop at any moment of time?
c. Determine the mean time spent (waiting time plus repair time) by the dump truck in the workshop.

Solution

Given data:

$$\text{Mean arrival rate}(\lambda) = \frac{10}{8} \text{ per hour}$$

$$\text{Mean service rate}(\mu) = \frac{1}{\left(18/60\right)} = \frac{10}{3} \text{ per hour}$$

a. Fraction of time that the service engineer man is busy is given by

$$p_{n>0} = \rho = \frac{\lambda}{\mu} = \frac{10/8}{10/3} = \frac{3}{8} = 0.375$$

That is, the probability that there will be at least one dump truck in the workshop is 0.375. Hence, the fraction of time that the service engineer is busy is 0.375.
On the other hand, the probability that there will be no dump truck in the workshop is given by

$$p_0 = 1 - 0.375 = 0.625$$

Therefore, the fraction of time the service engineer will be sitting idle is 0.675.

b. The average number of dump trucks at the workshop, L_s is given by

$$L_s = \frac{\lambda}{\mu - \lambda} = \frac{10/8}{(10/3) - (10/8)} = \frac{3}{5} \approx 1$$

That is, at any moment of time, the average number of dump trucks in the workshop is 1.

c. The mean time spent in the workshop, W_s is given by

$$W_s = \frac{1}{\mu - \lambda} = \frac{1}{(10/3) - (10/8)} = \frac{24}{50} \text{ hour} = \frac{24}{50} * 60 \text{ minutes} = 28.8 \text{ minutes}$$

Therefore, any dump truck arriving into the workshop has to spend 28.8 minutes for completion of repair works.

Example 10.3
In an opencast coal mine, dump trucks arrive at the loading point according to a Poisson process. The arrival rate is 10 dump trucks per hour. The dump trucks are served in order of arrival. The loading times are exponentially distributed. The mean loading time is 5 minutes.

a. Estimate the mean number of dump trucks at the loading point.
b. Estimate the mean number of dump trucks waiting in the queue.
c. Determine the waiting time of the dump truck before getting service.
d. Determine the probability that an arriving dump truck is finding two dump trucks at the loading point.

Solution

Given data:

$$\text{Mean arrival rate}(\lambda) = 10 \text{ trucks per hour}$$

$$\text{Mean service rate}(\mu) = \frac{60}{5} = 12 \text{ trucks per hour}$$

a. The average number of trucks at the loading point, L_s is given by

$$L_s = \frac{\lambda}{\mu - \lambda} = \frac{10}{12 - 10} = 5$$

That is, at any moment of time, the average number of dump trucks at the loading point is 5.

b. The average number of dump trucks at the loading point, L_s is given by

$$L_s = \frac{\lambda^2}{\mu(\mu-\lambda)} = \frac{10^2}{12(12-10)} = \frac{100}{24} \approx 4$$

That is, at any moment of time, the average number of dump trucks waiting in the queue is 4.

c. The mean time spent in the queue, W_q is given by

$$W_q = \frac{\lambda}{\mu(\mu-\lambda)} = \frac{10}{12(12-10)} = \frac{10}{24} \text{ hour} = \frac{10}{24} * 60 \text{ minutes} = 25 \text{ minutes}$$

The average waiting time in the queue before getting service is 25 minutes.

d. The probability that an arriving dump truck finds two dump trucks at the loading point is given by

$$P(L_s = 2) = p_2 = \rho^2(1-\rho) = \left(\frac{\lambda}{\mu}\right)^2\left(1-\frac{\lambda}{\mu}\right) = \left(\frac{10}{12}\right)^2\left(1-\frac{10}{12}\right) = 0.115$$

10.6.2 M/M/s QUEUING SYSTEM

The only difference between the M/M/s queuing model, and the M/M/1 model is the number of servers (loaders) deployed to provide service. All other characteristics remain the same as that of the M/M/1 model. In M/M/s model, the number of servers is more than 1. In an open-pit mine, if two shovels are deployed in the same face for loading the trucks one after another, then M/M/s queuing system should be followed for optimizing the system. A typical queuing characteristics M/M/s model is shown in Figure 10.5. In this type of model also a single-channel queue is followed. Thus, the characteristics of the M/M/ s queuing model is described below:

- Inter-arrival times of dump truck follow exponential distribution function with mean arrival rate is λ per unit time.
- Service times of the loader follow exponential distribution function with mean service rate is μ per unit time.
- The number of loaders deployed in the face for loading operation is *s*.
- Queuing capacity is infinite, that is, all the dump truck arrives in the loading point can wait in the queue for being loaded.

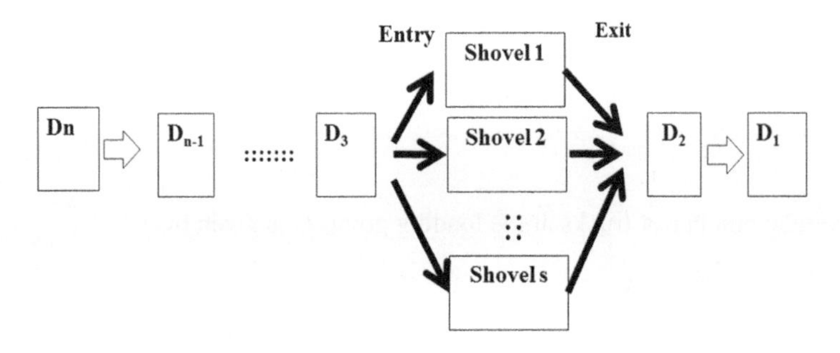

FIGURE 10.5 Queue discipline in M/M/s queuing model.

- Dump trucks are assumed to be coming from an infinite population.
- Dump trucks are served on a FIFO basis.

The basic assumption of the model is that the service rate of s identical servers is more than that of the arrival rate of customers in order to avoid the infinite queue length. The condition can be represented mathematically as follows:

$$\rho = \frac{\lambda}{s\mu} < 1 \tag{10.27}$$

In the above equation, ρ is the fraction of time the server is busy.

Let p_n denote the probability of n dump trucks in the queuing system (loading point). The flow behaviour of dump trucks at the loading point is shown in Figure 10.6. In the M/M/s queuing model, the arrival rate is constant with time and calling population, but the service rate varies and depends on the number of dump trucks at the loading point. Assuming the number of servers (loaders) deployed in the system is s, the service rate is $n\mu$ if n is less than s and $n\mu$ if n is greater and equal to s (shown in Figure 10.6). Mathematically, these can be represented as

$$\text{Mean arrival rate} = \lambda_1 = \lambda_2 \ldots = \lambda_N = \lambda$$

$$\text{And, Mean service rate} = \mu_n = \begin{cases} N\mu & \text{for } N < s \\ s\mu & N \geq s \end{cases} \quad N = 1,2,\ldots$$

At equilibrium conditions, the probability equations can be derived similarly to that of the M/M/1 model. Instead of equating the flow into and out of a single state n, the equilibrium equations can be derived by equating the flow in and out of different states $\{0, 1, \ldots, N\}$. The flow that occurs in state n depends on the two neighboring states $(n - 1)$ and $(n+1)$. At equilibrium condition,

Expected rate of inflow to state $N = \lambda_{N-1}p_{N-1} + \mu_{N+1}p_{N+1}$

Similarly, the expected rate of outflow from state $N = \lambda_N p_N + \mu_N p_N$

In the equilibrium state, the rate of inflow should be equal to the rate of outflow. Mathematically, this can be written as:

$$\text{Rate of inflow} = \text{Rate of outflow}$$

$$\Rightarrow \lambda_{N-1}p_{N-1} + \mu_{N+1}p_{N+1} = \lambda_N p_N + \mu_N p_N \tag{10.28}$$

From Figure 10.6, the equilibrium equation for $N = 0$ can be written as

$$\lambda p_0 = \mu_1 p_1$$

$$\Rightarrow p_1 = \frac{\lambda}{\mu_1} p_0 \tag{10.29}$$

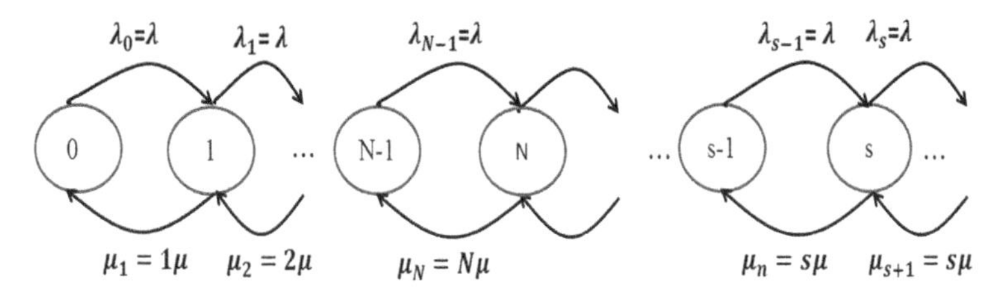

FIGURE 10.6 Flows in M/M/s queuing model.

For $N = 1$, the equilibrium equation can be derived from Eq. (10.28) as

$$\lambda p_0 + \mu_2 p_2 = \lambda p_1 + \mu_1 p_1$$

Substituting the value of p_1 in the above equation from Eq. (10.29), we get

$$\mu_2 p_2 = \lambda * \frac{\lambda}{\mu_1} p_0 + \mu_1 * \frac{\lambda}{\mu_1} p_0 - \lambda p_0$$

$$\Rightarrow p_2 = \frac{\lambda * \lambda}{\mu_1 * \mu_2} p_0 \tag{10.30}$$

Similarly, we can derive the value of p_N as

$$p_N = \frac{\lambda^N}{\mu_1 * \mu_2 * \ldots * \mu_N} p_0, \qquad N = 1,2,\ldots \tag{10.31}$$

Putting, $\mu_n = \begin{cases} N\mu & \text{for } N < s \\ s\mu & N \geq s \end{cases}$ $N = 1,2,\ldots$ in Eq. (10.31), we have

$$p_N = \begin{cases} \dfrac{\lambda^N}{\mu(2\mu)(3\mu)\ldots(N\mu)} p_0 = \dfrac{\lambda^N}{N!\mu^N} p_0 = \dfrac{(\lambda/\mu)^N}{N!} p_0 & N < s \\ \dfrac{\lambda^n}{\mu(2\mu)(3\mu)\ldots\{(s-1)\mu\}(s\mu)^{N-s+1}} p_0 = \dfrac{\lambda^N}{s!s^{N-s}\mu^N} p_0 = \dfrac{(\lambda/\mu)^N}{s!s^{N-s}} p_0 & N \geq s \end{cases}$$

Putting $\frac{\lambda}{\mu} = s\rho$, the above equation becomes

$$p_N = \begin{cases} \dfrac{(s\rho)^N}{N!} p_0 & N < s \\ \dfrac{(s\rho)^n}{s!s^{N-s}} p_0 & N \geq s \end{cases} \tag{10.32}$$

Again, we know the sum of the total probability for all the possible events is equal to 1. Therefore, we have,

$$\sum_{N=0}^{\infty} p_N = 1$$

$$\Rightarrow \sum_{N=0}^{s-1} p_N + \sum_{N=s}^{\infty} p_N = 1 \tag{10.33}$$

Putting the value of p_N from Eq. (10.32) in Eq. (10.33), we have

$$\Rightarrow \sum_{N=0}^{s-1} \frac{(s\rho)^N}{N!} p_0 + \sum_{N=s}^{\infty} \frac{(s\rho)^N}{s!s^{N-s}} p_0 = 1$$

$$\Rightarrow p_0 = \left\{ \sum_{N=0}^{s-1} \frac{(s\rho)^N}{N!} + \sum_{N=s}^{\infty} \frac{(s\rho)^N}{s!s^{N-s}} \right\}^{-1} = \left\{ \sum_{N=0}^{s-1} \frac{(s\rho)^N}{N!} + \frac{\rho^s}{s!} \sum_{N=s}^{\infty} \frac{(\rho)^{N-s} s^N}{s^{N-s}} \right\}^{-1}$$

$$\Rightarrow p_0 = \left\{ \sum_{N=0}^{s-1} \frac{(s\rho)^N}{N!} + \frac{(s\rho)^s}{s!} \sum_{N=s}^{\infty} (\rho)^{N-s} \right\}^{-1}$$

For $\rho = \dfrac{\lambda}{s\mu} < 1$, the value of p_0 can be rewritten as

$$p_0 = \left\{ \sum_{N=0}^{s-1} \frac{(s\rho)^N}{N!} + \frac{(s\rho)^s}{s!} \left(\frac{1}{1-\rho} \right) \right\}^{-1} \tag{10.34}$$

The performance measures for M/M/s model can be summarized as follows:

- **Average number of customers in the queue**

The average number of customers in the queue, L_q can be obtained by the expected number of dump trucks standing in the queue.

The average number of dump trucks at the loading point, L_s(including one being loaded) in the equilibrium state is given by the sum of all the expected number of dump trucks. This is given by

$$L_q = \text{Expected number of dump trucks in the queue} = \sum_{N=0}^{\infty} N p_{s+N}$$

$$= \sum_{N=0}^{\infty} N \frac{(s\rho)^{N+s}}{s^{\{(N+s)-s\}} s!} p_0 = \sum_{N=0}^{\infty} N \frac{(\rho)^{N+s} s^{N+s}}{s^N s!} p_0 = \sum_{N=0}^{\infty} N \frac{(\rho)^{N+s} s^s}{s!} p_0$$

$$= \frac{\rho^{s+1} s^s}{s!} p_0 \sum_{N=0}^{\infty} N \rho^{N+1} = \frac{\rho^{s+1} s^s}{s!} p_0 \frac{d}{d\rho} \left(\sum_{N=0}^{\infty} \rho^N \right) = \frac{\rho^{s+1} s^s}{s!} p_0 \frac{d}{d\rho} \left(\frac{1}{1-\rho} \right)$$

$$\Rightarrow L_q = \frac{\rho^{s+1} s^s}{s!} p_0 \frac{1}{(1-\rho)^2} = \frac{\rho^{s+1} s^s}{s!(1-\rho)^2} p_0 \tag{10.35}$$

- **Average number of customers in the system**

The average number of customers in the system, L_s can be obtained by adding the average number of dump trucks in service with the average number of dump trucks waiting in the queue (L_q). This is given by

$$\Rightarrow L_s = L_q + s\rho = \frac{\rho^{s+1} s^s}{s!(1-\rho)^2} p_0 + \frac{\lambda}{\mu} \tag{10.36}$$

- **Average time spent in the system**:

The average time spent by a dump truck in the system is given by

$$\Rightarrow W_s = \frac{L_s}{\lambda} = \frac{\dfrac{\rho^{s+1} s^s}{s!(1-\rho)^2} p_0 + \dfrac{\lambda}{\mu}}{\lambda} \tag{10.37}$$

- **Average time spent in the queue**

The mean waiting time by the dump truck in the queue, Wq, can be obtained as

$$W_q = \frac{L_q}{\lambda} = \frac{\frac{\rho^{s+1} s^s}{s!(1-\rho)^2} P_0}{\lambda} \tag{10.38}$$

Example 10.4
The mine manager has to make a decision on the number of loading machines needed to deploy in an opencast mine face in the following condition. It is observed that approximately 16 dump trucks per hour arrive in the loading face. The duration of loading a dump truck is approximately exponentially distributed with a mean of 5 minutes. How many loaders are needed such that the mean waiting time in the queue is less than 5 minutes?

Solution

Given data:

$$\text{Mean arrival rate}(\lambda) = 16 \text{ dump trucks per hour}$$

$$\text{Mean service rate}(\mu) = \frac{1}{5} \text{dump truck per minute}$$

$$= \frac{60}{5} \text{dump trucks per hour} = 12 \text{ dump trucks per hour}$$

In order to have $\rho = \dfrac{\lambda}{s\mu} < 1$, the value of $s \geq 2$

For $s = 2$, $\rho = 2/3$

$$W_q = \frac{\frac{\rho^{s+1} s^s}{s!(1-\rho)^2} P_0}{\lambda} = \frac{(2/3)^{2+1} 2^2}{2!(1-2/3)^2 * 16} P_0 = \frac{8*9*4}{27*2*16} P_0 = \frac{1}{3} P_0$$

We have,

$$P_0 = \left\{ \sum_{N=0}^{s-1} \frac{(s\rho)^N}{N!} + \frac{(s\rho)^s}{s!} \left(\frac{1}{1-\rho} \right) \right\}^{-1}$$

$$\Rightarrow P_0 = \left[\left(\frac{(4/3)^0}{0!} + \frac{(4/3)^1}{1!} \right) + \frac{(4/3)^2}{2!} \left(\frac{1}{1-\frac{2}{3}} \right) \right]^{-1} = \left(1 + \frac{4}{3} + \frac{8}{3} \right)^{-1} = \frac{3}{15}$$

$$\therefore W_q = \frac{1}{3} * \frac{3}{15} \text{hour} = \frac{1}{15} * 60 \text{ minutes} = 4 \text{ minutes}$$

With two shovels, the average waiting time is 4 minutes, which is less than desired upper bound of waiting time of 5 minutes. Hence, the deployment of two shovels is sufficient for loading operations.

Example 10.5

In a production bench, two shovels are deployed for loading the blasted material into dump trucks. Each shovel loads one dump truck at a time separately. When both shovels are occupied, the new arriving dump truck waits for their turn. The loading time is exponentially distributed with an average loading time of 3 minutes. Dump trucks arrive at the loading point according to a Poisson process with a rate of 20 dump trucks per hour.

a. Calculate the probability of N dump trucks at the loading point (waiting plus under loading), $N = 0, 1, 2, \ldots$.
b. Calculate the mean waiting time and mean number of dump trucks waiting at the loading point.
c. Determine the proportion of dump trucks who find both loaders are busy on arrival.
d. Determine the number of shovels for restricting the proportion of dump trucks who find all loaders are busy on arrival to at most 20%.

Solution

Given data:

Arrival rate $(\lambda) = 3$ per minutes $= 20$ per hour
Service rate $(\mu) = 20$ per hour
Number of server (s) = 2

$$\therefore \rho = \frac{\lambda}{s\mu} = \frac{20}{2 * 20} = \frac{1}{2}$$

a. The probability of zero dump truck at the loading point is given by

$$p_0 = \left\{ \sum_{N=0}^{s-1} \frac{(s\rho)^N}{N!} + \frac{(s\rho)^s}{s!}\left(\frac{1}{1-\rho}\right) \right\}^{-1}$$

$$= \left[\left\{ \frac{(2*1/2)^0}{0!} + \frac{(2*1/2)^1}{1!} \right\} + \frac{(2*1/2)^2}{2!}\left(\frac{1}{1-1/2}\right) \right]^{-1} \left(\frac{1}{1-1/2}\right) = (1+1+1)^{-1} = \frac{1}{3}$$

For $N < s$ (=2), the probability of N dump truck at the loading point is given by

$$p_N = \frac{(s\rho)^N}{N!} p_0$$

$$p_1 = \frac{(2*1/2)^1}{1!} * \frac{1}{3} = \frac{1}{3}$$

For $N \geq 2$, the probability of N dump truck at the loading point is given by

$$p_N = \frac{(s\rho)^N}{s!\,s^{N-s}} p_0$$

$$p_2 = \frac{(2*1/2)^2}{2!\,2^{2-2}} * \frac{1}{3} = \frac{1}{6}$$

The probability that there are zero dump trucks at the loading point is

$$P(N = 0) = 1/3$$

The probability that there is one dump truck at the loading point is

$$P(N = 1) = 1/3$$

The probability that there are two dump trucks at the loading point is

$$P(N = 2) = 1/6$$

b. The average number of dump trucks waiting at the loading point is given by

$$L_q = \frac{\rho^{s+1} s^s}{s!(1-\rho)^2} p_0 = \frac{\left(\frac{1}{2}\right)^{2+1} 2^2}{2!\left(1-\frac{1}{2}\right)^2} * \frac{1}{3} = \frac{1/2}{1/2} * \frac{1}{3} = \frac{1}{3} \approx 1$$

Therefore, the average number of dump truck presence at the loading point (waiting and under loading) is approximately 1.

The mean waiting time in the queue, Wq, can be obtained as

$$W_q = \frac{\frac{\rho^{s+1} s^s}{s!(1-\rho)^2} p_0}{\lambda} = \frac{\frac{\left(\frac{1}{2}\right)^{2+1} 2^2}{2!\left(1-\frac{1}{2}\right)^2} * \frac{1}{3}}{20} = \frac{1}{60} \text{hour} = 1 \text{ minute}$$

The average waiting time of a dump truck before start loading is 1 minute.

c. Both the shovels will be busy if the number of dump trucks in the system is two or more (≥ number of the loader). The probability is given by

$$P(N \geq s) = p_s + p_{s+1} + p_{s+2} + \cdots$$

We know,

$$p_{N+1} = \frac{(s\rho)^{N+1}}{s!s^{N+1-s}} p_0 = \frac{(s\rho)^N}{s!s^{N-s}} p_0 * \frac{s\rho}{s} = \rho * p_N$$

Therefore, the probability equation $P(N \geq s)$ can be rewritten as

$$P(N \geq s) = p_s + \rho p_s + \rho^2 p_s + \cdots$$

$$\Rightarrow P(N \geq s) = \frac{p_s}{1-\rho}$$

For $s = 2$, we have

$$P(N \geq 2) = \frac{p_2}{1-\rho} = \frac{1/6}{1-1/2} = \frac{1}{3}$$ [Refer solution (a) for p_2]

The desired fraction of dump trucks finding both loaders are busy is the probability that the system has at least two dump trucks in the system.

$$\therefore \text{Fraction of dump truck finding both loaders are busy} = P(N \geq 2) = \frac{1}{3}$$

d. For $s = 2$, the probability that the dump truck in the system find both the shovel busy is 1/3, which is more than 10%

Therefore, for $s = 3$,

$$\rho = \frac{\lambda}{s\mu} = \frac{20}{3*20} = \frac{1}{3}$$

The probability that the dump truck in the system find all the three shovels busy is given by

$$P(N \geq 3) = \frac{p_3}{1-\rho} = \frac{\frac{(s\rho)^3}{s!s^{N-s}} p_0}{1-\rho} = \frac{\frac{(3*1/3)^3}{3!3^{3-3}} * \frac{1}{3}}{1-\frac{1}{3}} = \frac{3}{18*2} = \frac{1}{12}$$

The probability is less than 10% $\left(= \frac{1}{10}\right)$. Therefore, three shovels are required.

Example 10.6

Three shovels are deployed in a production face for loading operation. The shovels are loading coal into dump trucks, which are arriving according to a Poisson process with a rate of 15 dump trucks per hour. The loading time is exponentially distributed with a mean of 10 minutes.

a. What is the average number of shovels busy with loading operations?
b. What is the probability that a dump truck has to wait in a queue before loading starts?
c. Determine the average waiting time of a dump truck.

Solution

Given data: Arrival rate $(\lambda) = 15$ per hour

Service rate $(\mu) = \frac{60}{10}$ per hour $= 6$ per hour

Number of server (s) = 3

$$\therefore \rho = \frac{\lambda}{s\mu} = \frac{15}{3*6} = \frac{5}{6}$$

a. The average number of shovels busy with loading operation $= s\rho = 3*\frac{5}{6} = 2.5$.

b. We know,

the probability that no dump truck is at the loading point is given by

$$P(N=0)=\left\{\sum_{n=0}^{s-1}\frac{(s\rho)^n}{n!}+\frac{(s\rho)^s}{s!}\left(\frac{1}{1-\rho}\right)\right\}^{-1}$$

$$=\left|\left\{\frac{\left(3*\frac{5}{6}\right)^0}{0!}+\frac{\left(3*\frac{5}{6}\right)^1}{1!}+\frac{\left(3*\frac{5}{6}\right)^2}{2!}\right\}+\frac{\left(3*\frac{5}{6}\right)^3}{3!}\left(\frac{1}{1-\frac{5}{6}}\right)\right|^{-1}$$

$$=\left(1+\frac{5}{2}+\frac{25}{8}+\frac{125*6}{8*6}\right)^{-1}=\left(\frac{8+20+25+125}{8}\right)^{-1}=\frac{8}{178}$$

$$=0.044$$

The probability that there will be no dump truck at the loading point is 0.044.

If all shovels are busy, the dump truck has to wait on arriving. All the shovels will be busy if the number of dump trucks in the system is equal to or more than three (≥ number of servers). Thus, the probability that the number of dump trucks at the loading point is three or more is given by

$$P(N\geq 3)=p_3+p_{3+1}+p_{3+2}+\cdots$$

$$\Rightarrow P(N\geq 3)=p_3+\rho p_3+\rho^2 p_3+\cdots$$

$$\Rightarrow P(N\geq 3)=\frac{p_3}{1-\rho}$$

The probability that $N(N\geq s)$ dump trucks are at the loading point is given by

$$P(N=3)=p_3=\frac{(s\rho)^3}{s!s^{3-s}}p_0$$

$$\Rightarrow P(N=3)=\frac{(3*5/6)^3}{3!3^{3-3}}*\frac{8}{178}=\frac{(3*5/6)^3}{3!3^{3-3}}*\frac{8}{178}=\frac{125}{8*6}*\frac{8}{178}=\frac{125}{6*178}$$

$$\therefore P(N\geq 3)=\frac{p_3}{1-\rho}=\frac{125}{6*178*(1-5/6)}=\frac{125}{178}=0.70$$

Therefore, the probability that a dump truck arriving at the loading point has to wait before getting service is 0.70.

c. The average waiting time of a dump truck is given by

$$W_s=\frac{\frac{\rho^{s+1}s^s}{s!(1-\rho)^2}p_0+\frac{\lambda}{\mu}}{\lambda}=\frac{\frac{(5/6)^{3+1}3^3}{3!(1-5/6)^2}*\frac{8}{178}+\frac{15}{6}}{15}$$

$$=\frac{\frac{625*27*8}{6*36*1*178}+\frac{15}{6}}{15}=0.40\text{ hour}=24\text{ minutes}$$

Therefore, the average time a dump truck has to spend at the loading point is 24 minutes.

Example 10.7

In a power plant, coals are purchased from two different mines. The coals from mine are transported through dump truck and fed at two different unloading points. Arrival and unloading rates at the first unloading platform are 12 and 18 per hour, respectively and that for the other unloading platform are 18 and 24 per hour, respectively. It is given that arrival pattern is a Poisson processes and unloading times are exponentially distributed.

a. If the two unloading platforms operate independently, determine the mean number of dump trucks waiting in the queue and their mean waiting time at each unloading point.
b. Analyse the effect of operating the two unloading points with a single queue with an arrival rate of 30 per hour and a service rate of 30 per hour.

Solution

a. The queuing characteristics of the two loading platforms are summarized below.

Parameter	Platform 1 (M/M/1 Model)	Platform 2 (M/M/1 Model)
λ	12	18
μ	20	24
$\rho = \lambda / \mu$	0.60	0.75
$L_q = \dfrac{\lambda^2}{\mu(\mu - \lambda)}$	$0.90 \approx 1$	$2.25 \approx 3$
$W_q = \dfrac{\lambda}{\mu(\mu - \lambda)}$	0.075 hour = 4.5 minutes	0.125 hour = 7.5 minutes

The average waiting time in the queue at the first unloading platform is 4.5 minutes, and the average number of dump trucks waiting in the queue at any moment of time is 1. Similarly, the average waiting time in the queue at the first unloading platform is 7.5 minutes, and the average number of dump trucks waiting in the queue at any moment of time is 3.

b. The effect of operating the two servers with a single queue with an arrival rate of 30 (= 12 + 18) per hour and a service rate of 44 (= 20+24) per hour is given below.

Parameter	Two Server Single Queue (M/M/2 model)
λ	30
μ	44
s	2
$\rho = \lambda / s\mu$	0.34
$P_0 = \left\{ \sum_{n=0}^{s-1} \dfrac{(s\rho)^n}{n!} + \dfrac{(s\rho)^s}{s!} \left(\dfrac{1}{1-\rho} \right) \right\}^{-1}$	0.5
$L_q = \dfrac{\rho^{s+1} s^s}{s!(1-\rho)^2} P_0$	$0.09 \approx 0$
$W_q = \dfrac{\dfrac{\rho^{s+1} s^s}{s!(1-\rho)^2} P_0}{\lambda}$	0.003 hour = 0.18 minute

Thus, the average waiting time in the new queuing system is 0.18 minutes, and the average number of dump trucks waiting in the queue at any moment of time is approximately zero.

It is clear from the results that two platforms with a single queue significantly reduce, the queue's length, and waiting time in the queue from two independent platforms with two different queues. Thus, in this case, the second option (M/M/s) queuing system offers more efficient and economical operation than the two independent M/M/1 queuing systems.

10.6.3 Infinite server queue model (M/M/∞)

The number of servers in the M/M/∞ queuing model is unlimited, unlike **M/M/1 and M/M/s** queuing models. This type of model is also called a self-service queuing model. In mines, this type of model can be observed when the dump truck operator does the self-checking in the workshop, dump trucks unloading the overburden at the dumpsite without any space restriction, etc. In this type of model also a single-channel queue is followed, but dump trucks do not have to wait for getting the service as there is an infinite number of servers that exist. A typical queuing characteristic M/M/∞ model is shown in Figure 10.7. The model is explained with the example that the dump trucks are unloading the overburdened material in the external dumpsite, where no restriction of space for unloading the material exists.

The characteristics of the M/M/∞ model are as follows:

- Inter-arrival times of dump truck follow exponential distribution function with mean arrival rate is λ per unit time.
- Unloading times follow exponential distribution function with mean service rate is μ per unit time.
- The number of servers is infinity.
- Queuing capacity is infinite, that is, all the dump truck arrives in the loading point can get a space.
- Dump trucks are assumed to be coming from an infinite population.

The flow of customers in this type of system is represented in Figure 10.8.

In **M/M/∞** type queuing model,

$$\text{Mean arrival rate for } N^{th} \text{ state} = \lambda_N = \lambda, \qquad N = 0,1,2,\ldots \tag{10.39}$$

$$\text{Mean service rate for } N^{th} \text{ state} = \mu_N = N\mu, \qquad N = 0,1,2,\ldots \tag{10.40}$$

As the number of servers is infinity, the service rate is equal to the number of dump trucks in the system. Any dump truck will immediately get the service on arrival, and no need to wait for getting the service.

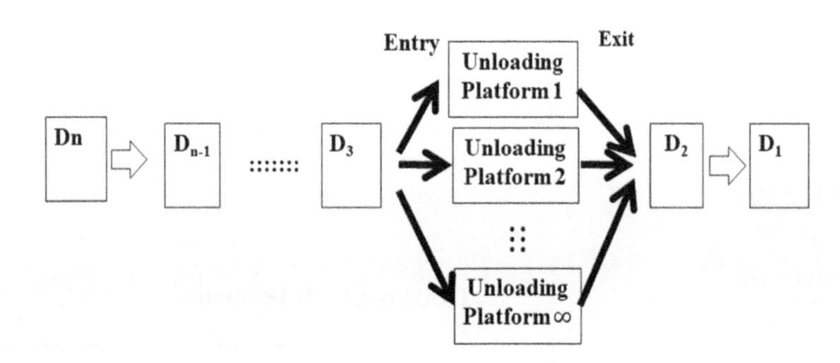

FIGURE 10.7 Arrival and Service Pattern of M/M/∞ Queuing Model.

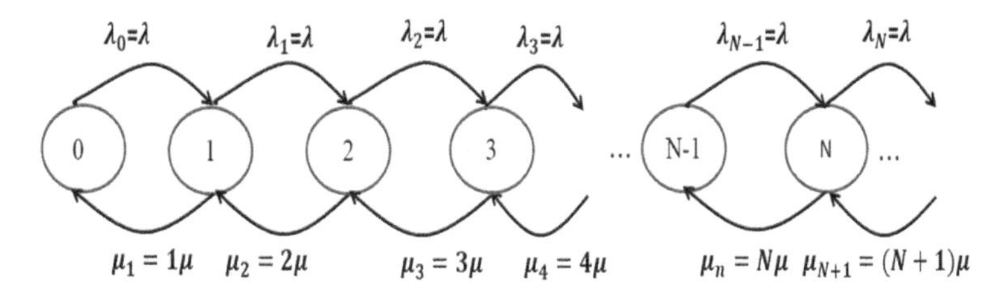

FIGURE 10.8 Flows in M/M/∞ queuing system.

Thus, the probability equation (Eq. 10.31) for p_n can be derived in a similar method as done for M/M/s queuing model.

$$p_N = \frac{\lambda^N}{\mu_1 * \mu_2 * \ldots * \mu_N} p_0, \qquad N = 1, 2, \ldots \tag{10.41}$$

Putting the value of μ_N from Eq. (10.40) in Eq. (10.41), we have

$$p_N = \frac{\lambda^N}{1\mu * 2\mu * \ldots * N\mu} p_0, \qquad N = 1, 2, \ldots$$

$$\Rightarrow p_N = \frac{1}{N!}\left(\frac{\lambda}{\mu}\right)^N p_0, \qquad N = 0,1,2,\ldots$$

$$\Rightarrow p_n = \frac{1}{N!}(\rho)^N p_0, \qquad N = 0,1,2,\ldots \tag{10.42}$$

We know, the probabilities p_N also satisfy

$$\sum_{N=0}^{\infty} p_N = 1$$

$$\Rightarrow p_0 + p_1 + p_2 \cdots p_N + \cdots \infty = 1$$

Putting the values of $p_0, p_1, \ldots, p_N$ from Eq. (10.42) in the above equation, we get

$$p_0 + \frac{1}{1!}(\rho)^1 p_0 + \frac{1}{2!}(\rho)^2 p_0 \cdots \frac{1}{N!}(\rho)^N p_0 + \cdots \infty = 1$$

$$\Rightarrow p_0 = \frac{1}{1 + \frac{1}{1!}(\rho)^1 + \frac{1}{2!}(\rho)^2 \cdots \frac{1}{N!}(\rho)^N + \cdots \infty} = \frac{1}{1 + \frac{\rho^1}{1!} + \frac{\rho^2}{2!} \cdots \frac{\rho^N}{N!} + \cdots \infty} = \frac{1}{e^\rho} = e^{-\rho}$$

$$\therefore p_N = \frac{\rho^N}{N!} e^{-\rho}, \qquad N = 0, 1, 2, \ldots \tag{10.43}$$

The average performance measures for M/M/∞ model can be determined as follows:

- **The average number of dump trucks in the queue:** The average number of customers in the queue, L_q is zero as there is an infinite number of servers exist in the system. Any customer gets service immediately after arrival into the system.

  $L_q = 0$ [Number of servers are infinite and thus waiting time in the queue is zero]

- **The average number of dump trucks in the system:** The average number of dump trucks in the system, L_s is given by

$$L_s = \frac{\lambda}{\mu} \tag{10.44}$$

- **The average time spent by the dump trucks in the system**: The average time spent in the system is the average service time to each customer. It is given by

$$W_s = \frac{1}{\mu} \tag{10.45}$$

- **The average time spent in the queue:** The average waiting time in the queue, Wq, is zero.

$$W_q = 0$$

10.6.4 (M/M/s): (FCFS)/K/K QUEUING SYSTEM

This model is a limited source model, where only K dump trucks can accommodate the service stations (loading point, workshop, etc.). This type of queuing model is called the machine repair model or the cyclic queue model. The model is demonstrated with the example of a limited repair facility available for the dump trucks at the mine workshop, as shown in Figure 10.9. This type of model considers a finite source model, in which arrivals are drawn from a small population. In the finite source model, the arrival of dump trucks in the repair facility depends on the number of dump trucks deployed in the system, unlike an infinite source model, where the arrivals are independent of the number of dump trucks in the system.

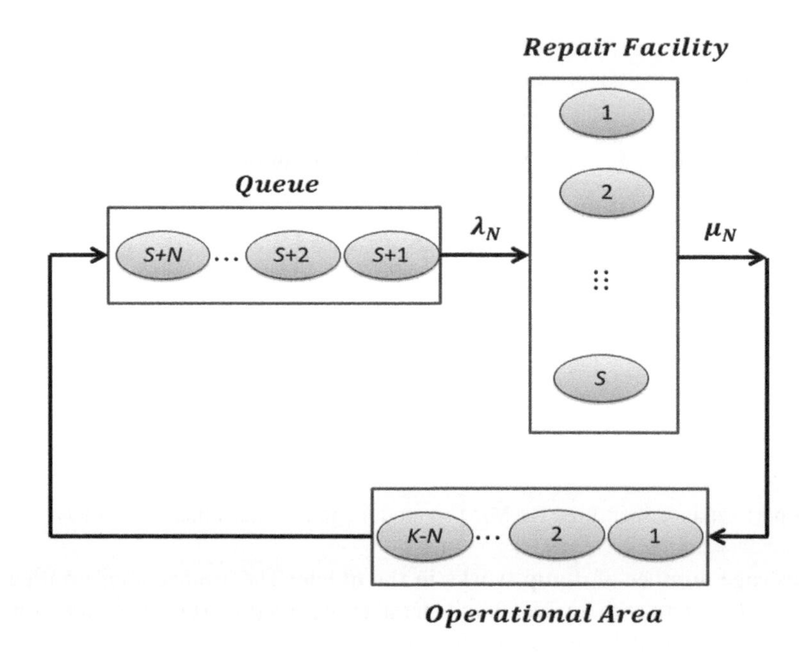

FIGURE 10.9 Flow of customer in machine repairing model.

This model can be represented as M/M/s/G/K/K queue system. The characteristics of the queuing system are as follows:

- The breakdown occurrence of dump trucks is a Poisson process, and the distribution of time interval of the breakdown follows an exponential distribution with a mean breakdown rate is λ per unit time. Moreover, once the dump truck gets breakdown, the same arrives at the repair facility, and hence the inter-arrival times of dump trucks at the repair facility also follow exponential distribution function with a mean arrival rate is λ per unit time.
- Service times of the repair follow exponential distribution function with mean service rate is μ per unit time. It is assumed that all the service persons are equally efficient.
- The number of repair-persons employed in the workshop is s.
- Queuing capacity is finite, N. That is, a limited number of dump truck arrives in the workshop can wait in the queue for repair.
- Dump trucks are assumed to be coming from a finite population, K.
- Dump trucks are served on a FIFO basis.

Let there be N dump trucks in the repair facility (system), and hence K-N dump trucks are in the operational area of the mine.

Let T is the time until the breakdown of the dump truck occurs among the operating dump truck population (K–N). T is exponentially distributed with parameter (K-N).

The arrival occurs in the Poisson process with mean $(K-N)\lambda$ when N dump trucks are in the repair facility.

$$\therefore \text{Mean arrival rate} = \lambda_N = \begin{cases} (K-N)\lambda, & N = 0,1,\ldots,K-1 \\ 0, & N = K \end{cases} \tag{10.46}$$

The value of μ_N can be determined by looking at the number of repair workers currently in repair service. If a truck breaks down occurs when all servers are busy, it has to wait in the queue to be repaired. Each occupied repair worker completes repairs with mean service rate μ. For N number of dump trucks at the repair facility, the service rate μ_N can be determined as follows:

$$\mu_N = \begin{cases} N\mu, & 0 \le N < s \\ s\mu, & s \le N \le K \end{cases} \tag{10.47}$$

Thus, the probability equation (Eq. 10.31) for p_N can be derived in a similar method as done for M/M/s queuing model.

$$p_N = \left(\frac{\lambda_0 \lambda_1 \ldots \lambda_N}{\mu_0 \mu_1 \ldots \mu_N} \right) p_0$$

In this case, the mean arrival rate (λ_N) depends on the number of dump trucks in the repair facility, and hence the value is not constant, unlike the infinite calling population. Let the mean arrival rates are λ_0, $\lambda_1,\ldots,\lambda_n$ for $0,1,\ldots,N$ number of dump trucks in the repair facility, respectively. Putting the values of λ_0, $\lambda_1,\ldots,\lambda_n$ and $\mu_0,\mu_1,\ldots,\mu_N$ from Eq. (10.46) and Eq. (10.47) in the above equation, we get

$$p_N = \begin{cases} \dfrac{K\lambda * (K-1)\lambda * \cdots * (K-N)\lambda}{\mu * 2\mu * \cdots N\mu} p_0 & N = 1,2,\ldots,s-1 \\[2ex] \dfrac{(K-1)\lambda * (K-2)\lambda * \cdots * (K-N)\lambda}{1\mu * 2\mu * \cdots * (s-1)\mu * s\mu * s\mu * \cdots s\mu} p_0 & N = s, s+1, \ldots, K \end{cases}$$

$$=\begin{cases}\dfrac{K*(K-1)*\ldots\{K-(N-1)\}}{1*2*\ldots N}\left(\dfrac{\lambda}{\mu}\right)^{N}*\dfrac{(K-N)*(K-N-1)*\ldots*1}{(K-N)*(K-N-1)*\ldots*1}p_0 & N=1,2,\ldots,s-1\\ \dfrac{K*(K-1)*(K-2)*\ldots*\{K-(N-1)\}}{(1*2*\ldots s)*s^{N-S}}*\dfrac{(K-N)*(K-N-1)*\ldots*1}{(K-N)*(K-N-1)*\ldots*1}\left(\dfrac{\lambda}{\mu}\right)^{N}p_0 & N=s,s+1,\ldots,K\end{cases}$$

$$=\begin{cases}\dfrac{K*(K-1)*\ldots\{K-(N-1)\}*(K-N)*(K-N-1)*\ldots*1}{(1*2*\ldots N)*\{(K-N)*(K-N-1)*\ldots*1\}}\left(\dfrac{\lambda}{\mu}\right)^{N}p_0 & N=1,2,\ldots,s-1\\ \dfrac{K*(K-1)*(K-2)*\ldots*\{K-(N-1)\}*(K-N)*(K-N-1)*\ldots*1}{(1*2*\ldots s)*s^{N-S}*\{(K-N)*(K-N-1)*\ldots*1\}}\left(\dfrac{\lambda}{\mu}\right)^{N}p_0 & N=s,s+1,\ldots,K\end{cases}$$

$$\Rightarrow p_N=\begin{cases}\dfrac{K!}{N!(K-N)!}\left(\dfrac{\lambda}{\mu}\right)^{N}p_0 & N=1,2,\ldots,s-1\\ \dfrac{K!}{s!s^{N-s}(K-N)!}s^{N}p_0 & N=s,s+1,\ldots,K\end{cases} \tag{10.48}$$

We know,

$$\sum_{N=0}^{K}p_N=1$$

Expanding the above equation, we get

$$p_0+p_1+p_2+\cdots+p_K=1$$

After substituting the value of $p_0,p_1,p_2,\ldots,p_K$ from Eq. (10.48) and simplifying the equation, we get

$$p_0\left[\sum_{N=0}^{s-1}\left(\frac{K!}{N!(K-N)!}\right)\left(\frac{\lambda}{\mu}\right)^{N}+\sum_{N=s}^{K}\left(\frac{K!}{s!s^{N-s}}\right)\left(\frac{\lambda}{\mu}\right)^{N}\right]=1$$

$$\Rightarrow p_0=\left[\sum_{n=0}^{s-1}\left(\frac{K!}{n!(K-n)!}\right)\left(\frac{\lambda}{\mu}\right)^{N}+\sum_{n=s}^{K}\left(\frac{K!}{s!s^{n-s}}\right)\left(\frac{\lambda}{\mu}\right)^{N}\right]^{-1} \tag{10.49}$$

The mean performance measure of the M/M/s/G/K/K queuing model can be summarized as follows:

- **Average number of failed dump trucks in the system:** The average number of failed dump trucks in the system, L_s is given by

$$L_s=\sum_{N=0}^{K}Np_N \tag{10.50}$$

- **Average number of failed dump trucks waiting in the queue for repair:** The average number of customers in the queue, L_q is given by.

$$L_q=\sum_{N=s}^{K}(N-s)p_N \tag{10.51}$$

- **Average time spent in the workshop:** The average time spent in the system is the average service time to each customer. It is given by

$$W_s = \frac{L_s}{(K - L_s)\lambda} \tag{10.52}$$

- **Average waiting time in the queue:** The average waiting time in the queue, W_q, is given by

$$W_q = \frac{L_q}{(K - L_s)\lambda} \tag{10.53}$$

Example 10.8

A mine has deployed 20 dump trucks for transportation of blasted materials from the working face. A dump truck requires service once every 16 days. The mine has two repair-persons at the workshop in each shift, who take an average of 12 hours to repair a dump truck. The times between two consecutive breakdown and repair times follow exponential distributions.

a. Find the average number of dump trucks in good condition.
b. Find the average downtime for a shovel that needs repair.
c. Find the fraction of the time a particular service engineer is idle.

Solution

Given data:
Number of dump trucks (K) = 20
Number of repair persons (s) = 2
Rate of the arrival of dump trucks at repair facility (λ) = 20/16 per day = 1.25 per day
Rate of repair of dump trucks (μ) = 24/12 per day = 2 per day

$$\therefore \rho = \frac{\lambda}{\mu} = \frac{1.25}{2} = \frac{5}{8}$$

We have,

$$p_N = \begin{cases} \dfrac{K!}{N!(K-N)!}\rho^N p_0 & N = 1 \\ \dfrac{K!}{s!s^{N-s}(K-N)!}\rho^N p_0 & N = 2,3,\ldots,20 \end{cases}$$

$$\therefore p_1 = \frac{K!}{N!(K-N)!}\rho^N p_0 = \frac{20!}{1!(20-1)!}(0.625)^1 p_0 = 12.5 p_0$$

For $N = 2,3,\ldots,20$,

$$p_N = \frac{K!}{s!s^{N-s}(K-N)!}\rho^N p_0$$

And,

$$p_{N+1} = \frac{K!}{s!s^{N+1-s}(K-N-1)!}\rho^{N+1} p_0 = \frac{K!}{s!s^{N-s}(K-N)!}\rho^N p_0 * \frac{1}{(K-N-1)s}\rho$$

$$\Rightarrow p_{N+1} = \frac{1}{(K-N-1)s}\rho * p_N$$

Putting the values of K, s, and ρ, we have

$$p_{N+1} = \frac{1}{2(20-N+1)} * \frac{5}{8} * p_N = \frac{5}{16(20-N+1)} * p_N \quad (10.50)$$

$$p_2 = \frac{K!}{s!s^{N-s}(K-N)!}\rho^N p_0 = \frac{20!}{2!2^{2-2}(20-2)!}\left(\frac{5}{8}\right)^2 p_0 = 74.21875 p_0$$

Similarly, the probabilities $p_3, p_4, \ldots, p_{20}$ are determined using Eq. (10.25). The values of $p_0, p_1, p_2, p_3, p_4, \ldots, p_{20}$ are listed in Table 10.2.

TABLE 10.2
Probabilities in Different Conditions

Probability of N dump truck at the repair facility	Probability value	$N * p_N$
$P(N=0) = p_0$	5.97345E-10	0
$P(N=1) = p_1$	1.38384E-09	1.38E-09
$P(N=2) = p_2$	8.21656E-09	1.64E-08
$P(N=3) = p_3$	4.62181E-08	1.39E-07
$P(N=4) = p_4$	2.45534E-07	9.82E-07
$P(N=5) = p_5$	1.22767E-06	6.14E-06
$P(N=6) = p_6$	5.7547E-06	3.45E-05
$P(N=7) = p_7$	2.51768E-05	0.000176
$P(N=8) = p_8$	0.000102281	0.000818
$P(N=9) = p_9$	0.000383553	0.003452
$P(N=10) = p_{10}$	0.001318464	0.013185
$P(N=11) = p_{11}$	0.004120199	0.045322
$P(N=12) = p_{12}$	0.011588058	0.139057
$P(N=13) = p_{13}$	0.028970146	0.376612
$P(N=14) = p_{14}$	0.063372195	0.887211
$P(N=15) = p_{15}$	0.118822865	1.782343
$P(N=16) = p_{16}$	0.185660727	2.970572
$P(N=17) = p_{17}$	0.232075909	3.94529
$P(N=18) = p_{18}$	0.217571165	3.916281
$P(N=19) = p_{19}$	0.135981978	2.583658
$P(N=20) = p_{20}$	0.042494368	0.849887
	$\sum_{N=0}^{K} p_N = 1$	$L_s = \sum_{N=0}^{K} Np_N = 16.66 \approx 17$

(a) Therefore, the average number of dump trucks in good condition, which are in an operational area is given by

$$L = K - L_s = 20 - 17 = 3$$

(b) The average downtime for a dump truck for repair is given by

$$W_s = \frac{L_s}{\bar{\lambda}}$$

And, $\bar{\lambda} = \lambda(K - L_s) = 1.25 * (20 - 16.66) = 4.175$ dump truck per day

$$\therefore W_s = \frac{L_s}{\bar{\lambda}} = \frac{16.66}{4.175} = 3.99 \text{ days} \approx 4 \text{ days}$$

Therefore, the downtime of a dump truck for repair work is approximately four days, which is very high, and thus a greater number of repair-persons should be employed.

(c) The fraction of the time that a repair-person will be idle is

$$= p_0 + 0.5 * p_1 = (5.97345)^{-10} + 0.5 * (1.38384)^{-9} \approx 0$$

The idle time of the repair-persons are approximately zero.

10.7 COST MODELS

There are mainly two types of the cost involves in a queuing system. These are the cost for providing service and not providing any service (i.e., waiting time cost). These two types of costs are conflicting. Increasing the cost of one will automatically decrease the cost of another. The trend of the cost curve is demonstrated in Figure 10.10. The objective is to minimize the total expected system cost

$$\text{Minimum } ETC = EOC + EWC \tag{10.54}$$

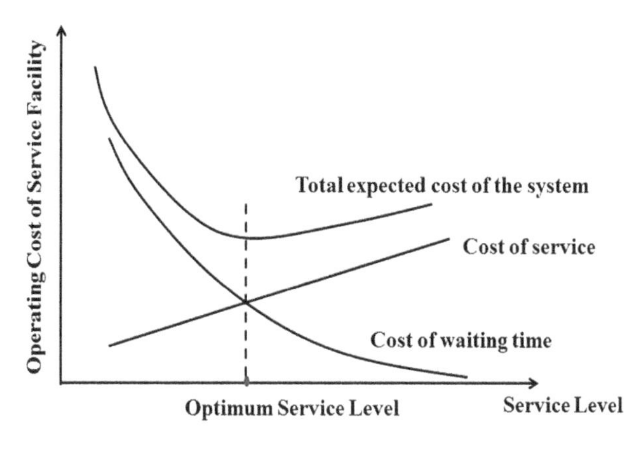

FIGURE 10.10 Cost characteristics of a queuing system.

where

ETC = Expected total cost per unit time
ESC = Expected cost of operating the facility per unit time
EWC = Expected cost of waiting per unit time

The expected waiting cost depends on the number of customers in the system, L_s. If C_w is the waiting cost per dump truck per unit time, the total expected waiting cost (EWC) can be determined as

$$EWC = C_w L_s$$

On the other hand, the expected service cost depends on the number of servers and their service rate. Let the number of servers is s and the service rate is μ. If the service cost per server per unit time is C_s, the total expected service cost (ESC) can be determined as

$$ESC = sC_s(\mu)$$

The service cost of a server, C_s is a function of service rate, μ

$$\therefore ETC = C_w L_s + sC_s(\mu) \tag{10.55}$$

The following examples illustrate the use of the cost model.

Example 10.9
A shovel deployed in a production face for loading operation where dump trucks arrive according to a Poisson distribution with a rate of 10 dump trucks per hour. The loading times are exponentially distributed with a mean of 1/μ hours. The shovel operator charges US\$0.5μ per hour, and the waiting cost of the dump truck is US\$2 per dump truck per hour. Determine the loading rate μ for which the average cost per hour is minimum.

Solution

Given data:

Mean arrival rate of dump truck (λ)= 10 per hour
Mean loading time = $1/\mu$ hours
$\therefore$ Mean loading rate (μ) = μ per hour
Service cost per unit of the shovel (C_s) = US\$0.5μ per hour
Cost of waiting per dump truck (C_w) = US\$2 per hour

The queuing system is an M/M/1 system.
We have,

$$\text{Average cost per hour} = ETC = C_w L_s + sC_s(\mu)$$

For one shovel, s = 1

$$\therefore ETC = C_w L_s + sC_s(\mu) = 2 * \frac{\lambda^2}{\mu(\mu-\lambda)} + 2*0.5\mu$$

$$\Rightarrow ETC = 2 * \frac{10^2}{\mu(\mu-10)} + 1\mu = \frac{200}{\mu(\mu-10)} + \mu$$

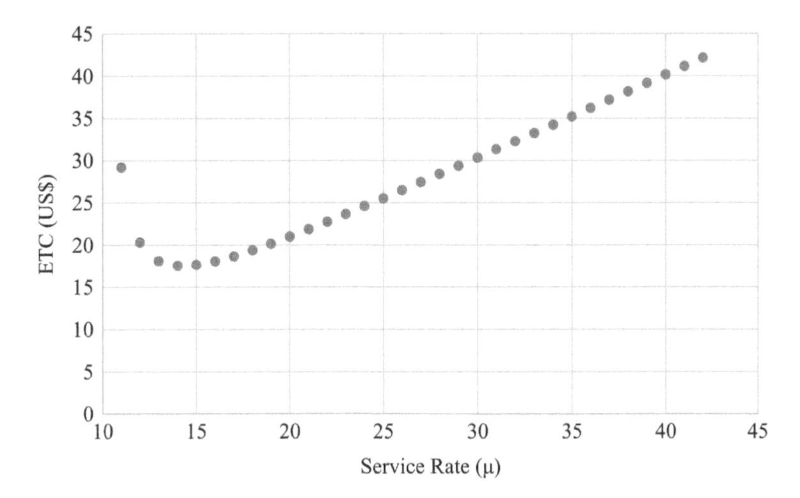

FIGURE 10.11 Expected total cost incurred with different loading rate.

For minimizing the ETC, we have

$$\frac{dETC}{d\mu}=0$$

$$\Rightarrow -\frac{200(2\mu-10)}{(\mu^2-10\mu)^2}+1=0$$

$$\Rightarrow \frac{200(2\mu-10)}{(\mu^2-10\mu)^2}=1$$

The above equation holds true with $\mu = 6.4$ and 14.2.

Since μ should be greater than λ, therefore $\mu = 14.2$ is acceptable. Thus, the minimum cost (ETC_{min}) is achieved with a loading rate of 14.2 dump trucks per hour. The cost of ETC with different service rates is also summarized in Figure 10.11. It is observed that the total expected cost is minimum for $\mu = 14.2$.

Example 10.10

A mine has six shovels for loading operations. The time until the breakdown of a shovel occurs exponential, with a mean of 100 hours. The time to repair a shovel is exponentially distributed with a mean of 10 hours. The repair cost is US$30 per hour. The cost due to loss of operating time is US$100 per hour per shovel. Analyse the expected total cost for employing one and two repair persons.

Solution

The given problem is a machine repair model with the following data:

Number of shovels deployed (K) = 4
The mean time to breakdown (λ) = 1/100 shovel per hour
The mean time to repair (μ) = 1/10 shovel per hour
Service cost (C_s) per repairman = US$30 per hour
Waiting cost (C_w) = US$100 per hour

Number of shovels in the system is N

Therefore, $\rho = \frac{\lambda}{\mu} = \frac{10}{120} = \frac{1}{12}$

Case 1: Number of repair person (s) = 1

$$\lambda_N = \begin{cases} (K - N)\lambda, & N = 0,1,2,3,4 \\ 0, & N > 4 \end{cases}$$

$$\mu_N = \mu,$$

The number of shovels at the repair facility is given by

$$L_s = \sum_{N=0}^{4} Np_N = 0 * p_0 + 1 * p_1 + 2 * p_2 + 3 * p_3 + 4 * p_4$$

Also,

$$p_N = \begin{cases} \frac{K!}{N!(K-N)!}\rho^N p_0 & N = 0 \\ \frac{K!}{s!s^{N-s}(K-N)!}\rho^N p_0 & N = 1,2,3,4 \end{cases}$$

$$\therefore p_1 = \frac{K!}{s!s^{N-s}(K-N)!}\rho^n p_0 = \frac{4!}{1!1^{1-1}(4-1)!}\left(\frac{1}{12}\right)^1 p_0 = \frac{1}{3}p_0$$

$$p_2 = \frac{4!}{1!1^{2-1}(4-2)!}\left(\frac{1}{12}\right)^2 p_0 = \frac{1}{12}p_0$$

$$p_3 = \frac{4!}{1!1^{3-1}(4-3)!}\left(\frac{1}{12}\right)^3 p_0 = \frac{1}{72}p_0$$

$$p_4 = \frac{4!}{1!1^{4-1}(4-4)!}\left(\frac{1}{12}\right)^4 p_0 = \frac{1}{864}p_0$$

Again,

$$p_0 + p_1 + p_2 + p_3 + p_4 = 1$$

$$\Rightarrow p_0 + \frac{1}{3}p_0 + \frac{1}{12}p_0 + \frac{1}{72}p_0 + \frac{1}{864}p_0 = 1$$

$$\Rightarrow p_0 = \frac{1}{1.431} = 0.698$$

$$\therefore L_s = 0 * 0.698 + 1 * \left(\frac{0.698}{3}\right) + 2 * \left(\frac{0.698}{12}\right) + 3 * \left(\frac{0.698}{72}\right) + 4 * \left(\frac{0.698}{864}\right)$$

$$\Rightarrow L_s = 0.497 \text{ shovel}$$

The expected cost of waiting and service per unit time is given by

$$ETC(s=1) = C_s s + C_w L_s = 30*1 + 100*0.497 = \text{US\$79.76 per hour}$$

Case 2: Number of repairman (s) = 2

$$\lambda_n = \begin{cases} (K-N)\lambda, & N = 0,1,2,3,4 \\ 0, & N > 4 \end{cases}$$

$$\mu_N = \begin{cases} N\mu, & N = 1 \\ s\mu, & N = 2,3,4 \end{cases}$$

We have,

$$L_s = \sum_{N=0}^{4} Np_N = 0*p_0 + 1*p_1 + 2*p_2 + 3*p_3 + 4*p_4$$

Also,

$$p_N = \begin{cases} \dfrac{K!}{N!(K-N)!}\rho^N p_0 & N = 0,\ 1 \\ \dfrac{K!}{s!s^{N-s}(K-N)!}\rho^N p_0 & N = 1,\ 2,\ 3,\ldots,4 \end{cases}$$

$$\therefore p_1 = \frac{K!}{N!(K-N)!}\rho^n p_0 = \frac{4!}{1!(4-1)!}\left(\frac{1}{12}\right)^1 p_0 = \frac{1}{3}p_0$$

$$p_2 = \frac{K!}{s!s^{N-s}(K-N)!}\rho^n p_0 = \frac{4!}{2!2^{2-2}(4-2)!}\left(\frac{1}{12}\right)^2 p_0 = 0.0416p_0$$

$$p_3 = \frac{4!}{2!2^{3-2}(4-3)!}\left(\frac{1}{12}\right)^3 p_0 = 0.00347p_0$$

$$p_4 = \frac{4!}{2!2^{4-2}(4-4)!}\left(\frac{1}{12}\right)^4 p_0 = 0.000144p_0$$

Again, we know,

$$p_0 + p_1 + p_2 + p_3 + p_4 = 1$$

$$\Rightarrow p_0 + 0.334p_0 + 0.0416p_0 + 0.00347p_0 + 0.000144p_0 = 1$$

$$\Rightarrow 1.379p_0 = 1$$

$$\Rightarrow p_0 = 0.725$$

$$\therefore L_s = \{0 + 1*(0.334) + 2*(0.0416) + 3*(0.00347) + 4*(0.000144)\}*0.725$$

$$\Rightarrow L_s = 0.310 \text{ dump trucks}$$

The expected cost of waiting and service per unit time is given by

$$ETC(s=2) = C_s s + C_w L_s = 30*2 + 100*0.310 = \text{US\$93.04 per hour}$$

Thus, the mine should employ one repair-person to minimize the cost.

Example 10.11

The owner of an iron ore mine is in the process of purchasing a loader (shovel) for loading operation in the opencast production benches. Vendors have proposed four models whose capacity are summarized below.

Model/Brand	Operating cost ($/hour)	Loading rate (dump trucks/hour)
1	10	12
2	12	14
3	14	16
4	15	18

Dump trucks arrive at the bench according to a Poisson distribution with a mean of eight dump trucks per hour. The waiting cost of the dump truck is $10 per dump truck per hour. If each model has equal failure probability, which shovel should be purchased for minimizing the operational and waiting cost?

Solution

Let the EOC_i is the expected operating cost of the loader of ith model/brand, and EWC_i is the corresponding expected waiting cost of dump trucks. The total expected cost per hour associated with i[th] type of loader is

$$ETC = EOC_i + EWC_i$$

$$\Rightarrow ETC = C_{1i} + C_{2i} * L_{1i}$$

$$\Rightarrow ETC = 24C_{1i} + 80C_{2i}$$

The values of C_1 are given by the data of the problem. We determine L by recognizing that each dump truck can be treated as an (M/M/1): (FCFS/∞/∞) model for all practical purposes. The arrival rate is λ = 10 dump trucks/hour. The expected number of dump trucks in the loading point with different loading rates are shown in Table 10.3.

The expected total costs (operating plus waiting) for different types of shovels are shown in Table 10.4.

TABLE 10.3
Expected Number of Dump Trucks for the Different Loading Rate

Model i	λ_i (dump trucks/hour)	μ_i (dump trucks/hour)	$L_{si} = \dfrac{\lambda_i^2}{\mu_i(\mu_i - i)}$
1	10	12	4.16
2	10	14	1.78
3	10	16	1.04
4	10	18	0.69

TABLE 10.4
Expected Total Costs (Operating Plus Waiting) for Different Types of Shovel

Model *i*	C_{1i} (US$/hour)	C_{2i} (US$/hour)	L_{si}	$ETC_i = C_{1i} * + C_{2i} * L_{1i}$ (US$/hour)
1	10	10	4.16	=10 + 10 * 4.16 = 51.6
2	12	10	1.78	=12 + 10 * 1.78 = 29.8
3	14	10	1.04	=14 + 10 * 1.04 = 24.4
4	18	10	0.69	=18 + 10 * 0.69 = 24.9

TABLE 10.5
Sample Observations of Service or Loading Time and Inter-Arrival Time for 30 Events

Sl. No.	Service time of shovel (minute)	Inter-arrival time of dump truck (minute)	Sl. No.	Service time of shovel (minute)	Inter-arrival time of dump truck (minute)	Sl. No.	Service time of shovel (minute)	Inter-arrival time of dump truck (minute)
1	5.25	1.22	11	3.98	4.21	21	4.66	3.25
2	4.26	0.56	12	4.56	6.12	22	5.52	2.64
3	6.24	4.56	13	6.85	5.22	23	3.12	1.24
4	3.58	3.26	14	7.88	4.56	24	5.78	1.86
5	4.25	5.62	15	7.56	3.26	25	6.22	1.62
6	6.58	1.22	16	5.86	2.13	26	4.55	4.56
7	7.56	0.88	17	4.56	2.22	27	5.22	4.23
8	4.86	2.56	18	4.87	5.62	28	3.89	3.48
9	5.65	3.21	19	5.88	4.56	29	4.56	2.82
10	4.68	2.44	20	6.42	1.88	30	4.22	1.26

Model 3 offers the lowest cost (= US$24.4).

10.8 CASE STUDY FOR THE APPLICATION OF QUEUING THEORY FOR SHOVEL-TRUCK OPTIMIZATION IN AN OPEN-PIT MINE

An open-pit iron ore mine located in Odisha, India, is planning to optimize the material haulage system for minimizing the cost. The loading and transportations of ores from face to the processing plant are one of the key operations. These activities incur a significant share of the total operating costs of the mine. This case study demonstrates the applications of queuing theory for optimizing the number of trucks required to deploy for a specified number of shovels used for loading operations at the production pit. The nature of these activities depends on the cycle time of shovel and trucks. Trucks generally arrive at the production pit in a random fashion. Additionally, the loading times of the shovels are also a random variable. The system deploys one shovel (10 m^3 capacity) and 22 trucks (60 m^3 capacity each) to load and transport ore from face to plant. The production capacity of ore in a shift from one production pit is 4000 tonnes. The distance between the production pit to the stockpile is 3 km. The loading (or service) time of the shovel and inter-arrival time of dump trucks at the production pit are also random variables. These data need to be monitored in the mines for a specific period for determining the average service time and inter-arrival time. The sample field data (arrival times and loading times) from the case study mine are reported in Table 10.5.

The average inter-arrival time and service time are determined from the observed data (Table 10.5) as 3.07 minutes and 5.30 minutes, respectively. The inter-arrival rate and service rate can be determined using the following equations.

$$\text{Inter-arrival rate}(\lambda) = \frac{1}{\text{Average inter-arrival time}} = \frac{1}{3.07} \text{ truck per minute}$$

$$= \frac{60}{3.07} \text{ truck per hour} = 19.54 \approx 20 \text{ trucks per hour}$$

$$\text{Loading rate or Service rate } (\mu) = \frac{1}{\text{Average service time}} = \frac{1}{5.30} \text{ truck per minute}$$

$$= \frac{60}{5.30} \text{ truck per hour} = 11.32 \approx 12 \text{ trucks per hour}$$

The distributions of the observed inter-arrival times and loading times (Table 10.5) should be examined for estimating the best-fit probability density function (PDF). The outliers in the data should be removed before estimating the best PDF. After removal of outliers, the best fit PDF can be estimated by using any of the software like SPSS, SYSTAT, MATLAB, etc. Here, the data is fitted with the exponential probability density function using MATLAB software with good accuracy, as shown in Figure 10.12. However, most mining applications are highly complex, and accurate fitting of the observed data is important for queuing analysis. The probability density function for inter-arrival distribution and service time distribution is estimated as follows:

$$f(t) = \frac{1}{\mu_1} e^{-\frac{t}{\mu_1}} = \frac{1}{5.59} e^{-\frac{t}{5.59}}$$

$$f(t) = \frac{1}{\mu_2} e^{-\frac{t}{\mu_2}} = \frac{1}{3.91} e^{-\frac{t}{3.91}}$$

Determining the Model Performance Measures

The model performance measures of the model are determined by considering exponential inter-arrival times and service times. The number of shovels deployed at the loading point is one. The calling population is limited with a maximum of 22 trucks, and thus maximum queue capacity at the loading point is 22 trucks. Therefore, the type of queuing model is (M/M/1): (FCFS/K/K). This is a

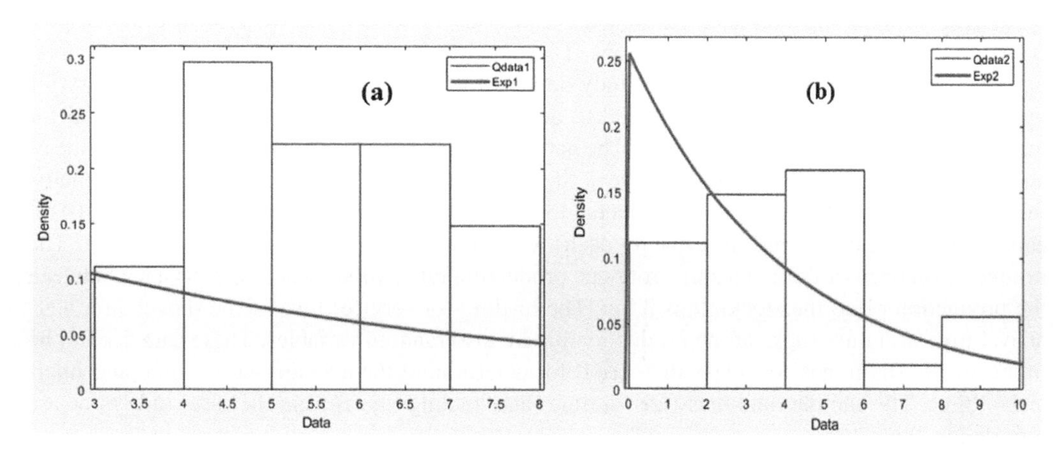

FIGURE 10.12 PDF of the observed inter-arrival times and service times.

machine repair queuing model with one service facility, as explained in Section 10.6.4. Therefore, the various model performance measures i.e., idle time of the shovel (P_0), the mean number of trucks waiting in the system (L_s), the mean number of trucks waiting in queue (L_q), waiting time in the queue (W_q), waiting time in the system (W_s), shovel utilization (η_s), truck utilization (η_t), cost incurred in per unit material handling (C_t), and total production (Q_n)) can be determined as follows:

- Probability of N – trucks at the loading points $= p_n$

$$\begin{cases} \dfrac{K!}{N!(K-N)!}\left(\dfrac{\lambda}{\mu}\right)^N p_0 & N = 1,2,\ldots,s-1 \\ \dfrac{K!}{s!s^{N-s}(K-N)!}\left(\dfrac{\lambda}{\mu}\right)^N p_0 & N = s,s+1,\ldots,K \end{cases}$$

- Probability of no trucks at the loading point (p_0)

$$= \frac{1}{\sum_{N=0}^{s-1}\left(\dfrac{K!}{N!(K-N)!}\right)\left(\dfrac{\lambda}{\mu}\right)^N + \sum_{N=s}^{K}\left(\dfrac{K!}{s!s^{N-s}}\right)\left(\dfrac{\lambda}{\mu}\right)^N}$$

- Mean number of trucks in the system $(L_s) = \sum_{N=0}^{K} Np_N$
- Mean number of trucks in the system $(L_q) = \sum_{N=s}^{K} (N-s)\, p_N$
- Mean time spent in the system $(W_s) = \dfrac{L_s}{(K-L_s)\lambda}$
- Mean time spent in the queue $(W_q) = \dfrac{L_q}{\lambda}$
- Utilization of the shovel $(\eta_s) = 1 - P_0$
- Utilization of the truck $(\eta_t) = 1 - \dfrac{W_q}{W_q + \text{Cycle time}} = 1 - \dfrac{W_q}{W_q + t_c}$

The cycle time of trucks includes service or loading time, including manoeuvring, loaded haul, unloading, empty haul, and waiting time at loading and unloading points, which is equal to 15 minutes. t_c is the cycle time of trucks

- Effective arrival rate $(\bar{\lambda}) = \lambda(K - L_s)$
- The total cost of unit production $(C_t) = \dfrac{\text{Hourly cost of operation}}{\text{Production per hour}} = \dfrac{C_S s + C_T N}{\mu * \eta_s * s * V}$

where

C_S = cost per hour of each shovel
C_T = cost per hour of each truck
μ = mean service or loading time
s = number of shovel
N = number of truck
V = capacity of truck

The hourly cost of shovel and truck are US$200 per unit and US$150 per unit, respectively at this mine. Here, the number of trucks needed to be optimized for one shovel and hence $S = 1$ and $N = 1, 2,\ldots,22$.

Results and discussion

The various performance measures like idle time of the shovel (P_0), the mean number of trucks waiting in the system (L_s), the mean number of trucks waiting in queue (L_q), waiting time in the queue (W_q), waiting time in the system (W_s), shovel utilization (η_s), truck utilization (η_t), cost incurred in per unit material handling (C_t), and total production (Q_n) of the defined queuing system are listed in Table 10.6. The graphical representation of different performance measures with the number of trucks deployed are represented in Figures 10.13–10.18.

The decision on the number of trucks needed to deploy for optimizing the cost of haulage system should be made based on the integration of the key performance measures like utilization of shovel and trucks, cost per unit haul of material, production in a specific duration, waiting time of the trucks, etc. It can be observed from Figure 10.18 that the production level is increasing significantly by increasing the number of trucks up to 5, and thereafter it becomes constant. Thus, deployment of more than 5-trucks unnecessarily increases the waiting time and cost per unit production. Figure 10.16 reveals that the probability of zero trucks at the loading point is almost equal to zero with five trucks. It can be observed from Figure 10.15 that the utilization of shovels is increasing with the increase in the number of trucks and reached 100% with the deployment of 5 trucks. At the same time, the utilizations of the trucks are continuously decreasing with increasing the number of trucks. It can be observed that increasing the number of trucks after five does not have any further scope of increasing the utilization of the shovel. On the other hand, the deployment of more than five trucks increases the number of trucks in the queue (Figure 10.14), and hence the waiting

TABLE 10.6
Performance Measures of the Defined Queuing System

λ	μ	K	$\bar{\lambda}$	P_0	L_s	L_q	W_q	W_s	η_s	η_t	C_t	Q_n
22	12	22	13.2	2.61E-17	21.40	20.40	1.70	1.78	1.00	0.94	4.86	5760.00
22	12	21	13.2	3.32E-16	20.40	19.40	1.62	1.70	1.00	0.94	4.65	5760.00
22	12	20	13.2	4.04E-15	19.40	18.40	1.53	1.62	1.00	0.94	4.44	5760.00
22	12	19	13.2	4.69E-14	18.40	17.40	1.45	1.53	1.00	0.94	4.24	5760.00
22	12	18	13.2	5.16E-13	17.40	16.40	1.37	1.45	1.00	0.94	4.03	5760.00
22	12	17	13.2	5.38E-12	16.40	15.40	1.28	1.37	1.00	0.94	3.82	5760.00
22	12	16	13.2	5.29E-11	15.40	14.40	1.20	1.28	1.00	0.94	3.61	5760.00
22	12	15	13.2	4.91E-10	14.40	13.40	1.12	1.20	1.00	0.94	3.40	5760.00
22	12	14	13.2	4.26E-09	13.40	12.40	1.03	1.12	1.00	0.94	3.19	5760.00
22	12	13	13.2	3.46E-08	12.40	11.40	0.95	1.03	1.00	0.94	2.99	5760.00
22	12	12	13.2	2.6E-07	11.40	10.40	0.87	0.95	1.00	0.94	2.78	5760.00
22	12	11	13.2	1.81E-06	10.40	9.40	0.78	0.87	1.00	0.94	2.57	5760.00
22	12	10	13.2	1.15E-05	9.40	8.40	0.70	0.78	1.00	0.94	2.36	5760.00
22	12	9	13.2	6.68E-05	8.40	7.40	0.62	0.70	1.00	0.94	2.15	5760.00
22	12	8	13.2	0.000348	7.40	6.40	0.53	0.62	1.00	0.94	1.94	5760.00
22	12	7	13.2	0.001614	6.40	5.40	0.45	0.53	1.00	0.94	1.74	5759.98
22	12	6	13.2	0.006556	5.40	4.40	0.37	0.45	1.00	0.93	1.53	5759.80
22	12	5	13.2	0.022935	4.40	3.40	0.28	0.37	1.00	0.93	1.32	5757.95
22	12	4	13.2	0.067983	3.40	2.40	0.20	0.28	1.00	0.93	1.11	5742.92
22	12	3	12.9	0.169005	2.41	1.43	0.12	0.21	0.98	0.93	0.92	5645.81
22	12	2	11.9	0.353414	1.46	0.56	0.05	0.14	0.90	0.93	0.77	5177.53
22	12	1	8.3	0.633214	0.63	0.00	0.00	0.08	0.63	0.93	0.78	3600.00

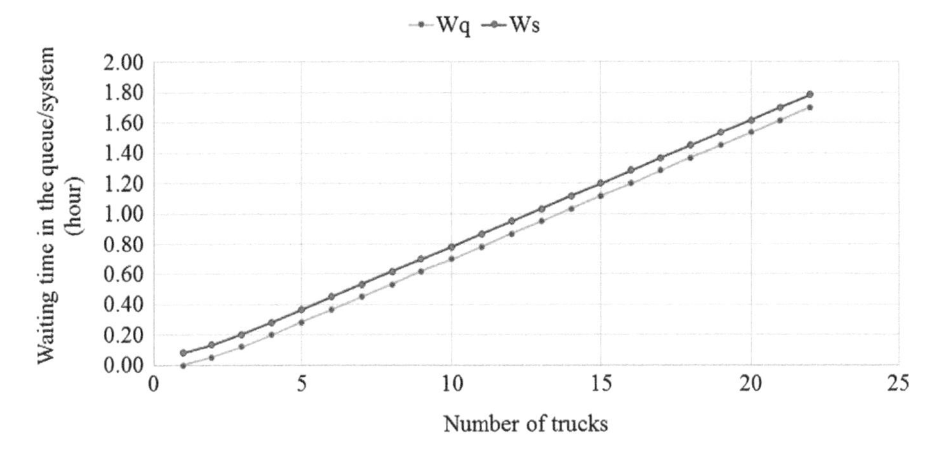

FIGURE 10.13 Waiting time in the queue/system versus the number of trucks deployed.

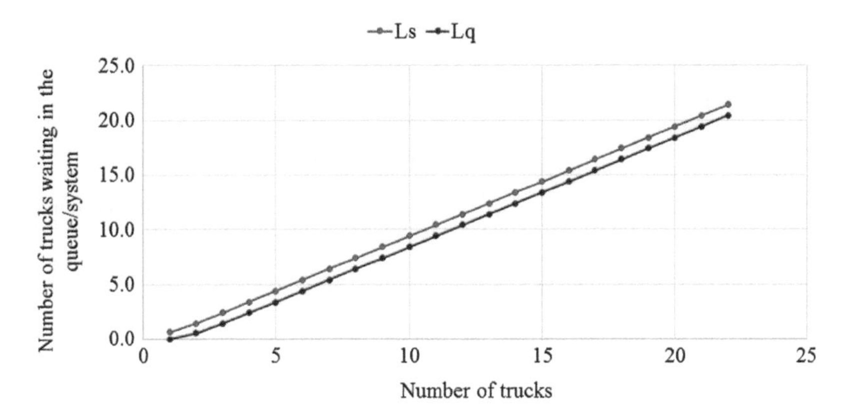

FIGURE 10.14 Number of trucks waiting in the queue/system versus the number of trucks deployed.

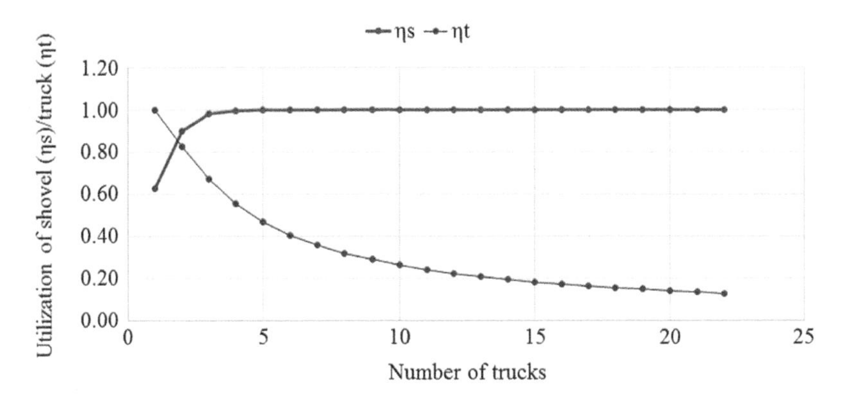

FIGURE 10.15 Utilization of the shovel and truck versus the number of trucks deployed.

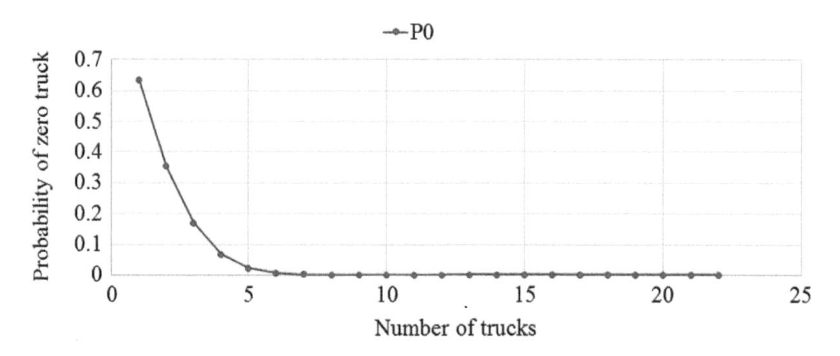

FIGURE 10.16 Probability of zero truck at the loading point versus the number of trucks deployed.

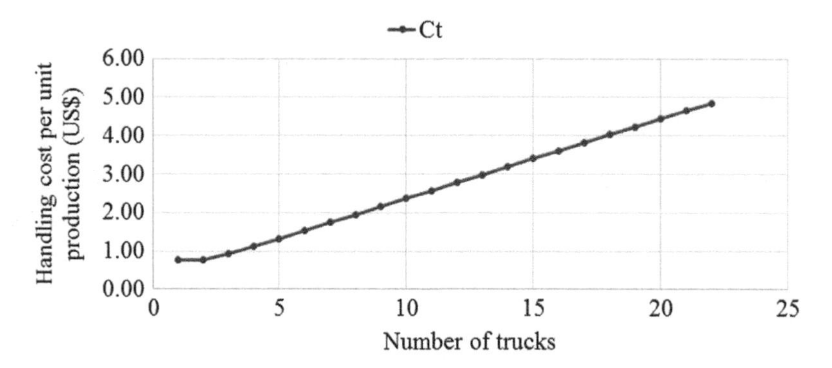

FIGURE 10.17 Material-handling cost (loading, transportation, unloading) versus the number of trucks deployed.

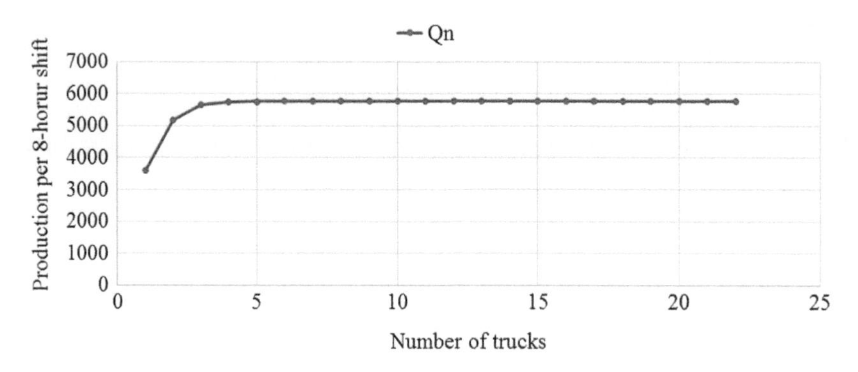

FIGURE 10.18 Materials handled in a shift versus the number of trucks deployed.

time (Figure 10.13) of each truck is also increased. On the other side, the deployment of fewer than five trucks may reduce the utilization of the shovel as the chance of zero trucks at the loading point is increased. Figure 10.17 indicates that the unit cost of material haul is also increasing with the increase in the number of trucks. Thus, the number of trucks should deploy as minimum as possible with maximum achievable production. It should be noted that the deployment of one shovel facilitates the target production of the mine. If the target production is higher than the maximum production capacity of one shovel, more number of shovel need to deploy and accordingly the model performance measure need to be analysed.

EXERCISE 10

Q1. It is observed that the arrival of mine trucks in the workshop for repair follows a Poisson distribution with mean arrival rate of 12 per 8-hour shift. The time spent to repair mine trucks by a serviceman in the mine workshop follows exponential distribution with a mean of 60 minutes. If the service offers in the order FCFS basis, determine the expected idle time of the repairman in a shift. Also determine the mean number of mine trucks in the workshop.

[**Ans.** 160 minutes; 2 units]

Q2. In the processing plant, arrivals of dump trucks follow Poisson distribution with a mean time of 12 minutes. There is a single unloading platform installed in the plant. The length of unloading the ores is found to be exponentially distributed with mean of 2 minutes.

a. Determine the probability that a dump truck arriving at the unloading point will have to wait.
b. Determine the mean number of trucks waiting in the queue.
c. Determine the probability that a dump truck has to spend (waiting plus unloading time) more than 2 minutes in the system.
d. The owner of the mine will install a second unloading platform when found that a dump truck on arrival has to wait an average of 4 minutes in the queue. Estimate the increase in arrival rate needed to install a second platform.

[**Ans.** a. 1/6; b. ≈ 0; c. 0.633; d. 200%]

Q3. A production bench in an opencast mine has space to accommodate only ten dump trucks. There is one loader (shovel) and can serve only one dump truck at a time. If a dump truck arrives in the production bench and finds no space, the dump truck moves to the alternative production bench. The inter-arrival time of dump truck follow exponential distribution with mean of 10 per hour and the loading time of shovel follow exponential distribution with mean of 5 minutes. Determine the probabilities of zero dump truck and n-dump truck at the production bench.

[**Ans.** 0.192; 0.111]

Q4. The general shift of a mine start at 8 AM and the arrival of workers at mine office follow an exponential distribution with mean of 2 minutes. Determine the followings:

a. Exponential distribution function for the defined inter-arrival time.
b. Probability that no worker arrives at the office by 8:04 AM
c. Probability that a workers arrived between 8:01 and 8:05 AM.
d. It is observed that 5 minutes past after arrival of last worker. If the current time is 8.25 AM, determine the probability that the next worker will arrive before 8:30 AM.

[**Ans.** a. $30e^{-30t}$; b. 0.135; c. 0.524; d. 0.918]

Q5. The time between breakdowns of a machine follow exponential with a mean of 18 hours. If the machine has been successfully worked without fail for 10 hours, determine the probability that the machine will continue to work without fail in the next 2 hours. Also, determine the probability that the break down will occur during the next 5 hours.

[**Ans.** 0.243]

Q6. The time between failures of a cap lamp is found to be exponential with a mean of 9000 hours, and the manufacturer issues a 1-year warranty on the cap lamp. What are the chances that the warranty will cover a fault repair?

[**Ans.** 0.623]

Q7. A mining company has just stored ten drill bits in the inventory system of the mine for regular replacement of wear out bit of drill machine. The inventory is replenished every 30 days with ten units of drill bit. The time between wear out of drill bit is exponential with a mean of three days. Determine the probability that the drill machine will remain non-functional for two days due to a shortage of drill bit in the inventory system.

[**Ans.** 0.487]

Q8. A mining company is screening the profiles of two service engineers for recruiting in a machine shop. The repair rate of first candidate is four machines per shift and charges US$200 per shift. On the other hand, the repair rate second candidate is six machines per shift and charges US$400 per shift. The shift duration is 8 hours. It is estimated that the breakdown cost of each machine is US$20 per breakdown hour due to production loss. It is assumed that machine breakdown occurrences follow Poisson distribution with a mean of 3 per shift and the repair times by each service engineers follow exponential distribution. Which service engineer should be recruited for more economical operation?

[**Ans.** Second]

Q9. A project officer must decide the number of shovels required to deploy in the production faces for loading operations. The officer uses the following criteria to determine the number of shovels in operation depending on the number of dump trucks in production faces with 1-shovel.

No. of dump trucks in production faces	No. of shovel in operation
1 to 4	1
4 to 7	2
>7	3

The arrival patterns of dump trucks in the production face follow Poisson distribution with a mean value of 10 dump trucks per hour. The loading times follows an exponential distribution with a mean of 5 minutes. Determine the number of shovel need to be deployed in the face for desirable service and the average number of dumpers in the loading face after deployment.

[**Ans.** 2 shovel; No. of dumper at the loading point ≈ 1]

Q10. In an opencast mine, production is done in one face. The loading face can accommodate at most six dump trucks, including those being loaded. Any dump truck on arrival if found six dump trucks at the loading point, go returns. The arrival distribution of dump truck is Poisson with a mean of 10 per hour. The loading times follow exponential distribution with a mean of 5 minutes. Estimate the followings:

a. Proportion of dump trucks return without being loaded.
b. Percentage utilization of the loaders.
c. The probability that an arriving dump truck will get space at the loading point.

[**Ans.** a. 8.1%; b. 76.9%; c. 0.919]

Q11. A mining company deploys one shovel for loading coal in an outgoing dump trucks. The mean arrival rate of dump truck at the loading point is 10 per hour. The mean loading rate of shovel is 12 dump trucks per hour. The inter-arrival times of dump trucks and loading times of shovel follow exponential distribution. The company planning to deploy an additional shovel to improve the mine's productivity. It is estimated that two shovels at the loading point will double the loading rate from 12 dump trucks per hour to 24 dump trucks per hour. Analyse the effect on the queuing performance measure parameters.

[**Ans.** Deployment of 2-loaders is a better choice for economical operation]

Q12. In Q11, if the salary of the dump truck operator is $25 per hour and that of the shovel operator is $30 per hour. Both the operators draw salary for waiting or idle time of the machines. Determine the hourly cost savings in deploying two shovels in respect to one shovel.

[**Ans.** US$77.25]

Q13. In Q11, the company considers to deploy the second shovel in an additional blasting faceto speed-up the process of loading. If the loading rate of each shovel is 12 dump trucks per hour and the dump trucks will continue to arrive at the rate of 10 per hour. Analyse the performance measures of the new queuing system.

[**Ans.** Though the deployment of shovel in a different face improves the performance of the system but not up-to the mark of deployment in the same face for simultaneous service to single dumper at a time]

Q14. There are two service engineers, who are responsible for 15 dump trucks. The dump trucks run on an average of 30 hours, then require an average 3-hour service period. The service time and inter-arrival time are exponentially distributed . Find out various measures of performance.

[**Ans.** Probability of zero dumper at the service point = 0.118; Average number of dump truck at the service point ≈ 3; average number of running dump truck ≈ 12; number of dumper waiting in the queue for service ≈ 1; Waiting time at the service point = 5.22 hours; waiting time in the queue = 1.32 hours]

11 Non-linear algorithms for mining systems

11.1 INTRODUCTION

A program is said to be non-linear if either the objective function or at least one constraint is non-linear. The similarity of a non-linear program with the linear program is that both consist of an objective function, generalized constraints, and the bound of the variables. Most of the real systems are inherently non-linear. This chapter demonstrates different characteristics of nonlinear programs and their solution algorithms. It is always difficult to obtain an optimal solution of a non-linear program due to the following reasons:

- It is difficult to distinguish the local optimum and global optimum solutions.
- It is not mandatory that extreme points always represent the optimal solution as in the linear programming.
- An optimal solution depends on the consideration of the initial solution.
- Different algorithms may provide different solutions.

11.2 STATIONARY POINTS

The function $Z = f(x)$ (shown in Figure 11.1) has five stationary points, which occur when the slope of $f(x)$ is zero. Stationary points are thus identified by solutions to Eq. (11.1).

$$\frac{df(x)}{dx} = 0 \qquad (11.1)$$

Stationary points may be local minima (x_2, x_5), maxima (x_1, x_4) or saddle points (x_3). The local optima are likely candidates for a global optimum.

Local Maximum and Global Maximum

The given function $f(x)$ has a local maximum at x_1, as the value of the function at x_1 is higher than at any other point of x in the neighbourhood of x_1. If the function $f(x)$ has a local maximum at $x_{1,}$ then the following condition should satisfy.

$$f(x_1 - \Delta x_1) < f(x_1) > f(x_1 + \Delta x_1)$$

A local maximum can be the candidate for global maximum.

The given function $f(x)$ has a global maximum at x_4 in the range of a< x <b because the value of the function is highest as compared to the value at any other point in the defined range.

Local Minimum and Global Minimum

The given function $f(x)$ has a local minimum at x_2, as the value of the function at x_2 is lower than at any other point of x in the neighbourhood of x_2. If the function $f(x)$ has a local maximum at $x_{2,}$ then the following condition should satisfy.

$$f(x_2 - \Delta x_2) > f(x_2) < f(x_2 + \Delta x_2)$$

DOI: 10.1201/9781003200703-11

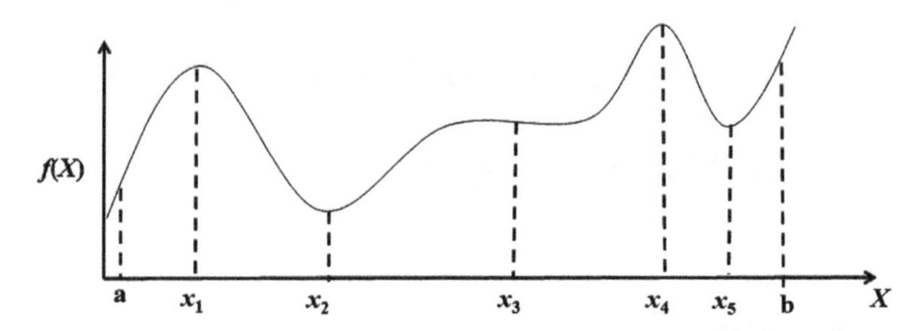

FIGURE 11.1 Stationary points of function f(x).

A local minimum can be the candidate for global minimum.

The given function $f(x)$ has a global minimum at x_5 in the range of a< x <b because the value of the function is lowest as compared to the value at any other point in the defined range.

Saddle point

The given function $f(x)$ has a saddle point at x_3, as the value of the function at x_3 is lower than any other neighbourhood point on one side and higher on the other. The slope of the function, $f(x)$, at any saddle point is equal to zero. If the function $f(x)$ has a saddle point at $x_{3,}$ then the following condition should satisfy.

$$f(x_3 - \Delta x_3) < f(x_3) < f(x_3 + \Delta x_3)$$

Convexity

If the slope of any function, $f(x)$ increases continuously, the function $f(x)$ is convex. The condition of convex function can be represented as

$$\frac{d^2 f}{dx^2} \geq 0$$

If all the eigenvalues of the function are positive, the function is strictly convex.

Concavity

If the slope of any function, $f(x)$ decreases continuously, the function $f(x)$ is concave. The condition of concave function can be represented as

$$\frac{\partial^2 f}{\partial x^2} \leq 0$$

If all the eigenvalues of the function are negative, the function is strictly concave.

11.3 CLASSIFICATIONS OF NON-LINEAR PROGRAMMING

11.3.1 Unconstrained optimization algorithm for solving non-linear problems

If the optimization problem has an objective function without any restrictions of the variables, the problem is referred to as an unconstrained optimization problem. Moreover, if the nature of the objective function is non-linear, the problem is called an unconstrained non-linear optimization problem.

Let $f(x)$ is a function of multiple variables, $X = (x_1, x_2, \ldots, x_n)$. A necessary condition for a stationary point, X is that each first partial derivative of $f(x)$ should be equal to zero.

$$\frac{\partial f}{\partial x_1} = 0, \frac{\partial f}{\partial x_2} = 0, \ldots, \frac{\partial f}{\partial x_n} = 0$$

The length of the first derivative indicates the rate of increase of the value of the function in the direction of the gradient. The next challenging task is to identify the stationary points which offer the local minimum or local maximum, or saddle point. In the case of single-variable function, the value of the double derivative of the function at different stationary points helps in identifying local maximum or local minimum. In the case of a multi-variable function, this is done by the Hessian matrix. The Hessian matrix is a square matrix, formed by the double derivatives of the function as follows:

$$H = \begin{bmatrix} \frac{\partial^2 f}{\partial x_1^2} & \frac{\partial^2 f}{\partial x_1 \partial x_2} & \cdots & \frac{\partial^2 f}{\partial x_1 \partial x_n} \\ \frac{\partial^2 f}{\partial x_2 \partial x_1} & \frac{\partial^2 f}{\partial x_2^2} & \cdots & \frac{\partial^2 f}{\partial x_2 \partial x_n} \\ \vdots & \vdots & \ddots & \vdots \\ \frac{\partial^2 f}{\partial x_n \partial x_1} & \frac{\partial^2 f}{\partial x_n \partial x_2} & \cdots & \frac{\partial^2 f}{\partial x_n^2} \end{bmatrix}$$

In the next step, the Hessian matrix is used to determine the eigenvalue of the matrix using the following Eq. 11.1:

$$H * \mathrm{V} = \lambda * \mathrm{V} \tag{11.2}$$

In Eq. (11.1), H is nxn Hessian matrix, V is non-zero n x 1 vector, and λ is a scalar quantity. The value of λ, which offers the solution of Eq. (11.2), is the eigenvalue of the Hessian matrix. The vector, V, which corresponds to this value, is called an eigenvector. The above equation can be rewritten as

$$H * \mathrm{V} - \lambda * \mathrm{V} = 0$$

$$\Rightarrow H * \mathrm{V} = \lambda * \mathrm{I} * \mathrm{V} = 0$$

Where, I is an identity matrix of the same order as that of the Hessian matrix.

$$\Rightarrow H = \lambda * \mathrm{I} = 0$$

$$\Rightarrow \begin{bmatrix} \frac{\partial^2 f}{\partial x_1^2} & \frac{\partial^2 f}{\partial x_1 \partial x_2} & \cdots & \frac{\partial^2 f}{\partial x_1 \partial x_n} \\ \frac{\partial^2 f}{\partial x_2 \partial x_1} & \frac{\partial^2 f}{\partial x_2^2} & \cdots & \frac{\partial^2 f}{\partial x_2 \partial x_n} \\ \vdots & \vdots & \ddots & \vdots \\ \frac{\partial^2 f}{\partial x_n \partial x_1} & \frac{\partial^2 f}{\partial x_n \partial x_2} & \cdots & \frac{\partial^2 f}{\partial x_n^2} \end{bmatrix} - \lambda * \begin{bmatrix} 1 & 0 & \cdots & 0 \\ 0 & 1 & \cdots & 0 \\ \vdots & \vdots & \ddots & \vdots \\ 0 & 0 & \cdots & 1 \end{bmatrix} = 0$$

$$\Rightarrow \begin{bmatrix} \frac{\partial^2 f}{\partial x_1^2} & \frac{\partial^2 f}{\partial x_1 \partial x_2} & \cdots & \frac{\partial^2 f}{\partial x_1 \partial x_n} \\ \frac{\partial^2 f}{\partial x_2 \partial x_1} & \frac{\partial^2 f}{\partial x_2^2} & \cdots & \frac{\partial^2 f}{\partial x_2 \partial x_n} \\ \vdots & \vdots & \ddots & \vdots \\ \frac{\partial^2 f}{\partial x_n \partial x_1} & \frac{\partial^2 f}{\partial x_n \partial x_2} & \cdots & \frac{\partial^2 f}{\partial x_n^2} \end{bmatrix} - \begin{bmatrix} \lambda & 0 & \cdots & 0 \\ 0 & \lambda & \cdots & 0 \\ \vdots & \vdots & \ddots & \vdots \\ 0 & 0 & \cdots & \lambda \end{bmatrix} = 0 \quad (11.3)$$

Solve Eq. (11.3) for determining the values of λ at different stationary points. These values represent the eigenvalues of the Hessian matrix. The criteria for examining the minimum and maximum of a non-linear function at X are as follows:

- If all eigenvalues are positive at $X = X_0$, the function has a local minimum at X_0. Moreover, if all the eigenvalues are positive for all possible values of $X = X_0$, then the function has a global minimum at X_0.
- If all eigenvalues are negative at $X = X_0$, the function has a local maximum at X_0. Moreover, if all the eigenvalues are negative for all possible values of $X = X_0$, then the function has a global maximum at X_0.
- If some eigenvalues are positive and some are negative, or some are zero at $X = X_0$, then the function has neither a local maximum local minimum at X_0.

The algorithm is applied to solve the problem demonstrated in Example 11.1.

Example 11.1

A mine workshop has an inventory size of unlimited storage for the HEMM accessories. The other inventory parameters are listed in Table 11.1.

Determine the EOQ level for minimizing the inventory cost.

Solution

Demand of tyres per day (d_1) = 730/365 =2
Demand of lubricant oil per day (d_2) = 200000/365 = 548
Ordering cost of tyres (K_1) = US$ 300 per order
Ordering cost of lubricant oil (K_2) = US$ 250 per order
Purchasing cost of tires (C_1) = US$ 600 per unit
Purchasing cost of lubricant oil (C_1) = US$ 5 per unit
Holding cost of tires (h_1) = US$ 0.01 per unit per day
Holding cost of lubricant oil (h_2) = US$ 0.005 per unit per day

TABLE 11.1
Inventory Parameters

Item	Demand (unit per year)	Ordering cost (US$ per order)	Purchasing cost (US$ per unit)	Holding cost (US$ Per unit time)
Tires	730	300	600	0.01
Lubricant oil	200000	250	5	0.005

We have,

$$Q_i^* = \sqrt{\frac{2K_i d_i}{h_i}}$$

$$\Rightarrow Q_1^* = \sqrt{\frac{2K_1 d_1}{h_1}} = \sqrt{\frac{2*30000*\left(\frac{730}{365}\right)}{1}} \cong 346$$

$$\Rightarrow Q_2^* = \sqrt{\frac{2K_2 d_2}{h_2}} = \sqrt{\frac{2*25000*\left(\frac{200000}{365}\right)}{0.5}} \cong 7402$$

The total cost of inventory per unit time is given by

$$TC(Q_1, Q_2) = \sum_{i=1}^{2} \frac{d_i}{Q_i^*} K_i + c_i d_i + \frac{Q_i^*}{2} h_i$$

$$TC(Q_1, Q_2) = \frac{d_1}{Q_1} K_1 + c_1 d_1 + \frac{Q_1}{2} h_1 + \frac{d_2}{Q_2} K_2 + c_2 d_2 + \frac{Q_2}{2} h_2$$

$$\frac{\partial TC}{\partial Q_1} = \frac{\partial}{\partial Q_1}\left(\frac{d_1}{Q_1} K_1 + c_1 d_1 + \frac{Q_1}{2} h_1 + \frac{d_2}{Q_2} K_2 + c_2 d_2 + \frac{Q_2}{2} h_2\right)$$

$$\Rightarrow \frac{\partial}{\partial Q_1}\left(\frac{d_1}{Q_1} K_1 + c_1 d_1 + \frac{Q_1}{2} h_1 + \frac{d_2}{Q_2} K_2 + c_2 d_2 + \frac{Q_2}{2} h_2\right) = 0$$

$$\Rightarrow \frac{\partial TC}{\partial Q_1} = -\frac{d_1}{Q_1^2} K_1 + 0 + \frac{h_1}{2} + 0 + 0 + 0$$

$$\Rightarrow \frac{\partial TC}{\partial Q_1} = -\frac{d_1}{Q_1^2} K_2 + \frac{h_1}{2}$$

Similarly,

$$\frac{\partial TC}{\partial Q_2} = -\frac{d_2}{Q_2^2} K_2 + \frac{h_2}{2}$$

The stationary points are given by solving

$$\frac{\partial TC}{\partial Q_1} = 0 \text{ and } \frac{\partial TC}{\partial Q_2} = 0$$

Therefore,

$$\frac{\partial TC}{\partial Q_1} = -\frac{d_1}{Q_1^2} K_1 + \frac{h_1}{2} = 0 \Rightarrow Q_1 = \sqrt{\frac{2K_1 d_1}{h_1}} = \sqrt{\frac{2*30000*\left(\frac{730}{365}\right)}{1}} \cong 346$$

$$\frac{\partial TC}{\partial Q_2} = -\frac{d_2}{Q_2{}^2}K_2 + \frac{h_2}{2} = 0 \Rightarrow Q_2 = \sqrt{\frac{2K_2 d_2}{h_2}} = \sqrt{\frac{2*25000*\left(\frac{200000}{365}\right)}{0.5}} \cong 7402$$

The double derivative of TC can be determined as

$$\frac{\partial^2 TC}{\partial Q_1^2} = \frac{2d_1}{Q_1^3}K_1$$

$$\frac{\partial^2 TC}{\partial Q_2^2} = \frac{2d_2}{Q_2^3}K_2$$

$$\frac{\partial^2 TC}{\partial Q_1 \partial Q_2} = 0$$

$$\frac{\partial^2 TC}{\partial Q_2 \partial Q_1} = 0$$

Therefore, $HTC(Q) = \begin{bmatrix} \frac{\partial^2 TC}{\partial Q_1^2} & \frac{\partial^2 TC}{\partial Q_1 \partial Q_2} \\ \frac{\partial^2 TC}{\partial Q_2 \partial Q_1} & \frac{\partial^2 TC}{\partial Q_2^2} \end{bmatrix} = \begin{bmatrix} \frac{2d_1}{Q_1^3}K_1 & 0 \\ 0 & \frac{2d_2}{Q_2^3}K_2 \end{bmatrix}$

Eigenvalues of matrix H are obtained by

$$|H - \lambda I| = 0$$

$$\Rightarrow \begin{bmatrix} \frac{2d_1}{Q_1^3}K_1 & 0 \\ 0 & \frac{2d_1}{Q_2^3}K_2 \end{bmatrix} - \lambda \begin{bmatrix} 1 & 0 \\ 0 & 1 \end{bmatrix} = 0$$

$$\Rightarrow \begin{bmatrix} \lambda - \frac{2d_1}{Q_1^3}K_1 & 0 \\ 0 & \lambda - \frac{2d_2}{Q_2^3}K_2 \end{bmatrix} = 0$$

$$\Rightarrow \lambda_1 = \frac{2d_1}{Q_1^3}K_1 \text{ and } \lambda_2 = \frac{2d_2}{Q_2^3}K_2$$

The eigenvalues (λ_1 and λ_2) at stationary points Q_1 (= 346) and Q_2 (7402) can be obtained by putting the values of d_1, d_2, K_1, and K_2. The eigenvalues are determined as $\lambda_1 = 2.89 * 10^{-5}$, $\lambda_2 = 6.75 * 10^{-7}$. As both the eigenvalues are positive, the function is a convex function. Also, as the eigenvalues depend on the value of Q_1 or Q_2, the function is not strictly convex. Therefore, the function $TC(Q)$ has a local minimum at $Q = (346, s7402)$. For all positive Q, the function $TC(Q)$ has a global minimum at $Q = (346, 7402)$.

The total inventory cost is

$$TC(Q)=\left(\frac{2}{346}*300+600*2+\frac{346}{2}*0.01\right)+\left(\frac{548}{7402}*250+5*548+\frac{7402}{2}*0.005\right)$$

$$\Rightarrow TC(Q)=1201.9+2760.3=\text{US\$3962.2 per day}$$

11.3.2 Constrained optimization algorithm for solving non-linear problems

If the optimization problem has an objective function with at least one restriction on the variables, the problem is referred to as a constrained optimization problem. Moreover, if the nature of the objective function is non-linear, the problem is called constrained non-linear optimization problem.

Let $f(x)$ is a function of n variables, $X=(x_1,x_2,\ldots x_n)$ need to be optimized subject to m number of constraints as follows:

$$\begin{aligned} &\mathit{Maximum\,/\,Minimum}\ f(x_1,\ x_2,\ \ldots,x_n)\\ &\text{s.t.}\ g_1(x_1,\ x_2,\ \ldots,x_n)=b_1\\ &\quad g_2(x_1,\ x_2,\ \ldots,x_n)=b_2\\ &\quad \ldots\ldots\ldots\ldots\ldots\ldots\\ &\quad g_m(x_1,\ x_2,\ \ldots,x_n)=b_m \end{aligned}$$

It should be noted that the objective function and constraints are continuous and differentiable. The constraint limits $b_1, b_2 \ldots . b_m$ are constants, and the decision variables $x_1, x_2, \ldots, x_n$ may be either positive or negative. It must also be assumed that solutions to the model are bounded (i.e., the optimal value of $f(x)$ is not infinite).

The method of Lagrange multipliers can be used to determine the stationary points. The method is used to the problem having equality constraints. The conditions for stationary points are derived by multiplying each equality constraint by a Lagrange multiplier, λ_i, and subtracting from the objective function as follows:

$$\begin{aligned} H(x_1,\ x_2,\ \ldots,x_n;\ \lambda_1,\ \lambda_2,\ \ldots,\lambda_m) &= f(x_1,\ x_2 \ldots x_n)-\lambda_1\left[g_1(x_1,\ x_2,\ \ldots,x_n)-b_1\right]\\ &-\lambda_2\left[g_2(x_1,\ x_2,\ \ldots,x_n)-b_2\right]\ldots-\lambda_m\left[g_m(x_1,\ x_2,\ \ldots,x_n)-b_m\right] \end{aligned}$$

$$\Rightarrow H=f(x_1,\ x_2,\ldots,x_n)-\sum_{i=1}^{m}\lambda_1(g_i-b_i)$$

The multiplier λ_i can be envisioned as a unit penalty cost incurred whenever $g_i(x_1,x_2,\ldots,x_n)$ exceeds b_i. If the net effects of these costs are subtracted from the objective function, a new objective function $H(x_1,x_2,\ldots,x_n;\lambda_1,\lambda_2,\ldots,\lambda_m)$ is obtained. The subtraction of constraint equations after multiplying with λ_i from the objective function does not affect the objective function as the value of the constraint in the revised objective function are zero.

The function $H(x_1,x_2,\ldots,x_n;\lambda_1,\lambda_2,\ldots,\lambda_m)$ contains the n original variables plus m unknown Lagrange's multipliers. If the function is treated as an unconstrained objective function, stationary points are determined by equating $(n+m)$ partial derivatives to zero.

$$\frac{\partial H}{\partial x_1} = 0, \frac{\partial H}{\partial x_2} = 0 \cdots \frac{\partial H}{\partial x_n} = 0$$

$$\frac{\partial H}{\partial \lambda_1} = 0, \frac{\partial H}{\partial \lambda_2} = 0 \cdots \frac{\partial H}{\partial \lambda_m} = 0$$

These equations may also be written as follows

$$\frac{\partial H}{\partial x_j} = \frac{\partial f}{\partial x_j} - \sum_{i=1}^{m} \lambda_1 \frac{\partial g}{\partial x_j} \, j = 1, 2, \ldots, n$$

$$\frac{\partial H}{\partial \lambda_i} = -g_i + b_i, \qquad i = 1, 2, \ldots, m$$

Solution of these $(n+m)$ equations will identify stationary points of the objective function, H. The values or roots of Lagrange's multiplier can be used to determine the minimum and maximum of the objective function at different stationary points. The conditions for examining the maximization and minimization are as follows:

- If each root of λ at $X = X_0$ is negative, the function has local maximum at X_0. Moreover, if each root of λ is negative and independent of X, the function has a global maximum at X_0.
- If each root of λ at $X = X_0$ is positive, the function has local minimum at X_0. Moreover, if each root of λ is positive and independent of X, the function has a global minimum at X_0.
- If some roots of λ are positive and some are negative at $X = X_0$, then the function has neither a local maximum nor a local minimum at X_0.

The algorithm is applied to solve the problem demonstrated in Example 11.2.

Example 11.2

$$\textit{Maximize } Z = 4x_1 - 0.1x_1^2 + 5x_2 - 0.2x_2^2$$

$$\text{Subject to, } x_1 + 2x_2 = 40$$

Determine the optimal solution to this nonlinear programming model using Lagrange's multiplier method.

Solution

The revised objective function is derived from the given objective function and constraint by considering the Lagrangian multiplier, λ, as follows:

$$H(x_1, x_2, \boldsymbol{\lambda}) = 4x_1 - 0.1x_1^2 + 5x_2 - 0.2x_2^2 - \boldsymbol{\lambda}(x_1 + 2x_2 - 40)$$

To determine the stationary points and the value of the Lagrangian multiplier, the partial derivatives of the revised objective function, H, are equated to 0.

$$\frac{\partial H}{\partial x_1} = x_1 - 0.1 * 2 * x_1 + 0 - 0 - \boldsymbol{\lambda}(1 + 0 - 0) = x_1 - 0.2x_1 - \boldsymbol{\lambda} = 0$$

$$\frac{\partial H}{\partial x_2} = 0 - 0 + 5 * x_2 - 0.2 * 2 * x_2 - \lambda(0 + 2 * 1 - 0) = 5x_2 - 0.4x_2 - 2\lambda = 0$$

$$\frac{\partial H}{\partial \lambda} = -1 * (x_1 + 2x_2 - 40) = 0$$

Solving the above three equations, we get

$$x_1 = 18.3, \quad x_2 = 10.8, \quad \lambda = 0.33$$

Since the value of λ is positive and independent of x, the objective function gives global maximum at $x_1 = 18.3$, $x_2 = 10.8$.

Substituting the values for and into the original objective function yields the value as follows:

$$Z = 4 * 18.3 - 0.1 * 18.3^2 + 5 * 10.8 - 0.2 * 10.8^2 = 70.42$$

This result can also be obtained from the revised objective function as

$$H = 4x_1 - 0.1x_1^2 + 5x_2 - 0.2x_2^2 - \lambda(x_1 + 2x_2 - 40) = 4 * 18.3 - 0.1 * 18.3^2$$
$$+ 5 * 10.8 - 0.2 * 10.8^2 - 0.33 * (18.3 + 2 * 10.8 - 40) = 70.42$$

The Lagrange multiplier approach can be extended to solve nonlinear programming problems with inequality constraints. The Karush-Kuhn-Tucker (KKT) conditions can be used to determine the stationary points. The KKT necessary conditions for maximization/minimization are explained in Table 11.2.

The method for solving a non-linear optimization problem with inequality constraints is explained in Example 11.3.

Example 11.3

The inventory in a mining workshop has an area of 2000 sq. ft. for storage of truck tires. The demand for tires in the mines is two units per day. The ordering cost of tires is US$300 per order, and the holding cost is US$0.01 per unit per day. The purchasing cost is US$600 per unit. The average space required to the tire is 12 sq. ft. per unit. Determine the optimal inventory plan.

Solution

Given data:

Demand of tyres (d) = 2 units per day

Ordering cost of tyres (K) = US$300 per order

TABLE 11.2
KKT Conditions for Maximization/Minimization

	Necessary conditions		
Type of optimization	**Objective function**	**Constraints**	**Lagrange's multiplier**
Maximization	Concave	Convex	≥ 0
		Concave	≤ 0
		Linear	Unrestricted
Minimization	Convex	Convex	≤ 0
		Concave	≥ 0
		Linear	Unrestricted

Purchasing cost of tyres (C) = US\$600 per unit
Holding cost of tyres (h) = US\$0.01 per unit per day
Space required for tyre (a) = 12 sq. ft. per unit
Total available area (A) = 2000 sq. ft.

The objective is to minimize the total inventory cost. Therefore, we have

$$\text{Minimum}\, TC(Q) = \frac{d}{Q}K + Cd + \frac{Q}{2}h$$

s.t., $aQ \le A$

$$\Rightarrow \quad \text{Minimum}\, TC(Q) = \left(\frac{2}{Q}*300 + 600*2 + \frac{Q}{2}*0.01\right)$$

s.t., $12Q \le 2000$

Before deriving the revised objective function, the inequality constraints should be converted into equality constraints by introducing a square of slack and surplus variables. In this case, the problem has only one constraint with ≤ sign. It is converted into equality constraint by introducing the square of a slack variable, s as follows:

$$12Q + s^2 = 2000$$

$$\Rightarrow 12Q + s^2 - 2000 = 0$$

The revised objective function is derived by subtracting the multiplications of Lagrange's multiplier, λ with the constraint equation from the original objective function as follows:

$$H(Q, \lambda, s) = \left(\frac{2}{Q}*300 + 600*2 + \frac{Q}{2}*0.01\right) - \lambda\left[12Q + s^2 - 2000\right]$$

To determine the stationary points and the value of the Lagrangian multiplier, the partial derivatives of the revised objective function, H with respect to $Q, \lambda,$ and s are equated to 0.

$$\frac{\partial H}{\partial Q} = -\frac{2}{Q^2}*300 + 0 + \frac{1}{2}*0.01 - 12\lambda = 0 \tag{11.4}$$

$$\frac{\partial H}{\partial \lambda} = 12Q + s^2 - 2000 = 0 \tag{11.5}$$

$$\frac{\partial H}{\partial s} = -2\lambda s = 0 \tag{11.6}$$

To determine the stationary points, the above three simultaneous equations need to be solved.

From Eq. (11.6), we have

$$\lambda = 0 \text{ or } s = 0$$

Putting $\lambda = 0$ in Eq. (11.4), we have

$$Q = \sqrt{\frac{600}{0.005}} = \sqrt{120000} \approx 346$$

Putting $s = 0$ in Eq. (11.5), we have

$$Q = \frac{2000}{12} \approx 166$$

$Q = 346$ is not feasible as this violating the space constraint and thus $Q = 166$.

$$\text{Therefore, the EOQ level of the tire}\left(Q^*\right) = 166 \text{ units}$$

KKT conditions for minimization of the objective function at the stationary point

The objective function is minimization, and thus the objective function should be convex type. It is clear that the double derivative of the objective function is positive for all positive Q, and hence the function is convex type.

$$\text{That is, } \frac{\partial^2 TC}{\partial Q^2} = \frac{1200}{Q^3} \geq 0 \qquad \text{for } Q \geq 0$$

There is only one constraint, which is linear type. Hence there is no restriction in the sign of Lagrange's multiplier, λ, for satisfying the condition for minimization. Therefore, the total cost function offers the minimum value at a stationary point, Q.

The total inventory cost at $Q = 166$, is given by

$$TC\left(Q = 166\right) = \left(\frac{2}{Q} * 300 + 600 * 2 + \frac{Q}{2} * 0.01\right)$$

$$= \left(\frac{2}{166} * 300 + 600 * 2 + \frac{166}{2} * 0.01\right) = \text{US\$1204.4 per day}$$

11.4 CASE STUDY ON THE APPLICATION OF NON-LINEAR OPTIMIZATION FOR OPEN-PIT PRODUCTION SCHEDULING

This case study presents a non-linear optimization application for open-pit production scheduling with cut-off grade optimization and stockpiling. An application of copper mine is selected for the production schedule. When mine has a stockpile, two key decisions need to be made: (a) how much materials and what quality should be sent to stockpile; and (b) how much materials and what quality should be sent from the stockpile to process plant.

The case study mine is located in Africa, and it has three distinct rock types: sulfide, oxide, and mixed. The resource model of the deposit was prepared by applying the ordinary kriging method and using 1-m composite data from exploration drilling (Chiles and Delfiner, 2012; Goovaerts, 1997). The $20 \times 20 \times 10$ metres blocks were selected for resource modelling. Each block has 10,800 tonnes of material available. A total number of 26,021 blocks are available within the resource model for mining optimization.

The production schedule with stockpiling and grade blending model maximizes the discounted cashflow over the mine life after satisfying slope, reserve, resource (mining, processing, and refining), and stockpile material handling constraints. However, the blending of materials in the stockpile makes the model non-linear (Bley et al., 2012). The formulation of the proposed mixed-integer quadratic programming problem used the following indices, parameters, variables, objective function, and constraints.

Indices

N = number of mining blocks within the orebody model;
D = number of material destinations (processing streams, stockpile, waste dump);
T = life of mining operation or the scheduling horizon;

i = set of blocks $\{1,\ldots,N\}$;
ξ_i = set of immediate predecessors of block i;
d = set of material destinations $\{1,\ldots,\mathrm{D}\}$;
t = set of periods or years $\{1,\ldots,T\}$.

Parameters

δ = discounted rate;

S_{dt} = discounted selling price per unit of metal produced at processing destination d during period t, where $S_{dt} = \frac{S_d}{(1+\delta)^t}$;

m_t = discounted mining cost during period t, where $m_t = \frac{m}{(1+\delta)^t}$;

h_t = discounted material re-handling cost in stockpile during period t, where $h_t = \frac{h}{(1+\delta)^t}$;

c_{dt} = discounted processing cost per unit of ore processed at processing destination d during period t, where $c_{dt} = \frac{c_d}{(1+\delta)^t}$;

r_{dt} = discounted refining cost per unit of metal at processing destination d during period t, where $r_{dt} = \frac{r_d}{(1+\delta)^t}$;

f_t = discounted fixed or administrative cost during period t, where $f_t = \frac{f}{(1+\delta)^t}$;

$\bar{M}, \underline{M}$ = lower and upper limit of the mining capacity;
$\bar{C}_d, \underline{C}_d$ = lower and upper limit of the processing capacity at a destination d;
$\bar{R}_d, \underline{R}_d$ = lower and upper limit of the refining capacity at a destination d;
$\bar{p}_t, \underline{p}_t$ = discounted penalty per unit lower and upper bound deviation in mining, processing and refining capacities, where $\bar{p}_t = \frac{\bar{p}}{(1+\delta)^t}$ and $\underline{p}_t = \frac{\underline{p}}{(1+\delta)^t}$;

p' = total penalty for deviations from mining, processing and refining capacities;
a_i = quantity of material in mining block i;
g_i = grade or metal content of mining block i;
μ_{id} = recovery in mining block i at processing destination d.

Variables

$x_{it} \in \{0,1\}$ = 1 if block i is extracted during period t and 0 otherwise;
$y_{idt} \in \{0,1\}$ = 1 if block i is sent to destination d during period t and 0 otherwise;
o_t = quantity of ore available in stockpile at the end of period t;
qo_t^d = quantity of ore moved from stockpile to processing destination during period t;
e_t = quantity of metal available in stockpile at the end of period t;
qe_t^d = quantity of metal moved from stockpile to processing destination during period t;

The objective function of the proposed non-linear mixed-integer model can be written as:

Objective function

$$\max z = \left[\left(\begin{array}{c}\sum_{t=1}^{T}\left\{\left(\sum_{i=1}^{N} -m_t\; a_i\; x_{it}\right) - h_t qo_t^d\right\} + \\ \sum_{t=1}^{T}\sum_{d=1}^{D}\left(\left(\left(s_{dt} - r_{dt}\right)\left(\sum_{i\in N}\left(a_i g_i \mu_{id}\right) y_{idt} + qe_t^d\right)\right) - c_{dt}\left(\sum_{i\in N} a_i y_{idt} + qo_t^d\right)\right)\end{array}\right)\right] - \sum_{t=1}^{T} f_t - \acute{p} \tag{11.7}$$

$$\acute{p} = \left[\begin{array}{l}\sum_{t=1}^{T}\bar{p}_t \max\left\{0,\left(\bar{M} - \left(\sum_{i\in N} a_i x_{it}\right)\right)\right\} + \\ \sum_{t=1}^{T}\underline{p}_t \max\left\{0,\left(\left(\sum_{i\in N} a_i x_{it}\right) - \underline{M}\right)\right\} + \\ \sum_{t=1}^{T}\bar{p}_t \max\left\{0,\left(\bar{C}_d - \left(\sum_{i\in N} a_i y_{idt} + qo_t^d\right)\right)\right\} + \\ \sum_{t=1}^{T}\underline{p}_t \max\left\{0,\left(\left(\sum_{i\in N} a_i y_{idt} + qo_t^d\right) - \underline{C}_d\right)\right\} + \\ \sum_{t=1}^{T}\bar{p}_t \max\left\{0,\left(\bar{R}_d - \left(\sum_{i\in N} a_i g_i \mu_{id} + qe_t^d\right)\right)\right\}\end{array}\right] \tag{11.8}$$

Constraints

$$x_{it} - \sum_{\tau=1}^{t} x_{j\tau} \le 0;\ \forall\ t,i \text{ and } j \in \xi_i \tag{11.9}$$

$$\sum_{t=1}^{T} x_{it} \le 1; \forall\ i \tag{11.10}$$

$$\bar{M} \le \sum_{i\in N} a_i x_{it} \le \underline{M};\ \forall\ t \tag{11.11}$$

$$\bar{C}_d \le \sum_{i\in N} a_i y_{idt} + qo_t^d \le \underline{C}_d;\ \forall\ d,t \tag{11.12}$$

$$\bar{R}_d \le \sum_{i\in N}\left(a_i g_i \mu_{id}\right) y_{idt} + qe_t^d \le \underline{R}_d;\ \forall\ d,t \tag{11.13}$$

$$o_t = \begin{cases}\sum_{i=1}^{N} a_i y_{idt}; & \text{if } t = 1 \\ \sum_{i=1}^{N} a_i y_{idt} - qo_t^d + o_{(t-1)}; & \text{if } t > 1\end{cases} \tag{11.14}$$

$$e_t = \begin{cases}\sum_{i=1}^{N}\left(a_i g_i \mu_{id}\right) y_{idt}; & \text{if } t = 1 \\ \sum_{i=1}^{N}\left(a_i g_i \mu_{id}\right) y_{idt} - qe_t^d + e_{(t-1)}; & \text{if } t > 1\end{cases} \tag{11.15}$$

$$qo_t^d \le o_{(t-1)};\ \forall\ d,t \ge 2 \tag{11.16}$$

$$qe_t^d \le e_{(t-1)};\ \forall\ d,t \ge 2 \tag{11.17}$$

$$qo_t^d\ e_{(t-1)} = o_{(t-1)} qe_t^d \tag{11.18}$$

The objective (Eq. 11.7) of this optimization problem is to maximize the discounted economic value (revenue minus cost) by minimizing the deviations (Eq. 11.8) from production targets over the mine life. The slope requirements for each mining block are respected by Eq. (11.9). The reserve constraints, i.e., a block can be mined once over the planning horizon, are respected by Eq. (11.10). The mining, processing, and refining capacity constraints are satisfied by Eqs. (11.11)–(11.13), respectively. Eqs. (11.14) and (11.15) calculate the quantity of ore and metal available at the stockpile; whereas, Eqs. (11.16) and (11.17) keep track of the supply of ore and metal from the stockpile to the processing destinations. The non-linear constraints (Eq. 11.18) are added in the optimization formulation to ensure constant ore to the metal ratio in stockpile over the mine life. This will ensure the consistent quality of blended materials sent from the stockpile to the process plant.

For solving the optimization, the algorithm proposed by Paithankar et al. (2020) is used in this case study. The solution strategy has three sequential steps: (a) a decision of extraction sequence is made; (b) a destination policy of the blocks is calculated, given the extraction decision as input; and finally (c) decision on the quantity of ore metal transport from the stockpile to plant is made, given the decisions from the first two steps as inputs. These three steps are integrated within a genetic algorithm (GA)-based optimization framework that maximizes the objective function iteratively, as discussed in Paithankar et al. (2020). Within the GA-based framework, the extraction sequence problem is solved using a maximum flow algorithm (Paithankar and Chatterjee, 2019; Chatterjee and Dimitrakopoulos, 2020). The economic value of block i is calculated using the below equation:

$$v_i = \left[(S-r)g_i\mu_i - c - m\right]a_i \tag{11.19}$$

The construction of the graph for the maximum flow algorithm was decided based on the v_i. In the graph, all the mining blocks are considered nodes in the graph. In the graph, there are two additional nodes: source and sink, and are connected with the mining block nodes. If $v_i > 0$, a connection is made using an arc from the source node, the block i, and the capacity of the arc is the value of economic block value. On the other hand, if $v_i \leq 0$, a connection is made from block i to the sink with an arc capacity of the absolute value of the block economic value. Using a maximum flow algorithm, this graph can produce an ultimate pit limit. This graph can be extended to a multiple-period graph using the discounted price and using additional nodes and arcs (Chatterjee et al., 2016; Paithankar and Chatterjee, 2019; Paithankar et al., 2020, 2021). To generate a variable pit size, a weight (w_{it}) was introduced in the graph, and these weights were optimized by GA (Paithankar and Chatterjee, 2019) to generate an optimal extraction sequence $(x_{it} = 1)$ that satisfies all physical and operational constraints.

After getting the extraction sequence $(x_{it} = 1)$, the grade and metal content of the material are used to decide destinations. A binary variable $y_{idt} \in \{0,1\}$ is used for the destination based on the optimal value for the cut-off grade $\left(\bar{\bar{g}}_{dt}\right)$, which is optimized within the same GA framework. In this case study, the destinations are considered: waste dump, stockpile, and processing stream. If the block i has the grade above the optimal cut-off grade $\left(\bar{\bar{g}}_{dt}\right)$, the materials are sent to the processing stream. The materials are sent to the stockpile if the block grade is higher than the break-even cut-off grade but below the optimal cut-off grade. The remaining extracted blocks are sent to the waste dump.

The determine the quantity of ore to be sent from the stockpile to the processing plant for meeting the processing requirements, a linear decision variable $z_t \in [0,1]$ is used. The quantity of ore from the stockpile to processing destination (qo_t^d) is calculated by $qo_t^d = z_t\left(o_{(t-1)}\right)$, where $o_{(t-1)}$ is the available ore quantity in the stockpile at the beginning of period t. Eq. (11.18) is then used to calculate the metal quantity sent to the process plant from the stockpile. It ensures that the non-linear mixing

TABLE 11.3
Economic Parameters Used for Non-linear Optimization of Copper Mine Production Scheduling

Parameters	Copper
Mining cost	$ 1/ t
Mining cost from stockpile	$ 0.25/t
Processing cost	$ 9 / t
Metal price	$ 4409.24 / t
Refining cost	$ 661.39 / t
Penalty coefficient for shortage cost of ore	$ 15 / t
Penalty coefficient for the surplus cost of ore	$ 15 / t
Penalty coefficient for shortage cost of metal	$ 440 / t
Penalty coefficient for the surplus cost of metal	$ 220 / t
Discount rate	10%
Fixed Cost per period	$ 2,000,000

TABLE 11.4
Solution of the Non-linear Optimization

	Year 1	Year 2	Year 3	Year 4	Year 5	Year 6	Year 7
Material produced (MT)	9.8	13.1	19.6	29.8	28.7	19.3	26.8
Total ore sent to plant (MT)	5.01	4.41	4.52	5.03	5.05	4.83	4.76
Metal produced (10^4 tonne)	3.7	2.52	2.58	3.31	2.98	2.41	2.62
Materials at stockpile (MT)	0.52	1.12	2.34	4.86	5.31	4.29	4.10
Ore sent from stockpile to plant (MT)	0.00	0.00	0.51	0.26	0.48	2.03	0.05
Optimal cut-off grade (%)	0.82	0.65	0.68	0.78	0.62	0.57	0.61

constraints are satisfied. These parameters are simultaneously optimized using GA to maximize the objective function (Eq. 11.8), as discussed in Paithankar et al. (2020).

The economic parameters used for this study are presented in Table 11.3. For GA, a set of parameters need to be selected. The initial population, crossover probability, mutation probability, and parametric diversification probability are selected as 200, 0.8, 0.01, and 0.2. The selection strategy for the GA parameters is discussed in Paithankar and Chatterjee (2019). The number of production periods for this deposit is seven years, and the maximum mining capacity per period is 30 million tonne (MT). The plant's processing capacity is between 4 MT to 5 MT; whereas, the refinery capacity is within 24,000 tonne to 26,000 tonne. The stockpile and waste dump capacities are considered unlimited.

The stochastic production scheduling with stockpiling and cut-off grade optimization was implemented in Matlab software. Since GA is a population-based optimization algorithm, the model was run five times, and average values were reported. Table 11.4 shows the solution of the non-linear optimization. It is noted here that only upper bound constraint is used for mining capacity, and no deviation was allowed. It can be observed from the table that the material productions from all seven years are within the upper bound limit (30 MT). It was also observed that the material productions were low in the initial periods and increased in the later periods. The ore sent to plant in different production periods are very consistent and marginally deviates from the target limits. It is noted here that the optimization formulation allows the deviation of ore sent to the process plant, but the penalty is imposed for the deviation. It was also observed from the table that metal production significantly deviated in the first period. The possible reason for such deviation could be that the over-production of metal generates a large amount of cashflow at the first period, which can compensate for the overproduction penalty. It can be seen from the table that at the beginning of the mine life, a small portion of the extracted ore is sent to stockpile, which is due to the destination of high-grade extracted ore to the process plant. It was also observed that the amount of ore sent from the stockpile to the process plant is consistently low over the mine life, except in Year 6 when 2.03 MT of ore was sent from the stockpile to the process plant. From the optimal cut-off grade values, it can be seen the cut-off grade value is high at the initial period of the mine life and decreases at the later periods. It is quite expected as a higher cut-off grade ensures high-grade ores are sent to the process plant at the initial periods, and thus, produce more metal and cashflow. The total net present value generated from this project is 239.5 million US dollars (M$). To show the effectiveness of the applied non-linear optimization approach, the solution was solved without stockpile. In that case, no cut-off grade needs to optimize, and the basic maximum flow with GA proposed by Paithankar and Chatterjee (2019) was used for solving the problem. It was observed from the solution that the NPV from the project is 236.9 M$, which is 2.6 M$ less than the non-linear optimization model. These results demonstrated the effectiveness of the non-linear optimization for production scheduling of open-pit optimization.

Bibliography

Al-Chalabi, H.S., Lundberg, J., Wijaya, A., and Ghodrati, B. (2014). Downtime analysis of drilling machines and suggestions for improvement. *Journal of Quality in Maintenance Engineering*, 20: 306–332.

Allahkarami, Z., Sayadi, A.R., and Lanke, A. (2016) Reliability Analysis of Motor System of Dump Truck for Maintenance Management. In Kumar, U., Ahmadi, A., Verma, A., and Varde, P. (Eds.) *Current Trends in Reliability, Availability, Maintainability and Safety. Lecture Notes in Mechanical Engineering*. Springer, Cham. https://doi.org/10.1007/978-3-319-23597-4_50.

Aven, T. (2006). On the precautionary principle, in the context of different perspectives on risk. *Risk Management*, 8(3): 192–205.

Avriel, M. (2003). *Nonlinear Programming: Analysis and Methods*. Dover Publication Inc. ISBN 0-486-43227-0.

Barabady, J., and Kumar, U. (2005). Availability allocation through importance measures. *International Journal of Quality & Reliability Management*, 24(6): 643–657.

Barabady, J., and Kumar, U. (2008). Reliability analysis of mining equipment: A case study of a crushing plant at Jajarm Bauxite Mine in Iran. *Reliability Engineering and System Safety*, 93(4): 647–653.

Bellman, R. (1953). *An Introduction to the Theory of Dynamic Programming*. R-245. The RAND Corporation.

Bley, A., Gleixner, A.M., Koch, T., and Vigerske, S. (2012). Comparing MIQCP solvers to a specialised algorithm for mine production scheduling. In *Modeling, Simulation and Optimization of Complex Processes* (pp. 25–39). Berlin, Heidelberg: Springer.

Blitzstein, J.K., and Hwang, J. (2014). *Introduction to Probability*, Chapman & Hall/CRC Texts in Statistical Science, ISBN-13: 978-1466575578.

Boulding, K.E. (1985). *The World as a Total System*. Beverly Hills, CA: Sage Publications.

Chakravarti, I.M., Laha, R.G., and Roy, J. (1967). *Handbook of Methods of Applied Statistics*, Volume I, John Wiley and Sons, pp. 392–394.

Chatterjee, S., and Bandopadhyay, S. (2012). Reliability estimation using a genetic algorithm-based artificial neural network: an application to a Load-Haul-Dump machine. *Expert Systems and Applications*, 39: 10943–10951.

Chatterjee, S., Dash, A., and Sukumar, B. (2014). Ensemble support vector machine algorithm for reliability estimation of a mining machine. *Quality and Reliability Engineering International*, 31: 1503–1516.

Chatterjee, S., and Dimitrakopoulos, R. (2020). Production scheduling under uncertainty of an open-pit mine using Lagrangian relaxation and branch-and-cut algorithm. *International Journal of Mining, Reclamation and Environment*, 34(5): 343–361.

Chatterjee, S., Sethi, M.R., and Asad, M.W.A. (2016). Production phase and ultimate pit limit design under commodity price uncertainty. *European Journal of Operational Research*, 248(2): 658–667.

Chilès, J-P., and Delfiner, P. (2012). *Geostatistics: Modeling Spatial Uncertainty, Second Edition*. John Wiley.

Choudhury, S., and Chatterjee, S. (2014). Pit optimisation and life of mine scheduling for a tenement in the Central African Copperbelt. *International Journal of Mining, Reclamation and Environment*, 28(3): 200–213.

Cococcioni, M., and Fiaschi, L. (2021). The Big-M method with the Numerical Infinite M. *Optimization Letters*, 15: 2455–2468. doi: 10.1007/s11590-020-01644-6.

Colorni, A., Dorigo, M., and Maniezzo, V. (1991) Distributed Optimization by Ant Colonies. In: Varela, F. and Bourgine, P. (Eds.), *Proceedings of the European Conference on Artificial Life, ECAL'91*. Paris: Elsevier Publishing, Amsterdam, pp. 134–142.

Cormen, T.H., Leiserson, C.E., Rivest, R.L., and Stein, C. (2001). *Introduction to Algorithms* (2nd ed.), MIT Press & McGraw–Hill, ISBN 0-262-03293-7 .

Dantzig, G.B. (1948). *Programming in a Linear Structure*. Comptroller, United Air Force, Washington, DC, Tech rep.

Dantzig, G.B., and Thapa, M.N. (1997). *Linear Programming 1: Introduction*. New York: Springer.

Demirel, N., and Gölbaşı, O. (2016). Preventive replacement decisions for dragline components using reliability analysis. *Minerals*, 6(2): 51.

Demirel, N., Gölbaşi, O., Düzgün, Ş., and Kestel, S. (2014). System reliability investigation of draglines using fault tree analysis. In *Mine Planning and Equipment Selection* (pp. 1151–1158). Cham: Springer.

Dikin, I.I. (1967). Iterative solution of problems of linear and quadratic programming. *Dokl. Akad. Nauk SSSR.* 174(1): 747–748.

Dooge, J.C.I. (1973). *Linear Theory of Hydrologic Systems. ARS Technical Bulletin No. 1468.* Washington, DC: US Department of Agriculture.

Dowd, P.A., and Elvan, L. (1987). Dynamic programming applied to grade control in sub-level Open stoping. Transactions of the Institution of Mining and Metallurgy, Section A: Mining Technology, A171–A177.

Dreyfus, S. (2002). Richard Bellman on the Birth of Dynamic Programming. *Operations Research*, 2002, 50(1): 48–51.

Eberhart, R.C., and Kennedy, J. (1995). A new optimizer using particle swarm theory. In Proceedings of the 6th international symposium on micro machine and human science (Nagoya, Japan) (pp. 39–43). Piscataway, NJ: IEEE Service Center.

Ezema, B.I., and Amakon, U. (2012). Optimizing profit with the linear programming model: A focus on Golden plastic industry limited, Enugu, Nigeria. *Interdisciplinary Journal of Research in Business*, 2: 37–49.

Frank, M., and Wolfe, P. (1956). An Algorithm for Quadratic Programming. *Naval Research Logistics Quarterly*, 3(1–2): 95–110. https://doi.org/10.1002/nav.3800030109

Furuly, S., Barabady, J., and Barabadi, A. (2013). Reliability analysis of mining equipment considering operational environments – A case study. *International Journal of Performability Engineering*, 9(3): 287–294.

Gautam, N. (2007). Queuing Theory. In *Operations Research and Management Science Handbook. Operations Research Series*. 20073432. Edited by Ravindran, A.R. https://doi:10.1201/9781420009712.ch9. CRC Press, Taylor & Francis Group, USA. ISBN 978-0-8493-9721-9.

Gharahasanlou, A.N., Mokhtarei, A., Khodayarei, A., and Ataei, M. (2014). Fault tree analysis of failure cause of crushing plant and mixing bed hall at Khoy cement factory in Iran. *Case Studies in Engineering Failure Analysis*, 2(1): 33–38.

Glover, F. (1986). Future Paths for Integer Programming and Links to Artificial Intelligence. *Computers and Operations Research*, 13(5): 533–549. https://doi:10.1016/0305-0548(86)90048-1.

Goldberg, D.E. (1989). *Genetic Algorithms in Search, Optimization, and Machine Learning*. Reading, MA: Addison-Wesley.

Goovaerts, P. (1997). *Geostatistics for Natural Resources Evaluation*. Applied Geostatistics Series. xiv + 483 pp. New York and Oxford: Oxford University Press. ISBN 0 19 511538 4.

Gupta, S., and Bhattacharya, J. (2011). Aspects of Reliability and Maintainability in Bulk Material Handling System Design and Factors of Performance Measure. In *Design and Selection of Bulk Material Handling Equipment and Systems*, Vol. 1, Edited by J. Bhattacharya, Wide Publishing, Kolkata, pp. 153–188.

Gustafson, A., Schunnesson, H., and Kumar, U. (2013). Reliability analysis and comparison between automatic and manual load haul dump machines. *Quality and Reliability Engineering International*, 31: 523–531.

Hadley, G., and Whitin, T.M. (1963). *Analysis of Inventory Systems*. Englewood Cliffs, NJ: Prentice Hall. ISBN: 0130329533.

Hoseinie, S.H., Ataei, M., Khalokakaie, R., and Kumar, U. (2011a). Reliability modeling of water system of longwall shearer machine. *Archives of Mining Sciences*, 56: 291–302.

Hoseinie, S.H., Khalokakaie, R., Ataei, M., and Kumar, U. (2011b). Reliability-based maintenance scheduling of haulage system of drum Shearer. *International Journal of Mining and Mineral Engineering*, 3: 26–37.

Kajal, S., Tiwari, P.C., and Saini, P. (2012). Availability optimization for coal handling system using genetic algorithm. *International Journal of Performability Engineering*, 9(1): 109–116.

Kalra, V.M., Thakur, T., and Pabla, B.S. (2015). Operational Analysis of Mining Equipment in Opencast Mineusing Overall Equipment Effectiveness (OEE). *IOSR-Journal of Mechanical and Civil Engineering, Special Issue of National Conference on Advances in Engineering, Technology & Management*, pp. 27–31.

Kantorovič, L.V. (1939). *Mathematical Methods of Organizing and Planning Production*. Publ. House Leningrad State Univ. https://doi.org/10.1287/mnsc.6.4.366

Kelley, J.E. (1961). Critical-path planning and scheduling: Mathematical basis. *Operations Research*, 9(3): 296–320.

Kelley, J.E., and Walker, M.R. (1959). Critical path planning and scheduling, *Proceedings of the Eastern Joint Computer Conference, Boston*, pp. 160–173.

Kendall, M. (1953) The Analysis of Economic Time Series, Part I: Prices. *Journal of the Royal Statistical Society*, 96 .

Kennedy, J.O.S. (1986). *Dynamic Programming Applications to Agriculture and Natural Resources.* London: Elsevier Applied Science Publishers. https://doi.org/10.2307/1242214

Kiefer, J. (1953). Sequential minimax search for a maximum. *Proceedings of the American Mathematical Society*, 4(3): 502–506, doi:10.2307/2032161, JSTOR 2032161, MR 0055639

Kleinrock, L. (1975). *Queueing Systems: Volume I – Theory.* New York: Wiley Interscience. p. 417. ISBN 978-0471491101.

Kuhn, H.W. (1955). The Hungarian method for the assignment problem. *Naval Research Logistics Quarterly*, 2(1–2): 83–97.

Kumar, U., Klefsjo, B., and Granholm, S. (1989). Reliability Investigation for a Fleet of Load Haul Dump Machines in a Swedish Mine. *Reliability Engineering and System Safety*, 26(4): 341–361.

Kurtz, L.F. (1992). Group Environments in Self-Help Groups for Families. *Small Group Research*, 23(2): 199–215.

Lamghari, A., and Dimitrakopoulos, R. (2012). A diversified Tabu search approach for the open-pit mine production scheduling problem with metal uncertainty. *European Journal of Operational Research*, 222(3): 642–652.

Lamghari, A., and Dimitrakopoulos, R. (2012). A diversified Tabu search approach for the open-pit mine production scheduling problem with metal uncertainty. *European Journal of Operational Research*, 222: 642–652.

Land, A., and Doig, A. (1960) An Automatic Method of Solving Discrete Programming Problems. *Ecometrics*, 28: 497–520. https://doi.org/10.2307/1910129.

Lerchs, H., and Grossmann, I.F. (1965). Optimum design of open pit mines. *Transactions CIM*, 58: 17–24.

Litterer, J.A. (1973). *The Analysis of Organizations*, Joseph August 2nd edition 1973.

Little, J.D.C. (1961). A Proof for the Queuing Formula: L = λW. *Operations Research* 9(3): 383–387. doi:10.1287/opre.9.3.383.

Malcolm, D.G., Roseboom, J.H., Clark, C.E., and Fazar, W. (1959). Application of a technique for a research and development program evaluation. *Operations Research*, 7: 646–669.

Mincom (2011). Optimizing inventory management for maximum asset performance. Available at: www.mining.com/optimizing-inventory-management-for-maximum-asset-performance/.

Morad, A.M., Pourgol-Mohammad, M., and Sattarvand, J. (2014). Application of reliability-centered maintenance for productivity improvement of open pit mining equipment: Case study of Sungun Copper Mine. *Journal of Central South University*, 21(6): 2372–2382.

Murty, K.G. (1983). *Linear Programming.* John Wiley. ISBN: 978-0-471-09725-9.

Paithankar, A., and Chatterjee , S. (2018). Forecasting time-to-failure of machine using hybrid Neuro-genetic algorithm – a case study in mining machinery. *International Journal of Mining, Reclamation and Environment*, 32(3): 182–195.

Paithankar, A., and Chatterjee, S. (2019). Open pit mine production schedule optimization using a hybrid of maximum-flow and genetic algorithms. *Applied Soft Computing*, 81: 105507.

Paithankar, A., Chatterjee, S., and Goodfellow, R. (2021). Open-pit mining complex optimization under uncertainty with integrated cut-off grade based destination policies. *Resources Policy*, 70: 101875.

Paithankar, A., Chatterjee, S., Goodfellow, R., and Asad, M.W.A. (2020). Simultaneous stochastic optimization of production sequence and dynamic cut-off grades in an open pit mining operation. *Resources Policy*, 66: 101634.

Palei, S., Das, K., and Chatterjee, S. (2020). Reliability-centered maintenance of Rapier Dragline for optimizing replacement interval of Dragline components. *Mining, Metallurgy & Exploration*, 37: 1121–1136. https://doi.org/10.1007/s42461-020-00226-5

Pelikan, M. (2005). Probabilistic Model-Building Genetic Algorithms. In *Hierarchical Bayesian Optimization Algorithm. Studies in Fuzziness and Soft Computing*, 170: 13–30. Berlin, Heidelberg: Springer. https://doi.org/10.1007/978-3-540-32373-0_2.

Press, W.H., Teukolsky, S.A., Vetterling, W.T., and Flannery, B.P. (2007). *Golden Section Search in One Dimension, Numerical Recipes: The Art of Scientific Computing (3rd ed.).* New York: Cambridge University Press. ISBN 978-0-521-880688.

Rahimdel, M.J., Ataei, M., Kakaei, R., and Hoseinie, S.H. (2013). Reliability Analysis of Drilling Operation in Open Pit Mines. *Archives of Mining Sciences*, 58: 569–578.

Rastrigin, L.A. (1963). The convergence of the random search method in the extremal control of a many parameter system. *Automation and Remote Control*, 24(10): 1337–1342.

Riddle, J. (1977). A Dynamic Programming Solution of a Block-Caving Mine Layout, In. Proceedings of the 14th International Symposium on the Application of Computers and Operations Research in the Mineral Industry, Society for Mining, Metallurgy and Exploration, Colorado, pp. 767–780.

Roy, S., Bhattacharyya, M., and Naikan, V.N. (2001). Maintainability and Reliability Analysis of Fleet Shovels. *Mining Technology*, 110: 163–171.

Samanta, B., Sarkar, B., and Mukherjee, S.K. (2002). Reliability assessment of hydraulic shovel system using fault trees. *Transactions of the Institution of Mining and Metallurgy, Section A: Mining Technology*, 111: 129–135.

Samanta, B., Sarkar, B., and Mukherjee, S.K. (2004). Reliability modelling and performance analyses of an LHD system in mining. *Journal of the Southern African Institute of Mining and Metallurgy*, 4: 1–8.

Sen, P.K. (1992) Introduction to Hoeffding (1948) A Class of Statistics with Asymptotically Normal Distribution. In Kotz, S., and Johnson, N. L., i, Vol I, pp. 299–307. Springer-Verlag. ISBN 0-387-94037-5.

Simchi-Levi, D., and Trick, M.A. (2013). Introduction to Little's Law as Viewed on Its 50th Anniversary. *Operations Research*, 59 (3): 535. doi:10.1287/opre.1110.0941.

Taha, H.M. (2008). *Operations Research: An Introduction*. 8th Edition, New Delhi: Prentice-Hall of India Private Limited.

Taha, H.M. (2016). *Operations Research: An Introduction*. 10th Edition. New Delhi: Prentice-Hall of India Private Limited.

Uzgören, N., Elevli, S., Elevli, B., and Uysal, Ö. (2010). Reliability Analysis of Draglines' Mechanical Failures. *Eksploatacja i Niezawodnosc*, 48: 23–28.

Vayenas, N., and Wu, X. (2009). Maintenance and reliability analysis of a fleet of load-haul-dump vehicles in an underground hard rock mine. *International Journal of Mining, Reclamation and Environment*, 23: 227–238.

Von Bertalanffy, L. (1968). *General System Theory: Foundations, Development*. New York: George Braziller.

Wang, W.H., Zhang, D.K., Cheng, G., and Shen, L.H. (2012). The dynamic fault tree analysis of not-cutting failure for MG550/1220 electrical haulage Shearer. *Applied Mechanics and Materials* (130–134): 646–649.

Weibull, W. (1951). A statistical distribution function of wide applicability. *Journal of Applied Mechanics*, 18: 293–297.

Wolff, R.W. (1982). Poisson Arrivals See Time Averages. *Operations Research*, 30(2): 223–231. doi:10.1287/opre.30.2.223

Yunusa-Kaltungo, A., Kermani, M.M., and Labib, A. (2017). Investigation of critical failures using root cause analysis methods: Case study of ASH cement PLC. *Engineering Failure Analysis*, 73: 25–45.

Index

A

B

C

D

E

P

Q

R

S

T

U

V

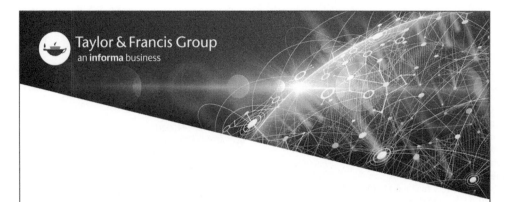